IEE MONOGRAPH SERIES 9

TELEVISION MEASUREMENT TECHNIQUES

Test card F

Fig. 10.13 (*see p.* 329)

TELEVISION MEASUREMENT TECHNIQUES

L. E. WEAVER, B.Sc., C.Eng., M.I.E.E.

Head of Measurements Laboratory
British Broadcasting Corporation

PETER PEREGRINUS LTD.

Published on behalf of

THE INSTITUTION OF ELECTRICAL ENGINEERS

Published by Peter Peregrinus Ltd.
2 Savoy Hill, London WC2R 0BP

ISBN: 0 901223 11 5

Printed in England
at the University Printing House, Cambridge
(Brooke Crutchley, University Printer)

CONTENTS

FOREWORD

It is hoped that this monograph will fill a gap that exists, at least in the English language, in the extensive and important field of video measurements. It includes a great deal of material until now accessible only to the specialist, and a certain amount which is believed to be original. It is inevitably written from the standpoint of the television broadcast engineer, since this is where the author's experience principally lies, but much of the material is applicable in related fields, such as television distribution systems for community and educational purposes. It is aimed at the professional engineer, and a sound knowledge of the basic principles of operational television is assumed. The techniques advocated are, wherever applicable, those recommended by the international bodies concerned, but a comparative approach is made to available methods, and local and national differences are pointed out. The views expressed, however, are those of the author, and must not necessarily be attributed to the British Broadcasting Corporation.

The author acknowledges the courtesy of the Institut für Rundfunktechnik, Munich, for permitting the reproduction of Figs. 5.21, 5.22, 15.1 and 15.2, and the Australian Broadcasting Commission for Fig. 5.10. His thanks are due to the Head of Designs Department of the BBC for allowing the resources of the department to be used in the preparation of the text, and not least to the Director of Engineering for permission to publish it. Finally, tribute must be paid to the invaluable assistance of the author's colleagues, unfortunately too numerous to mention individually.

1 INTRODUCTION

The primary role of measurements in television broadcasting may be defined broadly as the checking and standardisation of all the apparatus and circuits in the signal chain, starting with the signal source and ending with the radiated signal, to ensure that the predetermined standards of quality are maintained. One would like to extend this to the very end of the chain, the screen of the viewer's receiver; but, of course, this is quite impracticable, except in so far as the planning and siting of transmitters affect the received signal.

Measurements are all-pervasive in television: they affect every aspect of operational engineering, and markedly influence cost and efficiency. It is of paramount importance that correct measurements be carried out at every step, so as to maintain the system within the chosen tolerances, and that the most appropriate equipment and procedures be used.

The overall problem is threefold: the determination of the various types of distortion which can impair the final picture, the allocation of suitable tolerances to each of the components in the signal chain for each of these distortions, and the actual techniques and equipment employed for measuring the various distortions. These must not only ensure adequate accuracy with the lowest possible outlay in money and material, but be in all respects the most appropriate for the task for the conditions under which the measurements are made.

In the following Chapters, the techniques and equipment employed for measuring and testing the signal chain are reviewed. Video measurements are only a part of the complete problem, but it is unfortunately necessary to restrict this monograph to them, so that they can be dealt with in sufficient depth without unduly increasing the length of the text.

It would be pertinent here to consider what is meant by the distortion of a television picture. Television is a means of communication in which the original information is, as it were, coded, and subsequently decoded to yield, not the original information, but some acceptable substitute for it. For example, in television the original information is a visual scene in three dimensions and colour. The decoded output is the picture on the screen of the viewer's receiver which is quantised in the vertical direction, one dimension has been removed and, in many instances, only the luminance component of the information is retained. Further differences between input and output are brought about by psychophysical factors: a scene which may originally be virtually boundless becomes what is manifestly a small rectangle on the monitor or receiver; visual effects, such as a change of viewing angle or scene, are carried through independently of the volition of the observer; the information provided by the viewers' other senses may be totally unrelated to the scene, e.g. a torrid desert viewed in a cold, draughty room.

It is evident that the usual criterion for absence of distortion, i.e. an output identical with the input apart from a scale factor, cannot be applied here; and yet everyone has some notion of what constitutes a good or a bad picture. These purely subjective concepts vary from person to person; also the judgment of an individual varies with time and circumstances. Thus purely objective criteria cannot be applied to television-picture quality, and even subjective criteria can only be utilised on a statistical basis. One is constrained for practical reasons to apply objective criteria to the standardisation of the signal chain; yet it is all too often overlooked that they are not really fundamental.

A discussion of the standards of picture quality may seem irrelevant when the aim should be to distribute pictures of the highest standard that modern technology permits; and one would be inclined to agree if the economics had not to be considered. The second law of thermodynamics, in its broadest formulation, applies here as elsewhere, and every improvement in quality has to be paid for. Moreover, as the standard increases, the cost increases at a far greater rate—and not merely the capital cost of an improvement, but also the cost of maintaining the higher standard.

A broadcasting authority is thus forced implicitly to make a difficult decision (even though the problem may not be framed quite so crudely as is presented here): if perfection is not practicable, where should the line be drawn? If a high standard is intended, the cost will be high and the revenue needed correspondingly high. If the standard is set too low, there will be a risk of dissatisfaction on the part of the viewer.

In the formative years of television broadcasting, standards were necessarily set in a largely empirical manner based on the experience and judgment of individual engineers; although this could not be economically efficient, the relatively small scale of operations concealed the fact. Later, the increasing need for guidance and standardisation on such matters led to these being provided by the CCIR (Comité Consultatif International des Radiocommunications) and the CMTT (Commission Mixte CCIR/CCITT pour les Transmissions Télévisuelles), of the International Telecommunication Union, and the European Broadcasting Union. The consequent dissemination of the shared experience of engineers of many countries has been invaluable for the development of good standards of television engineering.

The efforts made in the mid-1960s towards the establishment of a common European colour-television system gave great impetus to the development of standardised methods of subjective testing, from which information on working tolerances may be derived. Some agreement on common methods of test and standards of performance has now been reached, but there is still far to go. The whole field of subjective testing and the determination of working tolerances is of great interest, but, regrettably, an adequate treatment would be far too lengthy for inclusion here. However, some notion may be gained from papers by Weaver (1959) and Prosser, Allnatt and Lewis (1964).

These subjective procedures, which derive their information from the statistical behaviour of observers to picture impairments, enable one to define a maximum permissible figure for each impairment which, although in the end must be based on some arbitrary decision, has at least the merit of having been derived in a consistent and scientific manner.

Once the global tolerance has been ascertained for each distortion, two further problems remain to be solved: how it is to be apportioned among the various elements of the signal chain, and how one is to estimate the combined effect of the simultaneous presence of a number of different types of picture impairment.

Until recently, the allocation of tolerances appears to have been performed very largely as though each distortion existed in isolation, whereas it is evident that most, if not all, of the distortions possible are likely to be present simultaneously, and moreover, the picture impairment increases with the number of distortions. This is attributable most likely to the lack of a valid general method of estimating the combined impairment of a number of distortions, and it seems curious that efforts were not made earlier to find a solution to this very important problem.

An answer has been supplied by Lewis and Allnatt (1965) as a result of a valuable series of subjective tests on a number of picture impairments. It makes possible the direct addition of impairments of different types in terms of a simple linear transform of a quantity determined during subjective testing by their method. They suggest that this quantity should be denoted 'impairment unit' or 'imp' for short. The number of imps corresponding to the various distortions can accordingly be summed arithmetically to obtain the 'imp value' corresponding to the overall picture impairment; moreover, one may then use this to determine the corresponding distribution of opinion.

The author has taken this somewhat further, in a suggestion (Weaver, 1968) that the specification of individual tolerances should be replaced by an overall rating factor, which may be the result of any combination of the individual distortions. This device is, strictly speaking, only applicable to a complete signal chain, because distortions of the same kind originating in successive links do not add in the Lewis and Allnatt way, but behave in a statistical manner, depending on the nature of the distortion (CCIR, 1960; Maurice, 1968).

When a decision has been taken on the maximum tolerable amount of a given distortion, the final part of the problem is to allocate to each individual link in the signal chain a suitable proportion of this value. This cannot be done simply by subdivision. First, the viewer's receiver, as an essential part of the signal chain, must take its share; and, since this must necessarily be manufactured to sell at a price sufficiently low for millions of people to be able to afford one, it deserves a very high proportion of the total tolerance, perhaps as high as 50% in certain instances. Cases also occur where the share cannot be restricted to what might seem a reasonable amount because of natural limitations; e.g. the signal/noise ratio of a camera may in some instances be very much worse than that of a very long distribution network because, using present techniques, the camera cannot be improved.

Finally, the magnitude of a given type of distortion measured between the ends of a given signal chain is itself statistical in nature. The individual values in some cases add and in others subtract, depending on the nature of the distortion. The exception to this is random noise, where no cancellation is possible. Certain empirical rules have been devised for the addition of distortions (Paddock, 1970) but Maurice (1968) points out that distortions are more correctly compounded by a process closely resembling the mathematical convolution of two functions of a single variable. Furthermore, no distortion is completely time-invariant; and, strictly speaking, one can never calculate the overall distortion of a long link, only the probability of its lying between certain limits. This applies in particular to long microwave links, where the probability of fading in one or more repeater sections is by no means negligible, so that there is a corresponding uncertainty in the overall signal/noise ratio (Siocos, 1969).

Once the individual tolerances have been decided, the next step is to ensure that they are maintained in service, and this is the chief role of measurement. It is customary to subject each component of the television chain to a severe acceptance test before it is put into service, after which it is maintained in an operable condition with a less rigorous maintenance test in between periods of service. Such tests may be comparatively leisurely, and carried out with individual line-repetitive or fullfield signals for each type of distortion.

However, the situation is changing quite rapidly, and for several reasons: the continuing increase in the number of channels and the hours of service, with the inevitable consequence that more and more work has to be carried out in an ever decreasing period of time; the development of automatic equalisation and monitoring methods; and the realisation that it is now feasible to construct electronic equipment which can carry out most, if not all, of the operational checking at present performed by engineers.

The first reaction to this trend of events has been a very considerable increase in the use of insertion test signals (Chapter 7), which are specially designed to yield the maximum information in the minimum time and can be carried on otherwise unused lines in the field-blanking interval of the video signal. Such signals do not offer quite the same facilities as line-repetitive test signals; they have a rather lower accuracy and certain inherent difficulties, but they provide the immense advantage of measurement (Shelley and Williamson-Noble, 1970), automatic control and automatic equalisation during the distribution and radiation of television signals. Despite the specialised equipment required, it is likely in the near future that a large proportion of all routine measurements will be carried out in this manner. Where a somewhat improved accuracy is desirable, and circuits can be made available for short out-of-service periods, and also under certain other special conditions, these insertion test signals can be employed in the line-repetitive form. Such 'compact' test signals (Chapter 6) take advantage of the highly condensed form of the insertion test signal to reduce the testing time considerably. Likewise, the evaluation of these signals can be carried out in time-multiplexed form, so as to make all the desired information available simultaneously.

Insertion test signals have for long been employed for the control of signal level at the input to a transmitter (Springer, 1959), and automatic control, not only of linear waveform errors, such as chrominance–luminance gain and delay inequalities, but also of nonlinear effects such as differential gain and phase, have been shown to be quite practicable. Automatic monitoring methods based on the use of such signals offer almost unlimited prospects for the overall control of a complex distribution network, apart from the invaluable statistical information which they can supply.

Nevertheless, it should not be supposed that line-repetitive or fullfield signals either have already lost or are soon likely to lose their importance. They will continue to be employed for acceptance testing and for general test purposes when the saving of time is of less importance than accuracy of measurement; and, of course, they remain the basic components of the special signals just mentioned, even if in a somewhat modified form. Therefore, the line-repetitive signal has been taken in the following Chapters as the fundamental type of test signal; when the use of each is thoroughly understood, any particular adaptation should present no great difficulty.

There is no doubt that video test and measurement techniques are of great interest and importance. Whether they are applied to research problems, to the design of television equipment or to the operational needs of a television service, they must be devised in a form applicable to the task in hand in the simplest and quickest manner consistent with the required standard of accuracy and with the least possibility of error. This means that an investigation into a new technique cannot be divorced from the design of the associated equipment, and by the same token, neither can any decision on the form of test signal or apparatus be taken without sufficiently extended practical tests.

Another factor which should be stressed is the importance of general standardisation of test and measurement techniques. This applies particularly where a broadcasting and a telecommunication authority are interested in the same distribution network or where more than one broadcasting authority use a network maintained by the same telecommunication administration. With many measurements, the significance of the result may well depend on the way they are carried out; and, if interpretations between interested parties are not to differ, it is essential that the techniques used be standardised as far as possible. In other words, they should be 'talking the same language'. This applies particularly to international programme exchanges, where the CCIR, CMTT and EBU have been striving for standardisation for many years, and a considerable measure of success has already been achieved. It is sincerely hoped that this monograph might also make some contribution towards achieving that very desirable end.

1.1 References

CCIR (1960): Documents of the XIth Plenary Assembly, Oslo, Recommendation 421–1, Annex IV

LEWIS, N. W., and ALLNATT, J. W. (1965): 'Subjective quality of television pictures with multiple impairments', *Electron. Lett.*, **1**, pp. 187–188

MAURICE, R. D. A. (1968): 'Tolerances for Pal colour television', *Roy. Televis. Soc. J.*, **12**, (4), pp. 86–93

PADDOCK, F. J. (1970): 'The relationship between individual link and chain distortions in a television network'. Proceedings of the IERE joint conference on television measuring techniques, London

PROSSER, R. D., ALLNATT, J. W., and LEWIS, N. W. (1964): 'Quality grading of impaired television pictures', *Proc. IEE*, **111**, (3), pp. 491–502

SHELLEY, I. J., and WILLIAMSON-NOBLE, G. (1970): 'Automatic measurement of insertion test signals', *IERE Conf. Proc.* 18, pp. 159–170

SIOCOS, C. A. (1969): Statistical principles in the supervision of technical performance in the color television network of the CBC', *IEEE Trans.*, **BC-15**, pp. 33–38

SPRINGER, H. (1959): 'Anwendung und Weiterentwicklung der Prüfzeilentechnik', *Rundfunktech. Mitt.*, **3**, pp. 40–50

WEAVER, L. E. (1959): 'Subjective impairment of television pictures', *Radio Electron. Engr.*, **36**, (5), pp. 170–179

WEAVER, L. E. (1968): 'The quality rating of colour television pictures', *J. Soc. Motion Picture Televis. Engrs.*, **77**, (6), pp. 610–612

2 MEASUREMENT OF LEVEL

2.1 Insertion gain

The signal path from the picture source to the input to the transmitter consists of a very long, complex chain of signal-handling equipment. The total length of this chain may be thousands of miles, and yet, when a 1 V pk–pk video signal is applied to the chain from the picture source, it is required that 1 V pk–pk, within close limits, appear at the input of the transmitter. Furthermore, any element of the chain must be replaceable with a similar unit with only a small and acceptable change in the overall gain; and if, at any point, a further item of equipment is inserted, the new characteristics of the signal path must be as planned.

The only way these requirements can be met in a flexible and manageable form is, as far as possible, for all equipment to be designed for insertion between fixed terminating impedances with unity insertion gain; i.e. the output level is ideally unchanged by the insertion of the equipment between the terminations, although the transmission characteristics may be modified as desired. If, in addition, each item of equipment has input and output impedances which are accurately matched to the terminations, it becomes possible to cascade any number of such items of equipment while maintaining the overall gain.

Gain, in the sense just used, refers to the ratios of the peak-to-peak video signal levels across the output termination before and after the insertion of equipment, and should really be called the 'video insertion gain' to make it clear that the signal is being considered as a waveform, because gain (or loss) in its general sense is a complex function of frequency.

In some instances, e.g. amplifiers and programme-distribution circuits, it is ideally required that the amplitudes of all the Fourier components of the waveform be unchanged by the insertion of another amplifier or circuit, whereas the change in the phase shift has to be strictly proportional to frequency, in order that all Fourier components shall be delayed by the same amount. In others, deliberate modifications of the amplitude and phase shift as a function of frequency may be required, e.g. to compensate for deficiencies elsewhere.

We have so far talked about insertion gain, although logically one might equally talk about insertion loss. There is no hard and fast rule, but it seems logical to use the term 'gain' when referring to networks containing active elements, and 'loss' when only passive elements are concerned.

Insertion gain is defined in practice as follows. Any signal source, no matter how complex, can be reduced by the application of Thévénin's theorem to a generator having an e.m.f. E in series with a source impedance Z_R. The power supplied by this generator, we will suppose, is normally delivered to a termination Z_S nominally equal to Z_R; then a voltage $e_o = EZ_S/(Z_R+Z_S)$ appears across Z_S.

A network is then inserted between the source and termination. As a result, the voltage across the termination Z_S becomes e_o'. The insertion gain is obtained as $20 \log_{10} (e_o'/e_o)$. The insertion loss would require the reciprocal e_o/e_o' to be taken.

In communication work, two main cases occur. The first is the insertion of a passive network, and is fairly complex, since reflections may occur at both input and output which are then modified by the transmission properties of the network. This is given detailed treatment in the literature (Shea, 1929), and will not be considered further here. Such networks are almost always wave filters for band limitation; equalisers are most usually constructed in constant-resistance form, and the possibility of reflections is much reduced.

The second type of transducer is much more common in television. It consists of equipment containing active devices, often highly complex, and is provided with a resistive termination at both input and output. With rare exceptions, these terminations are nominally 75 Ω, and the insertion gain is usually unity. As a result of the use of active elements between the input and output, the backward gain is so low as to be negligible, and it is not possible for terminal reflections to traverse the network. This results in a considerable simplification of the expression of the insertion gain.

Fig. 2.1 shows the insertion of such a network between terminations, which are now taken to be R_R and R_S, since they are nominally resistive. The input and output terminations of the network are R_i and R_o; and, as a result of the internal amplification, a generator of e.m.f. $2Ge_i$ appears in series with the output impedance of the network.

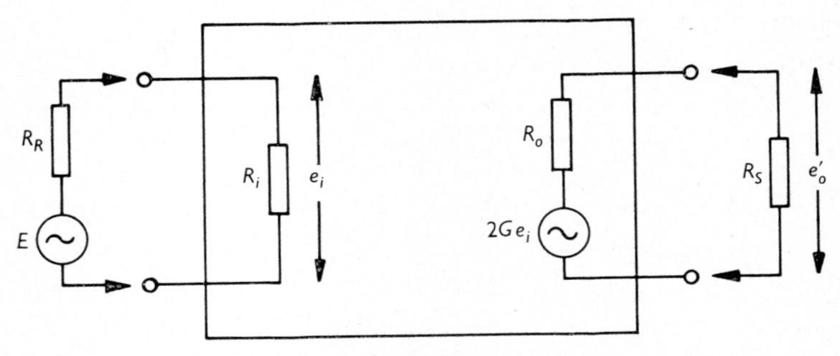

Fig. 2.1 Insertion gain of transducer with zero backward gain

The voltage across R_S before insertion is, as before,

$$e_o = \frac{ER_S}{R_R + R_S}$$

which, after insertion, becomes

$$e_o' = \frac{2GER_S R_i}{(R_R + R_b)(R_o + R_S)}$$

Hence the insertion gain is

$$20 \log_{10} \frac{2GR_i (R_R + R_S)}{(R_R + R_i)(R_o + R_S)}$$

A noteworthy feature of this expression is its tolerance to small changes in the terminations from their nominal values, which is one of the incidental advantages of the insertion-gain (-loss) principle. Where the attenuation in the reverse direction is low, as with cables and filters in the pass region, the output/input type of measurement can give rise to very serious errors, particularly when the electrical length is great. The reason is that the measured e_i includes any reflections returned from the output terminals, whereas, if the measurement is made as insertion loss, the corresponding term is a fraction of

E and, consequently, quite independent either of the network or of the terminal conditions. These reflections, by the way, are by no means necessarily unwanted; they may be the necessary consequence of the frequency-selective properties of the filter.

When the transducer is active, a further error may occur from reactive components of the terminating impedances, usually shunt capacitances, often aggravated by the shunt capacitance of the waveform-monitor input circuit and the connecting cable. The question of terminating impedances is considered in some detail in Chapter 8.

2.2 Measurement: general

The preferred circuit for the measurement of insertion gain or loss is given in Fig. 2.2*a*. It operates by comparing the level at the output of the transducer with that from a standard, calibrated attenuator B; consequently, when insertion gain is measured, a suitably calibrated attenuator A must be connected in front of the transducer; this would no doubt be required in any event to avoid overloading. In this case, it is possible to carry out the measurement entirely in terms of attenuator A by replacing attenuator B by a short length of coaxial cable.

The procedure consists in switching the output indicator to each of the paths in turn and adjusting the measuring attenuator until the two readings are as nearly as possible equal. Only in special cases will absolute equality be obtained; so the process is completed by interpolating between the readings at the two attenuator steps bridging the exact setting. Obvious precautions must be taken to ensure the quality of the terminations, including those of the attenuators. Short coaxial leads and good earthing are also required.

An important feature of the method is the use of a pair of precise 75 Ω resistors between the generator output and the two paths. Their role may seem obscure, until one realises that they ensure that the e.m.f. driving current into each path is the same for all conditions. The input circuit is thus equivalent to a constant-voltage generator feeding each path through 75 Ω, with the result that they are compared under true insertion-gain (or -loss) conditions. An advantage of this technique is the maintenance of equal levels in the two paths, which reduces the risk of errors due to crosstalk effects. It is therefore widely employed for the measurement of the insertion losses of filters and other networks, and indeed is sometimes called the 'filter test set' method.

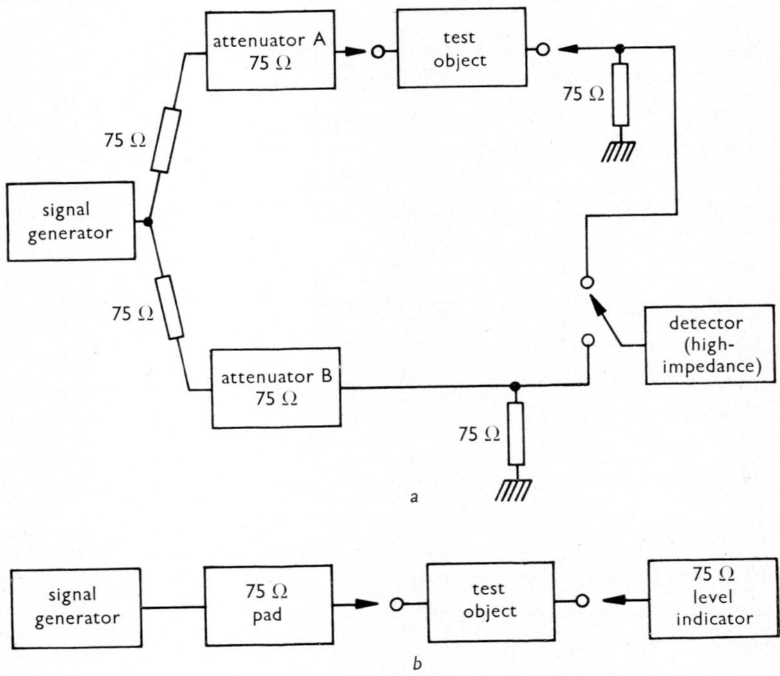

Fig. 2.2 Measurement of insertion gain or loss
a Changeover method
b Measurement from known impedance

Fig. 2.2*a* is drawn for an output-level indicator of high impedance, as a reminder that it may be used for waveform measurements by comparison of the output peak-to-peak levels, in which case the high-impedance input of a waveform monitor may be employed. In general, however, it is preferable, wherever possible, to use an indicator, whether for waveform or for sinusoidal measurements, which has a high-quality 75 Ω input impedance in its own right. The output switching then needs modification to ensure that both paths are terminated at any given instant, for which a coaxial switch is very desirable. The signal to be used for the measurement of video insertion gain is discussed in Section 2.3.

The comparison method of measuring insertion gain (or loss) just described has many advantages, but it is evidently not suitable for operational use, if only because the two ends of the test object must be available at the point of measurement. The alternative is to design the test generator so that the output impedance is precisely 75 Ω and to measure the received signal from the test object with a device having an accurate 75 Ω input impedance. The insertion of the equipment or network under test between these two will then give the true insertion gain (or loss).

The first necessity is to ensure that the generator has, in fact, the required 75 Ω impedance over the whole of the frequency range of interest, and also at all output levels if these should need to be varied. This is not unduly difficult with modern feedback output stages, which can be designed for an extremely low output impedance, perhaps of a few milliohms. The insertion in series with the output terminal of a precision 75 Ω resistor then enables the required condition to be fulfilled.

If, for some reason, it is not possible to provide a sufficiently good output impedance, the situation can be improved by adding an attenuator pad between the generator and the test object (Fig. 2.2*b*). The effect of a mismatch is the production of a reflection (Chapter 8), and it is evident that a reflection accompanying the test signal will have to traverse the pad twice, so that A decibels of attenuation will improve the generator output impedance by a factor of $2A$ decibels. This device must clearly be employed with caution to avoid overloading effects.

It is often assumed that the achievement of a good impedance is much easier with the level-indicating device than with the test generator, but this is not necessarily so; in cases of doubt, a measurement of the return loss should be made (Chapter 8). An instance where one may easily be deceived is the approximately r.m.s. levelmeter using a diode bridge, where the input impedance may be a function of input level. An example of this type of instrument which has been found very useful for sinusoidal measurements is given in Section 2.5.

The principal disadvantage of this second version of the insertion-loss method appears when one considers how to ensure sufficiently accurate readings, assuming that the terminations and any attenuators used are above suspicion. It is plain that, if all types of measurement are contemplated, the output level of the test generator into 75 Ω must be precisely known; likewise, the output indicator must be accurately calibrated. This problem is discussed in Section 2.4.

2.3 Measurement of video gain

The measurement of the insertion gain of the equipment used for the transmission or handling of video signals should be carried out with a standard video signal, and not with sinusoidal signals. One good reason is the need for much of this equipment for synchronising information to ensure correct operation; but, in addition, the gain at a single arbitrarily chosen frequency may be significantly different from the gain measured on a typical video signal, not only on account of variations in the gain–frequency product but, in certain instances, because of nonlinearity distortion.

The signal recommended by the CCIR for gain measurement on distribution networks (CCIR, 1966) is the monochrome sine-squared-pulse and bar signal, which has the virtue that its average picture level of very nearly exactly 50% is not far removed from the most probable level in programme signals (Quinn and Newman, 1965). For all gain measurements, only the picture component of the video signal is normally considered; so, in this instance, the level of the signal for gain measurements is taken as the height of the bar. However, this monochrome pulse-and-bar signal is, for convenience, frequently replaced by one of the composite chrominance and luminance sine-squared-pulse and bar signals. Also, if a measurement needs to be made during the transmission of a programme, one has no option but to utilise the luminance bar of the insertion test signal, which is then the only valid white-level reference; to make this possible, the bar amplitude is controlled within close limits on insertion.

Difficulty arises when the signal is degraded, particularly by relatively long-term distortions, such as line tilt, since various values of signal amplitude will be obtained unless the points at which the measurements are made on the bar top and at black level are standardised. Evidently, the safest point at which the bar top should be measured is its centre, since it is least likely to be affected by any of the distortions, but a similar choice cannot be made precisely at black level, owing to the presence of the other components of the signal, and an arbitrary decision must be taken.

With the CCIR signal, black level is measured at a point immediately following the sine-squared pulse, with the BBC chrominance–luminance pulse-and-bar signal (Section 6.3.6), at a point midway between the sine-squared pulse and the chrominance bar, and with the UK national insertion test signal, at a point slightly preceding the composite chrominance pulse. The first convention is a CCIR recommendation, and the latter two are conventions observed in the UK. With all such conventions, some common sense must be employed if the nominal point of measurement is obviously misplaced by the presence of an echo or a 'ring'.

It is common practice for a measurement of the synchronising-pulse amplitude to be made at the same time. This is defined by the level difference between the same blanking reference point and the centre of the base of the pulse, and its ratio to the difference between black level and white level is the 'picture/synchronising-pulse ratio'.

The preferred method of measuring the video gain is as follows. The level of the video signal from the test-signal generator, which must have a satisfactory 75 Ω output impedance, is set so that the amplitude of the picture-signal region between black and white levels exactly agrees with the reference value, i.e. 700 mV for system *I*. It is advantageous for this to be carried out by the method of Section 2.4, since this at the same time ensures not only the requisite accurate 75 Ω termination but also the precise measurement of the level from the generator. It is then connected to the apparatus under test; but if the gain is greater than unity, a calibrated 75 Ω attenuator must be inserted to reduce the overall gain appropriately; otherwise the reading may be vitiated by overloading effects. If the equipment contains some physical limitation on maximum amplitude, such as a clipper, particular care must be taken not to approach the threshold value.

The output level is measured, preferably by the same recommended method, but at least by some means having a sufficiently good 75 Ω impedance and adequate accuracy; the ratio of the two measured values, plus any attenuation which may have been inserted, gives the gain.

2.4 Measurement of level

The measurement of video level was discussed in Section 2.3 without specifying how this should be carried out; of course, in the end, the accuracy of the measurement depends entirely on the accuracy with which the various amplitudes can be measured.

Often the measurement is carried out with a master waveform monitor. The accuracy of amplitude measurement with this is discussed in Section 11.3, where it is shown that, if one takes into account all the sources of error, a typical overall uncertainty of ±2% is about the best one might expect. Hence, if the measurement of video insertion gain is made, in the case where the two ends of a circuit are not simultaneously available, with two waveform monitors whose errors are +2% and −2% respectively, the error of the measurement will amount to 4%, say 0·4 dB. This is serious, and may be greater, in fact, than the allowable tolerance for the circuit in question. If the waveform monitors are not of the very highest quality, or are poorly maintained, the possible uncertainty can be even greater.

The author conducted an investigation some years ago into complaints of errors in level measurement in a large studio complex, and it was found that errors of around 1 dB were actually occurring in certain instances, admittedly with equipment somewhat inferior to that in use today. As a result, the following method was devised to make the measurement as independent as possible of the characteristics of the waveform monitor used. It will be shown that the accuracy of the measurement is to all intents and purposes independent of both the frequency response and the deflectional linearity of the instrument. The method has been in operational service for a considerable period, and has proved to be extremely satisfactory (Weaver, 1961).

The basis of the method is a square wave with an impeccably flat top and bottom and a very accurately known peak-to-peak amplitude. The repetition frequency of the square wave is not at all critical; a frequency in the neighbourhood of 11 kHz has been found generally suitable for 625-line measurements.

The signal to be measured and the reference square wave are linearly added in precisely equal proportions in a device which presents an exact 75 Ω termination to the video signal to be measured, and the resultant signal is observed on a waveform monitor. A highly convenient method is to use the two inputs of a difference amplifier, which may well form part of the waveform monitor used, in which case the conditions that it should be correctly balanced and that its linearity should be faultless over its operating range ought to be fulfilled automatically. An alternative is to use a member of the well known family of constant-resistance mixing pads; here the delta connection of three 75 Ω resistors is the most convenient with its three output terminals connected to the input video signal, the square-wave generator and the waveform monitor, which in this case must be accurately terminated in 75 Ω. The source impedance of the square-wave generator must also be 75 Ω if the termination presented by the mixing pad to the input video signal is to be the 75 Ω required for a measurement of insertion gain. The equivalent star network can also be used, but the 25 Ω resistors required are generally less convenient. In either case, the loss of each input when the other pair is correctly terminated is exactly 6 dB.

The way the measurement is carried out is extremely simple to demonstrate, but unusually difficult to describe. An attempt will be made, however, with the aid of Fig. 2.3, in which it will be assumed, for simplicity, that the overall amplitude of the video signal is to be measured.

The linear addition of the video signal to the square wave results in two displays in sequence, vertically displaced by the peak-to-peak amplitude of the square wave. In practice, since the frequencies of repetition of the two waveforms are not synchronously related, one would not see both upper and lower displays presented as drawn on the diagram, but a less distinct picture, in which, however, the six horizontal lines corresponding to the three reference levels of the video signal in its two vertically separated positions are quite clear. These lines are labelled W_1, B_1, S_1 and W_2, B_2 and S_2 in the diagram. A somewhat similar effect is seen if a display corresponding to a few fields, instead of a few lines, is used.

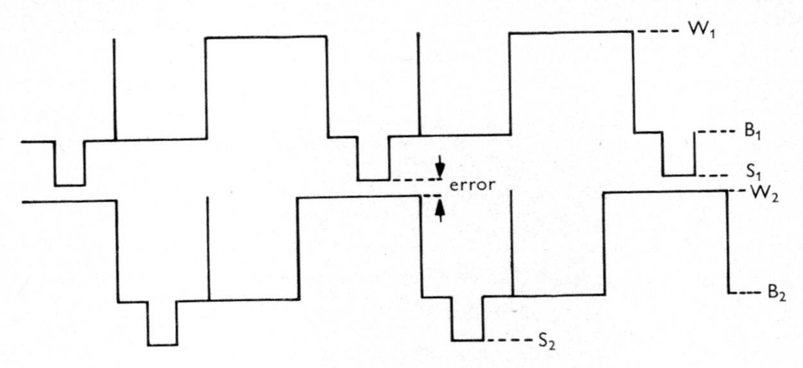

Fig. 2.3 Level measurement using standard square wave

The condition that the amplitude of the reference square wave should be exactly equal to the peak-to-peak amplitude of the video signal implies that S_1, the bottom of the synchronising pulse of the upper display, should exactly coincide with W_2, the white level of the lower display. It follows therefore that all that needs to be done to measure the amplitude is to vary the reference square-wave amplitude until these two lines, S_1 and W_2, coincide precisely. To improve the sensitivity, the vertical amplification may be increased as desired, since no reasonable nonlinearity distortion will affect the fact that the two levels being compared are set to the same potential. By the same token, no parallax error is possible. The amplitude of the square wave then gives the amplitude of the video signal.

In the preferred realisation of this method, the square-wave generator is provided with switchable attenuators, so that the calibrating amplitudes of 1 V, 0·7 V and 0·3 V pk–pk are available for the

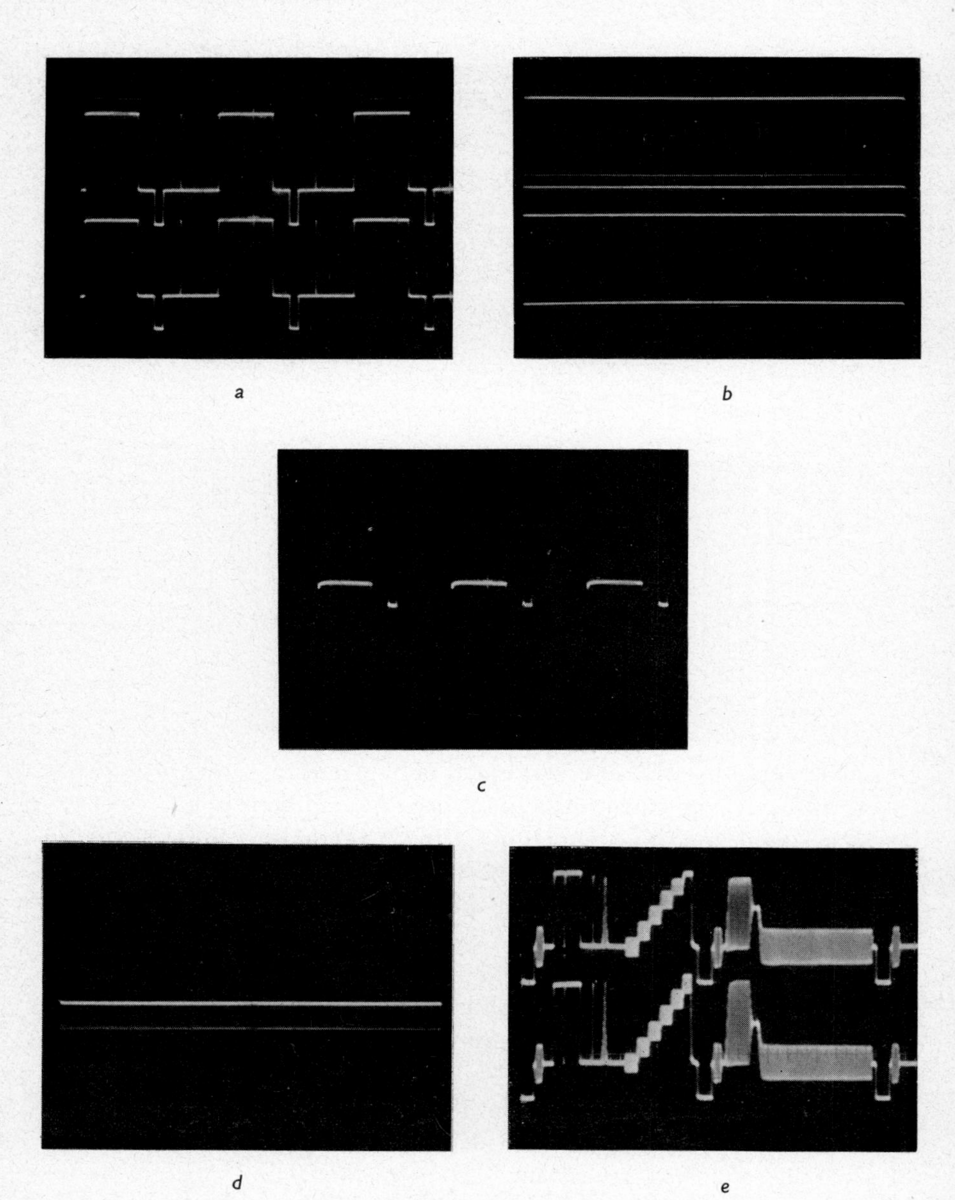

Fig. 2.4 Square-wave method: monochrome signals

a Sine-squared-pulse and bar: locked display
b As (*a*), but unlocked display
c As (*a*) with central region windowed
d As (*b*) with central region windowed
e Measurement using insertion test signal

measurement of the overall signal amplitude, the picture region and the synchronising-pulse amplitude, respectively. In addition, a vernier dial is incorporated to provide a total change within $\pm 0{\cdot}5$ dB with respect to each nominal value, and is thus calibrated directly in terms of the error from the nominal value, which is what is really required.

The measurement of a sine-squared-pulse and bar signal is shown in Figs. 2.4*a* and *b*, the former with a locked display and the latter with the display unlocked. In most instances, it is felt that the unlocked display will be preferred. Even a quite moderate gain in the waveform monitor permits a good sensitivity to be obtained, as is visible in Figs. 2.4*c* and *d*, in which the expanded central portion corresponds to an error of 0·25 dB. Fig. 2.4*e* shows how the same method can be applied to the measurement of video signals during programme time by utilising the bar of the insertion test signal to provide a white-level reference. Fig. 2.5 gives the application of the method to the measurement of colour signals.

Comparative tests have shown that, under these conditions, a number of engineers will obtain the same reading within typically $\pm 0{\cdot}02$ dB. When the generator is first calibrated against a standard cell or a digital voltmeter, its accuracy may be within, say $\pm 0{\cdot}05$ dB, or better, but subsequently it is likely to drift further away from the correct value as a result of aging of the reference Zener diode and the resistors, apart from any temperature effects.

The long-term accuracy of the generator is determined by the complexity of the generator circuitry and the quality of the components used to obtain the requisite stability; and this, in the end, comes down very much to a question of cost. One is therefore faced in the design of such equipment with a difficult choice: either to keep the cost to a minimum and to tolerate periodic recalibration, or to design an expensive but stable device.

A compromise solution which has been found satisfactory in practice is to assume that there is an advantage in keeping the level-measuring equipment cheap and simple, to encourage its use on the widest possible scale, and to provide in addition a highly accurate and stable version which may be kept in local maintenance areas as a means of ensuring correct calibration of the operational instruments.

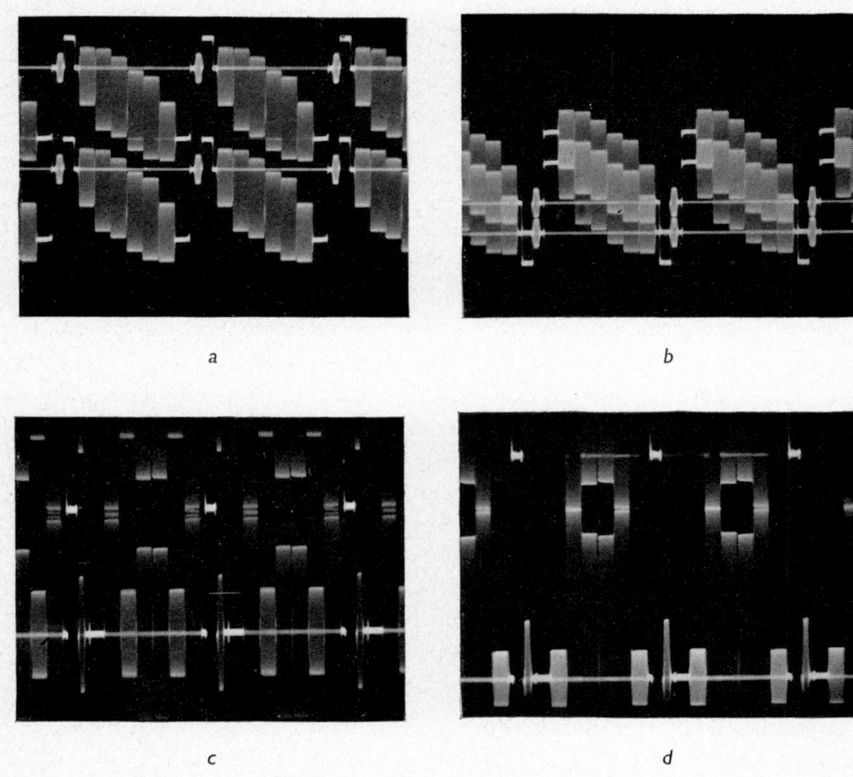

Fig. 2.5 Square-wave method: colour signals

a Measurement of colour-bar amplitude
b Measurement of synchronising-pulse amplitude
c Measurement of $B-Y$ signal: windowed display
d Measurement of $R-Y$ signal: windowed display

The latter also provides calibration facilities for the equipment used to check the chrominance–luminance errors of waveform monitors, as well as outputs of reference square waves and chrominance-subcarrier frequency for measurement purposes (Sections 2.5 and 11.4.1).

By such means, it should be possible to guarantee level-measurement uncertainty of within $\pm 0{\cdot}1$ dB under operational conditions, and better if sufficient care is taken. This naturally assumes optimum signal/noise ratio and low waveform distortion. It is also conditional on the use of short connecting leads during the measurement, for which reason the author strongly recommends apparatus small and portable enough to be taken directly to the point of utilisation, since the lengths of the leads used are thereby made more evident. When equipment is mounted in a bay, the total length of the cabling involved is often very much greater than one imagines, and may well grow insidiously. One is sometimes tempted to suppose that there must be a natural law which says that cable lengths, like entropy, always increase with time.

Where measurements have to be made on noisy signals, the errors can be decreased markedly by the application of sample-and-hold methods, in which the integration of a large number of samples makes possible the determination of the most probable value of the amplitude of the signal (Section 5.5). These incidentally lend themselves admirably to automatic measurement and control methods.

2.5 Standards of level

When one speaks of accuracy of voltage measurement, one thinks in terms of equipment which has an absolute calibration against a standard volt, naturally allowing of a certain possibility of error. The precise magnitude of this error is, in fact, very difficult to know, since the indirect nature of the calibration process involved introduces uncertainties which are difficult to deal with. It is furthermore surprising to find that the desirable accuracy in local standards of level for a high-quality colour service, at least as far as the measurement of subcarrier is concerned, approaches the limit at present possible. It seems realistic therefore to make one's principal objective the attainment of the greatest possible consistency over the whole of any given organisation, while still striving to reduce the absolute error to a minimum. In view of this, words below such as 'accuracy' must be interpreted as referring to the consistency of a measurement, rather than to its absolute value, so as to avoid the need to specify the latter even when the difference is small.

The only available standard of voltage is the Weston standard cell, and, fortunately, it can be readily obtained already aged and with a very low temperature coefficient. Incidentally, the modern versions are completely sealed and wholly indifferent to their mounting position; on the other hand, the maximum allowable drain is extremely small, and, if this is exceeded, they may be permanently damaged. Also their internal resistance is high, which puts a further limit on the minimum shunt resistance.

The Zener diode, while in some respects much more suitable than the standard cell for use in equipment, still has difficulty in matching at equivalent cost the stability and consistency of the inexpensive standard cells intended for instrument use. A compromise adopted to advantage by some manufacturers, for example, of digital voltmeters, is to use a Zener diode of slightly lower quality, which is then checked periodically against a standard cell.

The recommended technique for checking luminance (monochrome) amplitudes as well as the amplitudes of the various components of colour-test signals is the standard square wave, the method of use of which is described above. The fundamental principle of a successful standard-square-wave generator is given in Fig. 2.6*a*. The basis is a current generator which feeds 26·67 mA into an accurate 75 Ω resistor by way of a switch. Hence, when the switch is closed, an accurate 2 V pk–pk is developed across the resistor; and consequently, by Thévénin's theorem, it is equivalent, as far as the output terminals are concerned, to a generator with an e.m.f. of 2 V and an internal resistance of 75 Ω, as is required for insertion-loss measurements (Fig. 2.6*b*). It follows that, when the output terminals are closed with a 75 Ω resistor, 1 V pk–pk is developed across it, which becomes a rectangular wave when the switch is periodically operated.

As has already been stated, such generators need calibration at regular intervals to maintain their accuracy, apart from initial adjustment. This is most conveniently taken care of by arranging for the switch to be in the 'make' position and connecting across the output terminals a digital voltmeter of adequate accuracy. The current generator is then adjusted until the e.m.f. is precisely 2 V. Alternatively, a further switch can be added which transfers the current from the 75 Ω resistor to another resistor, the ratio of whose resistance to 75 Ω is numerically equal to the e.m.f. of the standard cell in use. This is shown dotted in Fig. 2.6*a*. The adjustment may now be made between the upper end of this resistor and the standard cell by measurement with a normal potentiometer.

There are two principal sources of error in this arrangement. The first is the resistance of the switch in the closed position, *r* in Fig. 2.6*a*.

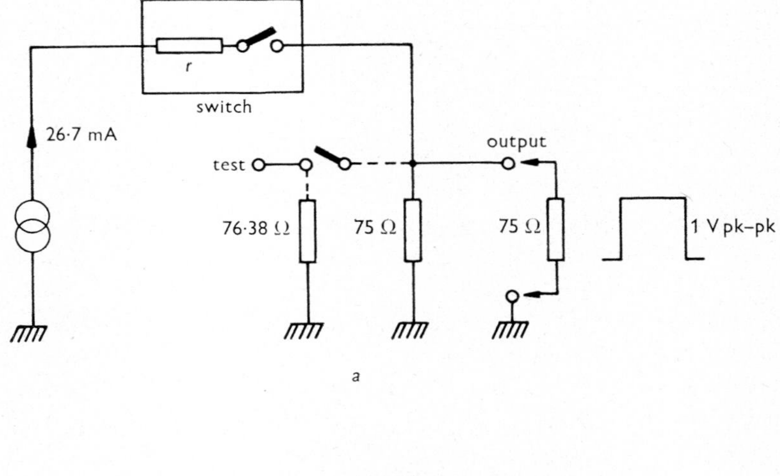

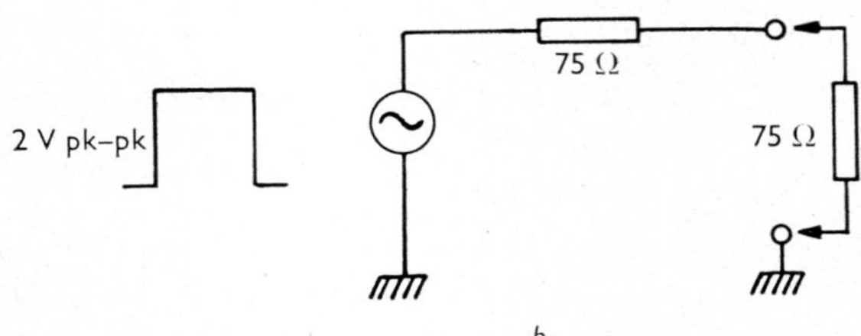

Fig. 2.6 Precision square-wave generator
a Basic circuit
b Equivalent circuit

This is of negligible importance, by virtue of the extremely high source resistance of the current generator, which effectively swamps *r*. The second error is the result of the switch presenting a finite resistance in the 'break' position, so that the output voltage over that region is not exactly zero. This is obviously more important with a semiconductor switch than with a relay, but, with careful design, it also can be made negligibly small. The former type of switch is greatly to be preferred, since more convenient repetition rates are achievable and since jitter is no longer a problem.

With this standard square wave, extremely accurate measurements of level can be carried out on waveforms, although it does not lend itself to the measurement of sinusoids such as might be required for the plotting of the insertion-gain/frequency-response of a circuit. An important application is the measurement of chrominance–luminance gain errors (Section 11.4.1); and, for this, a precise means of measuring alternating voltage is also required.

There is no available standard of alternating voltage, and it is again necessary to employ the Weston standard cell as a reference. For this to be practicable, a thermal transfer element of some kind must be introduced, resulting in a true r.m.s. measurement of the unknown alternating voltage. For this purpose, the high-frequency thermocouple is by far the most convenient (Hermach and Williams, 1960). Suitable types are those in which the heater and couple are in point thermal contact only via a miniature glass bead, and the whole thermocouple is sealed within an evacuated glass envelope. The use of very fine-gauge wire for the heater and a precise balance of the conductors make the couple e.m.f. virtually independent of frequency up to 20 or 30 MHz, and, with some types, the error is still small at 100 MHz or over.

The basic measurement principle consists in applying the unknown sinusoid to the heater of the thermocouple and noting the output current from the couple. The unknown is then removed and replaced by a direct voltage, which is varied until the same output current is obtained as in the previous case. The power dissipated in the heater is the same in each instance, from which the peak-to-peak voltage of the alternating input—for this is what is required in television applications—is derived on the assumption that its harmonic content is negligibly small; this is so in practice, since these test frequencies should always be filtered. The direct voltage used is then checked against a standard cell, so that the a.c. input is also compared with the cell, but, of course, at two removes.

A drawback of the thermocouple is the large variation in the heater resistance with input level. This is overcome by comparing the two inputs at the same couple current, but the problem still remains of adjusting the heater resistance precisely to 75 Ω. This is met first by operating at a fixed input level, and secondly by adjusting the heater to the required value at that level; this is a simple operation when a return-loss bridge is available (Chapter 8), and uses a series or shunt resistor as required. Large amounts of compensation must be avoided because of the very rapid loss of sensitivity which ensues.

The indirect nature of the thermocouple-transfer principle naturally introduces some uncertainty, but it appears that this can be kept small by a suitable design. According to Hermach and Williams (1960), it is possible under laboratory conditions to achieve an uncertainty of measurement within ±0·005% up to 20 MHz, although, under equipment-maintenance conditions, it is probably advisable to assume that this will be degraded by a factor of 4 or 5. At the risk of appearing to labour the point, it ought to be pointed out that hard-won accuracy can easily be lost by lack of attention to lead lengths. The loss, at 4·43 MHz, of a high-grade, double-screened coaxial cable is 0·005 dB/ft, so that an assumed error of ±0·02 dB may actually be doubled by the insertion of a 4 ft length of cable in series with the measurement equipment.

The utilisation of the output current from the couple to obtain a visual indication is worth a short discussion, since the use of a sensitive and fragile microammeter is inadvisable; besides, it is highly desirable for either a calibration or a measuring instrument to indicate directly the deviation from the correct value. The preferred technique is the use of an electronic chopper to convert the d.c. output into a square wave, which may be amplified stably to any appropriate extent.

The need for an amplified current, which is zero for the correct input to the thermocouple, and which varies in polarity with the sense of the deviation, can be met as follows. An opposing current is introduced in series with the couple, so that the resulting current is zero for the reference direct input voltage applied to the thermocouple heater. This resultant current is converted into a square wave by a form of modulator which is driven by a separately generated carrier wave of rectangular form. After suitable amplification, the square wave is synchronously demodulated with an identical modulator and the same carrier input. The amplified current thus regenerated is then applied to a meter which can now be relatively insensitive. In one practical realisation, the amplification was such that the whole of the meter scale, about 2·5 in long, corresponded to $\pm$0·5 dB with respect to the calibrated value, i.e. about 0·2 dB/in, which affords excellent discrimination.

Such a thermocouple transfer instrument is, however, not well adapted to operational work; and, in the development of a 'work horse' type of colour calibrator for checking the chrominance–luminance inequalities of waveform monitors (Section 11.4.1), the control of the colour subcarrier amplitude was effected by means of a precision Zener diode in the oscillator circuit. The resulting slight harmonic distortion was removed by a simple filter, the loss of which was allowed for. However, the thermocouple principle was utilised in a precision calibrator, with which the periodic checking of level-measuring equipment of all types can be carried out very expeditiously. Since the reference circuitry was already present, it was simple to increase the general usefulness of the instrument by also building into it the facility for generating precision level-measuring signals useful for special purposes.

In the development of the calibration standard, it was found to be necessary to construct a highly simplified version containing nothing but the thermocouple and its d.c. reference, with the wiring reduced to the absolute minimum. It could then serve as a 'standard' against which the calibrator itself could be checked, so that account could be taken, during all phases of development, of the small, but nevertheless cumulative, errors from the switching and extra wiring needed to provide a simple and convenient routine for calibration purposes. This instrument has also been found very useful in its own right for the measurement of a.c. level close to 1 V pk–pk. Estimates have suggested that its accuracy ought to be within $\pm$0·02 dB ($\pm$0·2%) at least to 30 MHz and probably beyond 100 MHz, for measurements on pure sinusoids, but it has not been possible so far to substantiate this figure beyond doubt by measurement.

The great drawback of the thermocouple instrument is its fragility. To maintain the sensitivity, thermocouples are most often operated not far from the burnout point of the fine wire employed for the heater, and a momentary inattention in setting the input level or even a switching surge can suddenly place one in the vexatious position of having to replace the thermocouple and entirely recalibrate the instrument.

In view of this, it is common to replace the thermocouple instrument for operational work by versions employing the square-law charac-

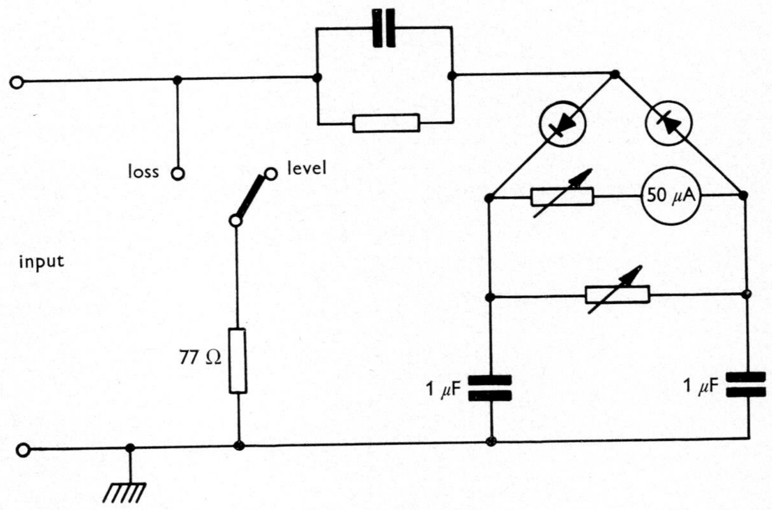

Fig. 2.7 Simple levelmeter

teristic of some semiconductors. An instance of an old design still performing yeoman service is shown in Fig. 2.7. It employs semiconductor diodes in a fullwave circuit at very low current, where the characteristic is very close to square law. The meter shunt provides a fine adjustment of sensitivity. The series *CR* circuit was originally included to equalise the response at the top end of the videoband; but with modern diodes, little or no correction is required. The input resistance of the rectifier circuit is about 3000 Ω, and requires a shunt 77 Ω resistor to provide a good 75 Ω input impedance; by itself, it is just about acceptable for what is often called, rather misleadingly, a 'level' measurement, i.e. one not terminated by the instrument.

Such instruments are usually restricted to a small range about the nominal value, say ± 2 dB, which helps with the maintenance of a good square-law characteristic, as well as avoiding too great a change in the impedance presented by the rectifier circuit. One notable fault is a sensitivity to temperature, which can be made negligible by maintaining the diode temperature constant in a miniature component oven. As an alternative to the rectifier diodes, field-effect transistors might be employed with some advantage.

2.6 References

CCIR (1966): Documents of the XIth Plenary Assembly, Oslo, Recommendation 421–1

HERMACH, F. L., and WILLIAMS, E. S. (1960): 'Thermal voltage converters for accurate voltage measurements to 30 MHz', *Trans. Amer. Inst. Elect. Engrs.*, **79**, Pt. I, pp. 200–206

QUINN, S. F., and NEWMAN, P. M. (1965): 'Distribution of average picture levels in television programmes', *Electron. Lett.*, **1**, p. 261

SHEA, T. E. (1929): 'Transmission networks and wave filters' (Van Nostrand)

WEAVER, L. E. (1961): 'The accurate measurement of video levels', *EBU Rev.*, Pt. A (Tech.), (69), pp. 2–4

3 RANDOM NOISE

3.1 General

The measurement of signal/random-noise ratio is one of the most important in television, and yet, paradoxically, it is the one which is all too often carried out to a much lower standard of accuracy than is required.

There are, of course, special difficulties associated with this measurement. The waveform of the random noise has no fixed levels which may easily be measured on a waveform monitor, since it is an ensemble of pulses whose amplitude distribution as a function of time is only known statistically. The final effect of the noise, i.e. the subjective impairment of the viewer's picture which it causes, is a function of the spectral distribution of the random noise associated with the video signal driving the picture tube; and this in turn depends on the picture source and the nature of the total transmission path between it and the receiver. With a colour signal, the high-frequency components of the noise which fall into the chrominance channel are demodulated in the receiver to provide comparatively low-frequency fluctuation effects in the colour picture, so that noise components which have comparatively little significance in a monochrome signal assume greater importance in a colour signal.

A further complicating factor is the frequent need to make a measurement in the presence of a video signal, either because it is necessary to carry out the measurement during programme time or because the circuit or apparatus concerned does not function correctly in the absence of synchronising information or blanking. Picture sources, e.g. cameras, must be blanked to remove spurious signals, and it may also be necessary to check the noise at various levels of illumination.

Not surprisingly, a great many methods have been devised for the measurement of random noise, and, in particular, a great deal of work has been carried out with the object of making the measurements as independent as possible of the nature of the picture source and the transmission path.

3.2 Definition of signal/noise ratio

The signal/noise ratio for a video signal in decibels is defined (CCIR, 1966) as $20 \log_{10} (E_v/E_m)$, where E_v is the peak-to-peak amplitude of the picture signal, i.e. between white level and blanking level, and E_m is the r.m.s. random-noise voltage. This assumes that frequency filtration is used to exclude noise below 10 kHz and above the nominal upper limit of the videoband. The latter is taken to be 5·0 MHz for system *I*, the 625-line system used in the UK. The lower-frequency filtration ensures that hum and other parasitic signals, such as invertor noise, which are measured separately, are not counted as random noise, and the upper limit removes noise which has no relevance to the video signal. It is also required in principle that a weighting filter be included in the measuring path.

3.3 Types of random noise

Random noise may be classified into a number of broad types which are distinguished by their differing spectra, 'spectrum' in this context being preferably interpreted as the distribution with frequency of the relative noise power in an infinitesimal frequency band, measured over a sufficiently long period. The latter restriction is required, since the distribution of the spectral components is statistical. However, the voltage spectrum is also used, i.e. the distribution with frequency of the r.m.s. voltage corresponding to the noise power.

The two most basic types of random-noise spectra encountered with video signals are 'flat', or 'white', noise in which the power spectrum is constant with frequency, and 'triangular' noise, where the power is directly proportional to the square of the frequency; this means, of course, that the r.m.s. voltage is directly proportional to the frequency; hence the name. However, these are seldom, or never, encountered in their pure form, since in practice the basic noise spectrum is modified, often substantially, by the characteristics of the apparatus or circuits concerned.

For example, the Johnson noise in a resistor is 'flat', and an image-orthicon tube generates substantially flat noise whose spectrum is subsequently lifted over the upper part of the band to a maximum of a few decibels owing to the effect of the aperture corrector. A photo-conductive tube such as the plumbicon or vidicon in a camera channel produces triangular noise which tends to be flattened over the lower part of the range by the preamplifier circuits, but which is lifted at the

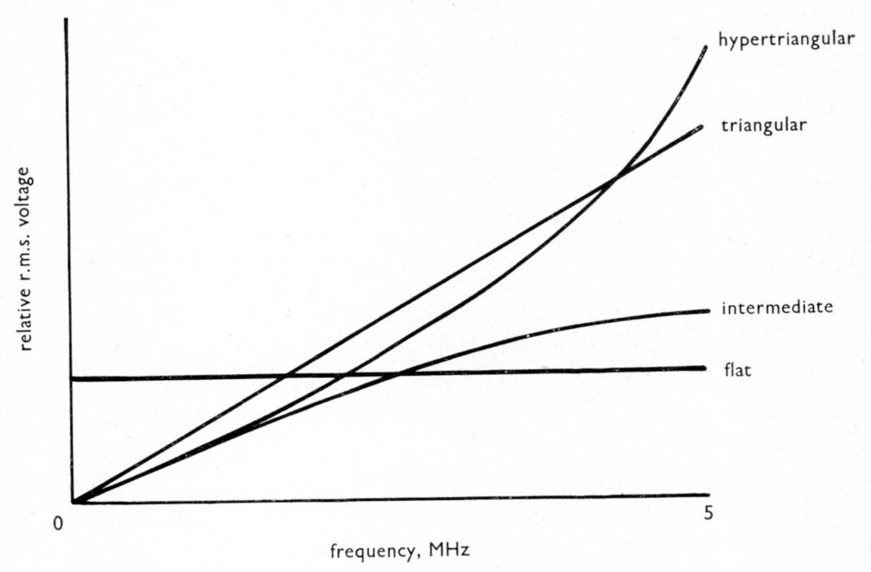

Fig. 3.1 Typical random-noise spectra

higher frequencies by the aperture corrector (Fix and Kaufmann, 1960). A spectrum occurs on short-haul links employing direct video transmission, as a result of the cable correction, in which the distribution is approximately triangular at the lower frequencies and approaches a parabola at the higher frequencies; random noise of this type is often termed 'hypertriangular'. Microwave links generate triangular noise whose spectrum is decreased proportionally with increasing frequency as a result of the emphasis and de-emphasis networks; such noise has been termed 'intermediate' (Allnatt and Prosser, 1966). Examples of these curves are given in Fig. 3.1.

3.4 Filtration

3.4.1 Band limitation

The bandwidth of the random noise accompanying a video signal may be considerably greater than that of the signal, since the noise bandwidth is determined by the apparatus and circuits handling the signal, whereas the video bandwidth is primarily determined by the resolution of the picture-generation equipment. Eventually, of course, the noise and signal bandwidths are equalised by the frequency limitation introduced by the transmitter and the viewer's receiver; they may even be equalised at a previous stage, but it is evident that a serious error is possible in the measurement of a signal/noise ratio if arrangements are not made to make the noise and signal bandwidths equal to the point of measurement.

For this reason, the use of a sharp-cutoff lowpass filter is recommended by the CCIR for use in front of the measurement equipment (CCIR, 1966). The attenuation/frequency curve of this filter and the element values for an effective bandwidth of 5 MHz (625-line: system *I*) are given in Fig. 3.2. The values can be scaled appropriately for other systems. The insertion loss of the filter over the major part of the passband is very small and can be considered to be negligible for most purposes. Note that the noise bandwidth for system *I* is taken to be 5 MHz and not 5·5 MHz, the nominal width of the videoband, to make the filter the same as in other 625-line systems.

3.4.2 Junction filter

During the measurement of random noise, it is advisable to remove the lowest frequencies by means of a highpass filter, to avoid including with the random noise any nonrandom forms of interference of relatively low frequency, such as supply hum with its higher harmonics, higher-frequency power-supply interference such as occurs with transistor invertors and which may have a frequency of several kilohertz, and also some types of quasi-aperiodic interference such as teleprinter clicks.

The CCIR has accordingly recommended the use of a highpass-lowpass filter combination (CCIR, 1966) with a crossover point at 10 kHz. This is illustrated in Fig. 3.3. The highpass section is employed for the measurement of random noise and the lowpass

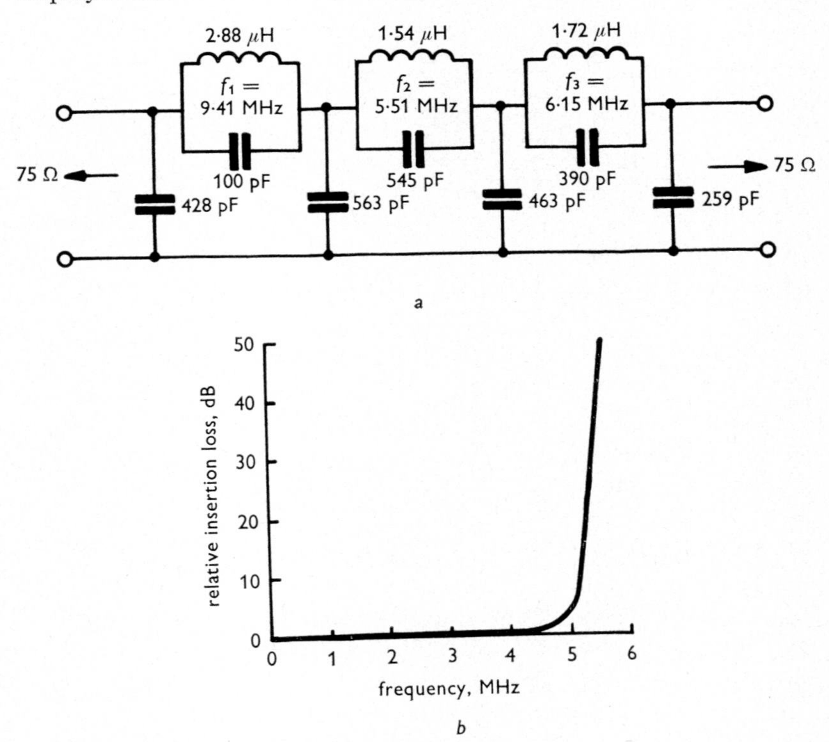

Fig. 3.2 CCIR band-limiting lowpass filter for 625-line systems
a Circuit of filter, *b* Insertion loss of filter

Capacitances, including strays, to be within ±2%
Inductances to be adjusted to make insertion loss maximum at indicated frequencies
Inductor Q factor = 80–125 at 5 MHz

section for the measurement of supply hum and other types of low-frequency interference. An advantage of this type of network is that the input terminals present a constant resistance at all frequencies when the output terminals are correctly terminated, and it may

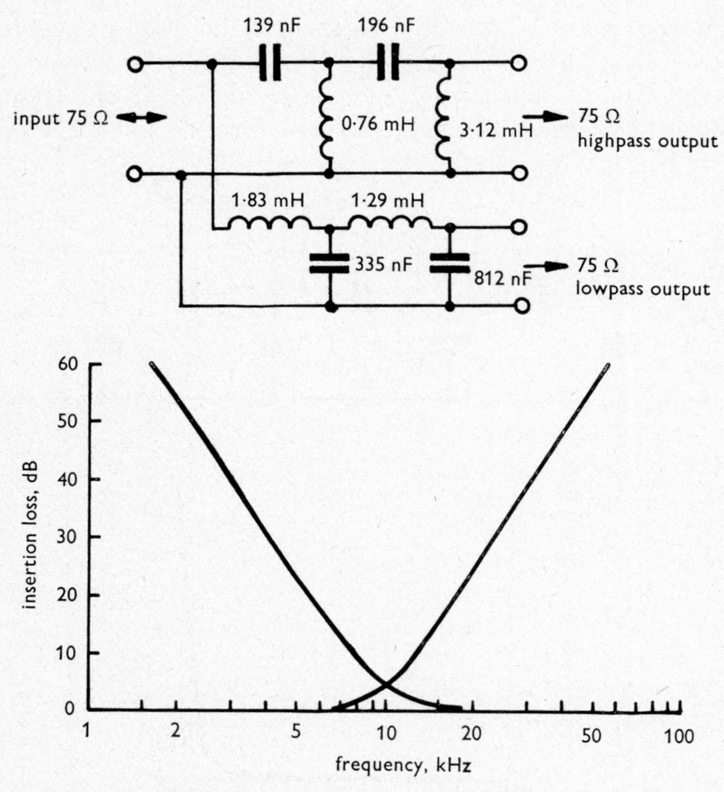

Fig. 3.3 CCIR junction filter for random-noise measurement
Inductances within $\pm 2\%$
Capacitances within $\pm 5\%$
Inductor Q factor $\geqslant$ 100 at 10 kHz

accordingly be used in series with the band-limitation filter, described in Section 3.4.1, without the risk of introducing irregularities due to reflections at the filter terminations.

3.5 Weighting networks

The precise shape of the spectrum of the random noise would mainly be of academic interest if it were not for its marked effect on the subjective degradation of the picture. The author found that the random-noise level with flat noise could be increased with triangular noise by about 4 dB with 405 lines and 6 dB with 625 lines for the same subjective picture degradation (Weaver, 1959). Prosser and Allnat (1966) found with 625 lines that, for a monochrome picture, triangular noise is less disturbing than flat noise by 5–8 dB, depending on the noise level.

The reason for this lies in the difference between the apparent visual structures of the two types of noise when they are viewed on the receiver screen. As might be expected from its very much higher frequency content, triangular noise has a much finer visual structure than flat noise and is less objectionable, because of the smaller 'grain size' which is less easily resolved by the eye. When the level of triangular noise is particularly high, areas of mean grey appear to 'sparkle' in a very characteristic manner.

On the other hand, flat noise possesses a larger apparent grain structure, and, with high levels, a random correlation takes place between the 'grains' in adjacent lines, causing a characteristic swirling effect comparable to the movement of solid particles in a boiling liquid.

The different subjective appearances of the same level of various types of random noise poses a considerable problem when long-distance links have to be measured, since the total random-noise power is the sum of a number of contributions from the signal source and from all the circuits which make up the link, each of which may affect the final picture by a different amount. It would therefore seem to be necessary to know the nature and the amount of the contribution of each type of noise, to arrive at a figure by means of which different circuit configurations can be compared on a reasonable basis. This would be extremely onerous, even if it were generally practicable.

The solution to the problem appears to have first been suggested by Mertz (1951); i.e. the use of weighting network analogous to that used for many years for noise measurement in speech and music transmission.

It operates as follows. It is known that a given level of sinusoidal interference causes a progressively smaller picture impairment as the frequency is increased; and, conversely, the permissible amplitude and delay distortion in a monochrome picture for a given degree of picture impairment increases progressively as the frequency at which the distortion occurs is increased. These effects can be explained in very broad terms by the decreasing size of the interference pattern or the contribution to the waveform distortion at the higher frequencies. One would then expect that the spectral components of the random noise would follow the same general rule, although not necessarily precisely the same law. The viewer's greater subjective tolerance of triangular noise, discussed earlier, appears to bear this out.

It then seems reasonable to suppose that, if random noise were weighted before measurement by predistorting its spectrum by the curve defining the relative sensitivity of the eye to the spectral components of the noise, equal powers of the different types of random noise would give rise to the same overall picture impairment, and there would no longer be any need for a knowledge of the spectrum of the noise. This is found to be true, and a great deal of work has gone into the design of a simple and easily reproducible noise-weighting network to provide the required amplitude characteristic.

A number of workers have accordingly attempted to find the most appropriate weighting curve by means of subjective tests using filtered bands of random noise, spread over the range of video frequencies (Barstow and Christopher, 1954; Gilbert, 1954; Kilvington *et al.* 1954; Maarleveld, 1957; Goussot, 1959; Müller and Demus, 1959; Yamaguchi, 1967). The agreement between the various measurements is not markedly high, probably owing to the differing experimental conditions, and in particular to the use of different bandwidths for the random noise. A recommendation was eventually issued by the CMTT (CCIR, 1966) based on the work of Müller and Demus (1959). The form of the proposed weighting network is shown in Fig. 3.4. It is a simple bridged-T constant-resistance equaliser with an insertion-loss characteristic given by

$$A = 10 \log (1 + \omega^2 \tau^2) \text{ decibels}$$

where τ is a time constant. The original recommendation proposed the same time constant $\tau = 0{\cdot}33\ \mu s$ for all 625-line systems as well as for 405 lines, and $\tau = 0{\cdot}166\ \mu s$ for 819 lines. This has now been changed for system I (CCIR, 1966), i.e. the 625-line system with a nominal video bandwidth of 5·5 MHz as used in the UK, to $\tau = 0{\cdot}20\ \mu s$. (Note: The nominal noise bandwidth is 5 MHz.)

This value for τ for system I, and incidentally the whole principle of the use of a weighting network, has now been checked by refined subjective testing methods (Prosser and Allnatt, 1966). It was found that the differences between the weighted flat and triangular signal/noise ratios for the same subjective impairment were not greater than

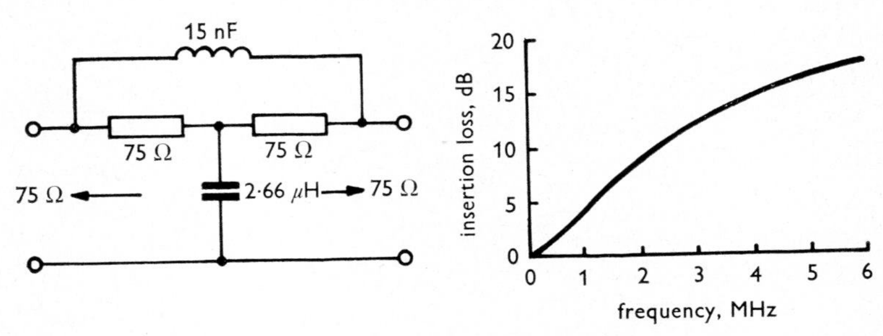

Fig. 3.4 CCIR random-noise-weighting network
Inductance, capacitance and resistance tolerances to be within ±1%
Q factor ⩾ 25 at 8 MHz
Insertion loss $= 10 \log_{10}\{1+(2\pi\tau f)^2\}$ decibels; $\tau = 200$ ns

±0·5 dB over the range of impairments except where the impairment was very small. The difference then increased to about 2 dB. It seems unlikely that a better performance could be obtained with a simple network.

The weighting network for the chrominance channel which has been proposed by the CCIR for system I is shown in Fig. 3.13. It is, of course, of bandpass form, and the shaping of the passband has been so chosen that the sidebands of the modulated subcarrier experience the same weighting up to about ±1 MHz from the carrier frequency, as is given by the monochrome network, which also serves as the luminance-weighting network with a colour signal. As with the monochrome measurement, the weighted signal/noise ratio is expressed as

$$20 \log_{10} \frac{\text{peak-to-peak picture amplitude}}{\text{r.m.s. weighted noise voltage}}$$

Subjective tests (Allnatt and Prosser, 1966) have confirmed the suitability of this weighting network, and, furthermore, have shown that the same picture impairment is produced by a luminance- (i.e. monochrome-) weighted signal/noise ratio of x decibels as by a chrominance-weighted signal/noise ratio of $x+6$ decibels. This offers the opportunity of deriving a single luminance–chrominance-weighted signal/noise ratio by combining the outputs of the luminance-weighting network and the chrominance-weighting network in series with a 6 dB pad and measuring the total weighted noise power.

On the other hand, the system I chrominance-weighting network is complex, and accordingly rather expensive, and the use of two separate networks makes it operationally cumbersome. There is a distinct need for a simple, combined luminance and chrominance network for routine-measurement purposes, perhaps on the lines of that proposed for the 525-line system M (CCIR, 1966).

3.5.1 Loss due to weighting network

The insertion-loss characteristic of the monochrome weighting network is given in Fig. 3.4. It is evident that the use of the network in front of a device which measures signal/noise ratio will result in a change of the measured signal/noise ratio as a function of the characteristics of the noise. The change in the reading can easily be calculated for flat and triangular noise whose power spectra are clearly defined.

Flat noise

Let the power in an infinitesimal band of frequencies dF be $K\,dF$, and let the nominal upper limit of the videoband be F_c megahertz. Then

$$\text{Loss (dB)} = 10 \log_{10} \frac{K\int_0^{F_c} dF}{K\int_0^{F_c} \frac{dF}{1+4\pi^2F^2\tau^2}}$$

$$= 10 \log_{10} \frac{2\pi F_c\tau}{\tan^{-1} 2\pi F_c\tau} = 10 \log_{10} \frac{\alpha F_c}{\tan^{-1}\alpha F_c}, \text{ say.}$$

Triangular noise

Let the noise power in the band dF be K_2F^2dF. Then loss due to the network is

$$10 \log_{10} \frac{K_2 \int_0^{F_c} F^2 dF}{K_2 \int_0^{F_c} \frac{F^2 dF}{1+4\pi^2 F_c^2 \tau^2}}$$

$$= 10 \log_{10} \frac{\frac{4}{3}\pi^2 F_c^2 \tau^2}{1 - \frac{\tan^{-1} 2\pi F_c \tau}{2\pi F_c \tau}}$$

$$= 10 \log_{10} \frac{\frac{1}{3}\alpha^2 F_c^2}{1 - \frac{\tan^{-1} \alpha F_c}{\alpha F_c}}$$

For system *I*, $\alpha = 0{\cdot}2\pi$ μs and $F_c = 5$ MHz, so that the losses of the network to flat and triangular noise are 6·5 and 12·3 dB, respectively.

3.6 Measurement

3.6.1 Visual method: simple form

The oldest, and probably still the most widely used, method of measurement, particularly under operational conditions, is the estimation of the quasi peak-to-peak value of the noise when the signal is observed on the screen of the waveform monitor. The popularity of the method is probably due, in the first place, to the fact that no extra equipment is needed; the waveform monitor is almost certain to be available at the point of measurement in any case, since it is required for other purposes. Secondly, measurements can be carried out in the presence of synchronising pulses or test signals, e.g. a sawtooth or a bar of variable height if measurements are required at various grey levels, or on lines in the field-blanking interval if the measurement must be made during programme time.

When a field- or line-repetitive signal is observed on the waveform monitor, the random-noise voltage gives rise to a bright, fuzzy band whose brightness appears to taper off fairly quickly at the top and bottom owing to the fact that the amplitudes of the individual pulses which make up the random noise have a Gaussian distribution, so that the average number of pulses per second falls exponentially with increasing pulse amplitude. The quasi peak-to-peak amplitude of the display is estimated, and its ratio to the amplitude of the picture component of the signal is calculated. The signal/noise ratio in the standard form is then derived by adding a constant, which may vary between 14 and 18 dB, and which represents the ratio between the quasi peak-to-peak voltage and the r.m.s. voltage of the noise (CMTT, 1968). The 14 dB corresponds to a choice of a quasi peak-to-peak amplitude, so that it is only exceeded in either direction for 0·5% of the time, and 18 dB to an amplitude exceeded only for 0·006% of the time.

The disadvantage of this simple method lies essentially in the difficulty of measuring the quasi peak-to-peak amplitude in a consistent and reliable manner, since the display brightness and contrast, the repetition rate, the judgment of the observer and many other factors affect the accuracy of measurement. The apparent change in amplitude with a change in the display brightness is clearly demonstrated in Fig. 3.5. Furthermore, to measure numerically large signal/noise ratios, the waveform-monitor amplifier will have to accept the simultaneous presence of a video-signal component of considerably greater amplitude, which may well affect the accuracy of measurement.

Claims are sometimes made that a very high precision is obtainable with this method, e.g. CMTT (1958*a*), but they are not supported by the author's experience. In a series of tests carried out by the author on ten skilled engineers, who were asked to measure signals presented to them with the same waveform monitor, but selecting their own display conditions, it was found that a given individual could almost invariably be relied on to repeat readings within about ± 1 dB, but that the total range of the readings might well be as much as 5 dB. Similar results have been reported by Putman (1966) from tests made on 45 engineers, where the maximum variation was found to be 6 dB. This wide disagreement between observers can only confirm what has been indicated above: namely that the readings given by the basic visual method are inevitably affected by many factors, personal, instrumental, experimental, environmental and so on, according to circumstances, all of which inevitably contribute to a greater or lesser extent to the statistical spread of error. Strict standardisation of operating conditions can

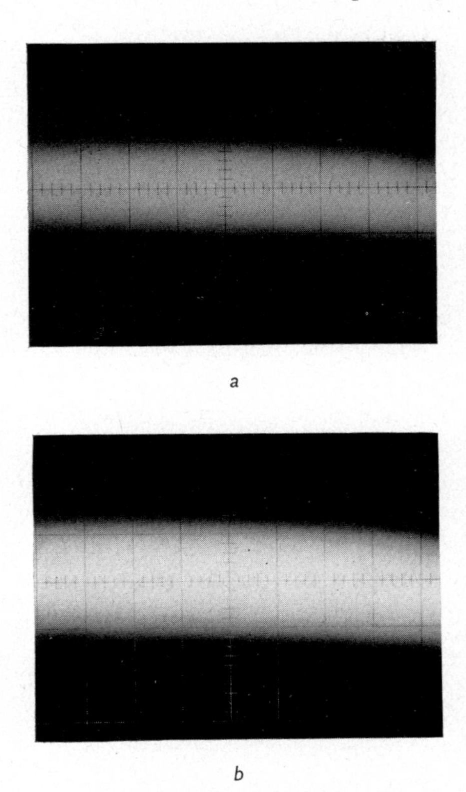

Fig. 3.5 Effect of exposure time on the apparent amplitude of white noise
a Normal exposure
b 4-times-normal exposure

reduce the variations, but one can never rely on the readings to the extent to which this is possible with suitable objective methods.

The only device the author has found to improve the accuracy of the simple visual method in any reliable manner is the use of a storage oscilloscope having a fairly long storage time, since the integration of the noise peaks results in a fairly well defined band of displayed noise from which the quasi peak-to-peak value may be obtained. However, this is not very practicable for operational purposes.

3.6.2 Tangential methods

A very useful improvement in the basic visual method can be obtained for comparatively little extra complication by a technique described by Garuts and Samuel (1969) under the name of 'tangential noise measurement', although the principle seems to have been anticipated by Krivocheev (CMTT, 1958).

As advocated by Garuts and Samuel, it is only applicable to instances where the noise of a circuit is available, in the absence of a video signal. Rather as in the BBC method of level measurement (Chapter 2), the signal to be measured is linearly added to a square wave, the amplitude of which becomes a criterion with which the unknown level is compared.

The procedure will be explained by means of the waveform photographs of Fig. 3.6 In (*a*) is shown the random noise to be measured, which we will assume corresponds to a video signal at the point of standard level, while the calibrating square wave appears in (*b*). The two waveforms are added via a constant-impedance splitting pad, with a calibrated attenuator in series with the square-wave path, so that, when the amplitude of the latter is sufficiently large, the combined signal has the appearance of (*c*) in which the two bands of noise are separated by a dark interval. As the amplitude of the square wave is progressively reduced, the central dark band narrows and at the same time becomes less intense, until eventually the critical condition shown in (*d*) is reached, at which point the brightness in the central region appears to be uniform. Any further decrease in the square-wave amplitude results in a brightening of the central area.

This is explained as follows. Assuming that the random noise has a Gaussian distribution of amplitudes, the brightness distribution of the corresponding display on a waveform monitor will also be of Gaussian form. Now it is not difficult to show that the sum of two Gaussian distributions with the same standard deviation is flat over the central region when, and only when, the two curves are separated by two standard deviations, so that, in the practical case, the critical square-wave amplitude corresponds to two standard deviations and hence, by a well known relation, to twice the r.m.s. voltage. From this

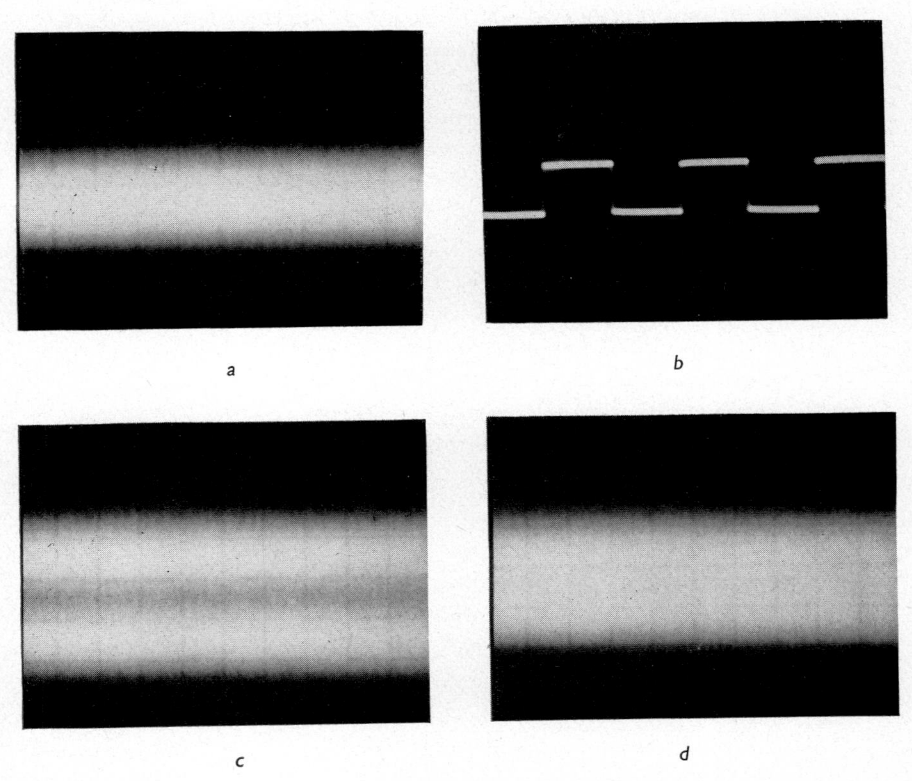

Fig. 3.6 Tangential method of random-noise measurement

a Noise level to be measured
b Calibrating square wave
c Addition of (*a*) and (*b*) at start of measurement
d Display for critical adjustment

square-wave amplitude, the signal/noise ratio can then obviously be calculated.

A useful practical improvement when the noise to be measured corresponds to standard video-signal level is to make the square-wave amplitude 1·40 V into 75 Ω, when the attenuator reading for the critical adjustment indicates the signal/noise ratio directly.

The important feature of this method is the provision of a definite criterion for the measurement which, in principle, should be independent of most factors which adversely affect the accuracy of the simplest visual method. At the same time, it avoids the need for an arbitrary factor to convert from quasi peak-to-peak to r.m.s. To ascertain whether the expected improvement is actually obtained in practice, a series of tests was carried out with the co-operation of a number of experienced engineers, who were asked to use the technique to measure a series of levels of random noise of three characteristic types: namely, flat noise, triangular noise and noise weighted with the standard CCIR network. The actual levels were, of course, known to the engineer in charge, but not to the observers.

It was not found possible at the time to carry out the total of the very large number of assays required for a really thorough examination; so the results are not regarded as conclusive, but are nevertheless of interest broadly. First of all, it became clear that satisfactory readings are not obtained unless a standardised procedure is adopted in which the critical adjustment is approached from a square-wave amplitude which is evidently too large; i.e. one commences with a definite dark central area. The reverse procedure gives very inconsistent results. Under these conditions, it was found that the spread of the measured values with any single observer is likely to be within ±0·5 dB, and the spread between different observers is not likely to exceed about 1·5 dB. This is not only an appreciable improvement over the simpler method, but is also slightly better than was claimed by Garuts and Samuel.

It has also been found by experiment that the tangential method can also very profitably be extended to the measurement of noise on a video signal containing a full-width bar ('lift') or black level only. The application of this to the measurement of noise on lines in the field-blanking interval is of especial importance, since it provides an improved visual technique for the estimation of signal/noise ratio under in-service conditions in the absence of more sophisticated equipment.

For this purpose, one or more lines in the field-blanking interval are required which must not have been tampered with, i.e. neither erased nor replaced, at any point downstream of the location on the distribution network to which the signal/noise ratio is required to correspond. The noise on such lines can only be observed on a waveform monitor by the use of a line-selection process, and here the simpler visual method is at a disadvantage. In its more usual applications, the display consists of the superposition of a large number of noise signals, which is of some assistance in the determination of the quasi peak-to-peak voltage; but, in a line-selected display, the noise takes on a quite different appearance, which can only be described crudely as a highly erratic sine wave. This is explained by Rice's theorem (Rice, 1945), which states that band-limited noise with a cutoff frequency of f_c passes through zero, on an average, at a rate of $1{\cdot}16 f_c$; in other words, each random-noise peak tends to take on the appearance of a halfcycle of a sinusoid of frequency $0{\cdot}58 f_c$. As a result, the determination of the quasi peak-to-peak voltage is even less satisfactory.

The basic measurement procedure is precisely as described above for noise in the absence of a video signal, except, of course, for the display, which must be arranged so as to show only the area of the signal over which the noise is to be measured. For this purpose the waveform monitor will require a suitable trigger derived from the original video signal before addition to the calibrating square wave. However, there is now one important difference in that the square-wave-repetition frequency is more critical because of the presence of the video signal, and it has been found that this should either be in the vicinity of 70 Hz or above, say, about 50 kHz. In the former, the precise value for optimum ease of adjustment is best found experimentally; but, in the latter, the precise value is relatively unimportant. Most users also find it beneficial to restrict the height of the display.

The sensitivity of adjustment in this measurement is every whit as good as with noise in the absence of a video signal, as is borne out by the photographs of Fig. 3.7, which illustrate, respectively, an experimental clean signal with added calibrating square wave, the same with superimposed random noise (note the well defined dark interval), and

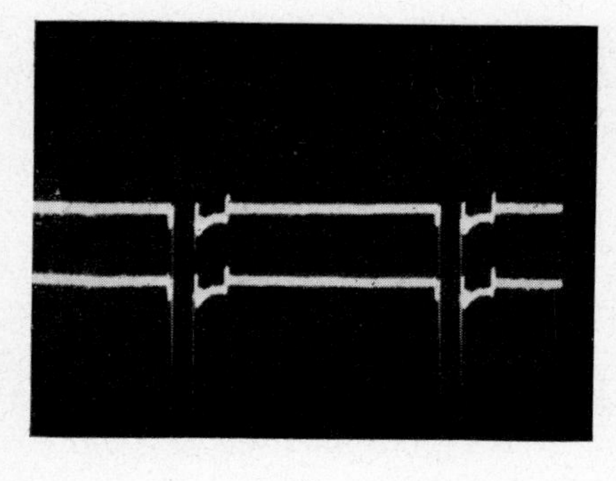

a

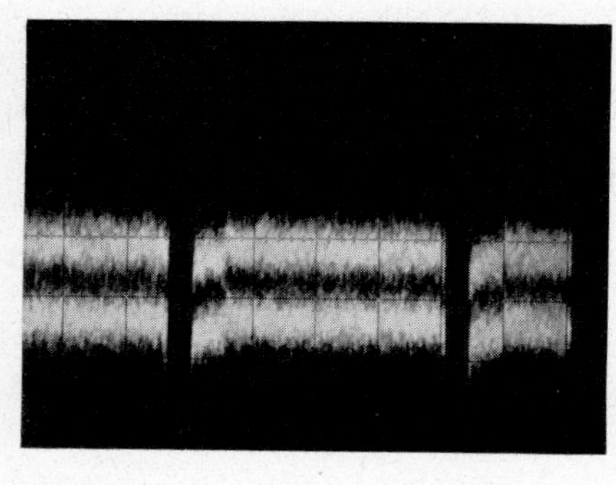

b

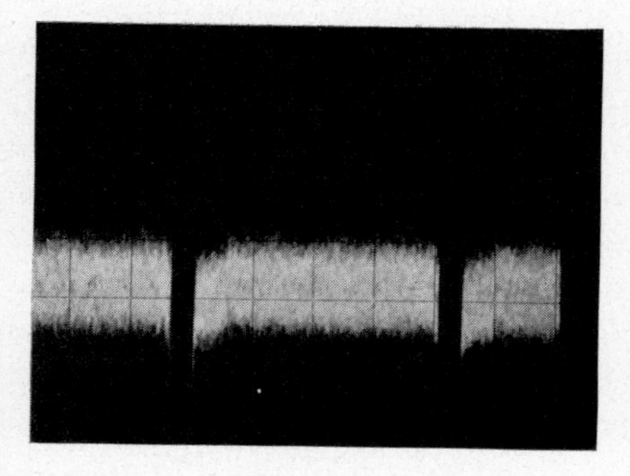

c

Fig. 3.7 Tangential method for noise measurements in field-blanking interval

- *a* Noisefree signal with added square wave
- *b* Signal shown in (*a*) with noise present
- *c* Display for critical adjustment

finally the waveform in the critical-adjustment condition. It may be worth pointing out, in view of the ever increasing demand for lines in the field-blanking interval for insertion purposes, that it has also been found perfectly practicable to carry out this measurement using, not an empty line in this region, but the equalising pulses.

Finally, in spite of the improved accuracy compared with the simplest visual method, it should not be supposed that the tangential method is able to replace objective methods of measurement where precision and repeatability are of importance. The accuracy of all such visual methods can only be expressed statistically, so that it is not possible to guarantee that considerable errors will not occasionally occur.

3.6.3 Comparison-patch method

A method for monochrome pictures, which at first sight appears to be very promising, makes use of a semiphotometric technique. It is believed to be due to Krivocheev (CCIR, 1959). It requires a picture monitor, a calibrated source of random noise, and equipment closely resembling the standard 'inlay' equipment, but with less extensive facilities.

A suitable region of approximately uniform tone is found on the picture monitor, with an area approximately one-hundredth of the useful picture area. The picture is removed from a rectangle within that area, and is replaced by a variable pedestal, which is then adjusted to match its immediate surroundings photometrically. If this area is too large, it is difficult to find a region of sufficiently even tone. A square with a side equal to one-tenth or one-twentieth of the picture height seems to be about right.

When the brightness of the patch has been matched to that of its surroundings, random noise is added to the patch, hitherto substantially free from random noise, until it is judged that the interfering effect of the noise within the patch is equal to that along its edges. The noise power added to the inlaid patch is then said to be equal to the random-noise power of the signal at the level of grey used; hence the signal/noise ratio can be calculated.

The method was investigated by the author. It was first found that the brightness match could be carried out with very good accuracy at a brightness not too close to picture black. At about 50% of white level, corresponding to the background level in a standard test card, ten engineers achieved the same reading within ±1%. Unfortunately, the spread of the results was very much wider for the random-noise measurement.

Using the originally recommended flat noise within the inserted patch, it was found that, with flat noise on the picture, the standard deviation of results was worse than 1 dB, depending on the actual noise level; but, with triangular noise, the standard deviation increased to 3·5 dB. The measured figures were roughly constant for the range of signal/noise ratio 25–35 dB; at numerically higher signal/noise ratios the error increased, but the reading became less dependent on the type of noise.

An attempt was then made to match the spectrum of the inserted noise to that of the picture noise. When a good match could be obtained, the improvement was significant; but in general, the spectrum of the noise was only known roughly, and an improvement to a standard deviation of 2 dB was found in an average instance. It might be mentioned that individuals were found to be very consistent on the whole; some could always be relied on to be within 1 dB or so, whereas two out of the ten subjects, in other respects very trustworthy observers, could be 5–6 dB in error.

It was concluded that

(*a*) the method is not sufficiently reliable

(*b*) four operational adjustments (position of patch, brightness, noise spectrum and noise level) are too many

(*c*) the amount of equipment required is considerable

(*d*) errors are introduced by hum and line and field tilt.

3.6.4 Oscilloscopic comparison

In a variant of the above method, the two random noise levels are compared, not on the screen of a picture monitor, but as waveforms on the screen of a waveform monitor. Although, at first glance, this may seem to be a retrograde step, in fact it has several advantages.

First, it is possible to pass the random-noise levels to be compared through identical highpass filters having a cutoff frequency of, perhaps, 100 kHz. This immediately removes the need for brightness matching and, incidentally, effectively avoids errors arising from low-frequency interference and distortion; the error introduced is very small and, in any case, can be taken into account. Secondly, it becomes possible to pass each random-noise sample through a weighting network, which minimises the differences between their spectra.

A recent instrument of this type has been described by Yamaguchi (1967). A special feature is the use of two weighting networks. The first is a network of the standard type already mentioned, which accounts for the variation in picture impairment with frequency. The second is unusual, in that it claims to take into account the variation of picture impairment with signal amplitude, and therefore contains amplitude nonlinear elements. An overall accuracy within $\pm 0{\cdot}5$ dB is claimed.

3.7 Objective methods

By 'objective methods' are meant those by which it is possible to read the random noise directly on a meter. In the simplest case, e.g. on a circuit which can operate correctly in the absence of signal information, it is basically only necessary to amplify the noise present at the output of the circuit to a level sufficient for it to be read conveniently on a meter capable of providing the r.m.s. voltage equivalent to the mean noise power.

More generally, signal information is present, and means have to be provided for separating the noise from the signal. This cannot be done when the picture content is purely arbitrary, since the apparatus is unable to discriminate between signal components and noise; but it can be achieved with certain signals of specialised form where, in effect, very simple criteria are applied to obtain this separation.

These objective methods are very much to be preferred to any using visual observation, since they are inherently independent of the observer, provided that the equipment is used in a standardised manner; and the accuracy can be made as good as is required over a very large range of signal/noise ratios. A further advantage is that weighting networks can be introduced with confidence that the measurement will be affected in the predicted manner.

3.8 Noise metering

Some points of importance arise in the design of the metering circuit. A thermocouple might seem to be indicated, in view of its wide bandwidth and true-r.m.s. measuring property, but thermocouples are much too fragile for routine operational work both from the mechanical point of view and on account of their intolerance of even quite small overloads. Other inherently power-measuring instruments are possible, but they also have practical drawbacks, e.g. the need for balancing arrangements and poor long-term stability.

The statistical nature of random noise ensures that, in principle, peak-reading and mean-reading detectors can also be used, provided the relationship between the reading provided by the detector and the r.m.s. voltage is clearly established. It is shown in standard text books that a true-mean-reading voltmeter reads 1·05 dB low on Gaussian noise, compared with a true-r.m.s. voltmeter; with a peak voltmeter, the error is a complicated function of the charge–discharge-time constants and therefore also a function of the signal level. However, peak-reading voltmeters are to be avoided, since the presence of even very small amounts of impulsive interference in the form of short pulses of high peak value can cause considerable errors, whereas the error with r.m.s. or mean-reading voltmeters under the same circumstances is likely to be negligible.

The voltage range of the metering device for a given error can be immediately determined from the probability distribution of amplitudes

$$p(v)\,dv = \frac{1}{\sigma\sqrt{(2\pi)}} \exp(-v^2/2\sigma^2)\,dv$$

where $p(v)\,dv$ is the probability of an instaneous amplitude lying between v and $v+dv$ and σ is the r.m.s. voltage. When this is evaluated for $v/\sigma = \pm 3{\cdot}29$, it appears that larger amplitudes occur for only 0·1 % of the time, and, for v/σ $\pm 3{\cdot}69$, they occur for 0·01 %

of the time. Evidently, even a total amplitude range of only 16 dB compared with the r.m.s. voltage would give no significant error for the measurement of signal/noise ratio.

Whatever type of voltmeter is used for the indicating instrument, a good practical scheme employs a calibrated attenuator in addition, so that the output-meter reading can be set to a fixed point on the scale in every instance, with the result that the ratio of the indicated voltage to the r.m.s. voltage is always constant; this allows an open scale to be used and a constant percentage error for a given change in meter reading. Significant additional advantage is that a simple protective circuit based on a Zener diode may be used to prevent damage to the meter as a result of accidental overloading, without affecting the accuracy of the higher readings, as a result of the reduction in the peak/r.m.s. ratio.

In principle, the time constant of the meter and its associated circuitry should also be taken into account, but, as has already been explained, the standard CCIR method of measurement involves the attenuation of frequencies below 10 kHz, so that it is improbable that the meter will not inherently possess an adequate time constant for this type of measurement. For special purposes, however, this factor might have to be taken into account.

The achievable accuracy when the amplified random-noise level is measured directly is quite high when suitable precautions are taken, but it is not economically sound to aim too high, since a high accuracy is likely to be associated with a high cost. In the author's experience, it is possible to achieve an overall accuracy of the signal/noise ratio within ±0·5 dB. This is adequate for practical purposes, and is considerably better than any visual method.

3.9 Direct measurement

An example of a modern design of instrument using direct measurement of the amplified random noise is a random-noise meter developed by the BBC. It will measure signal/noise ratios with an uncertainty of only ±0·5 dB from 20 to 50 dB and ±1 dB from 50 to 85 dB, taking into account all errors. It is intended to be used only in situations where the noise can be measured in the absence of a signal.

Fig. 3.8 gives the simplified block diagram. The instrument is essentially a high-gain high-stability amplifier with a bandwidth of

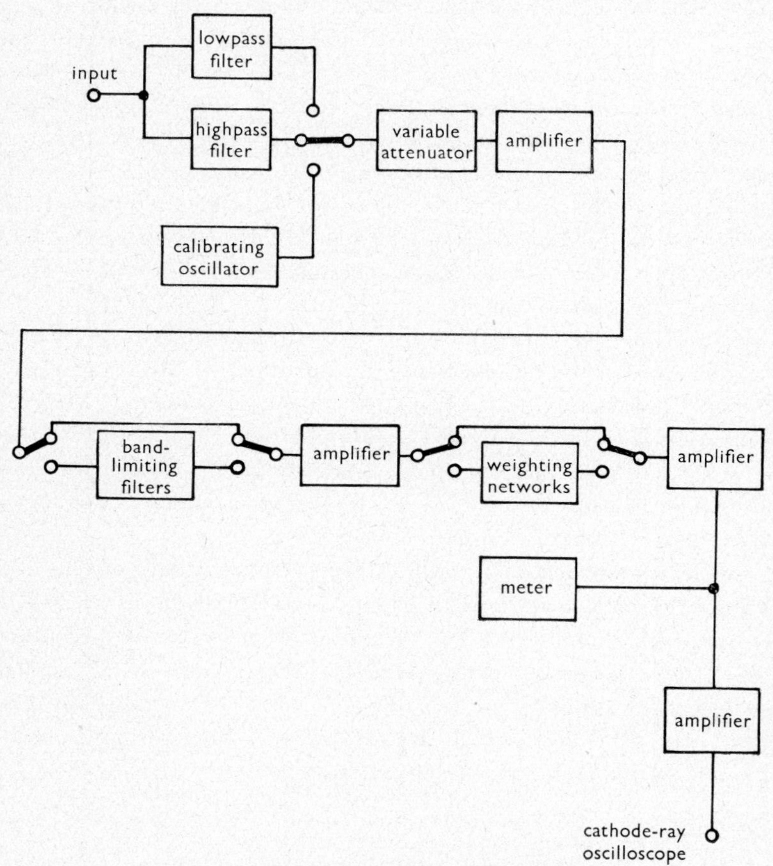

Fig. 3.8 Random-noise meter (simplified)

10 MHz, at the output of which is a detector of the preferred type described in Section 3.8. All the necessary band-limiting filters, weighting networks and junction filter are included, and appropriate combinations can be selected by means of a switch, so that measure-

ments can be made on either 405-line or 625-line systems, the latter including chrominance-noise measurement.

The measurement is carried out by connecting the output of the apparatus under test to the input of the random-noise meter, and then adjusting the attenuator provided until a standard reading is obtained on the meter. If the measurement is made at a point where the composite-signal level is 1 V pk–pk, the attenuator reading gives the signal/noise ratio directly in decibels. The amplifier gain is calibrated by the provision of a sinusoidal generator whose output is held constant against a reference Zener diode.

By the use of semiconductors throughout, the dimensions have been kept very small, the frontal area being only $5\frac{1}{4}$ in × 8 in, so that the instrument is very portable. Advantage has been taken of the very low power consumption of 6 W to make it completely autonomous when required, by replacing the normal mains power module by a mains/rechargeable accumulator module.

3.10 Frequency-discrimination methods

In a large number of instances, it is necessary to transmit some kind of television signal through the system under test, if only in a highly simplified form, to maintain the system in a state of normal operation. Examples are systems where black-level clamps are fitted or where the presence of synchronising pulses is necessary for control purposes. It may also be required to measure the noise at various signal levels. It then becomes necessary to find some reliable means of distinguishing between noise and signal components.

One possibility is to make the frequency spectrum of the transmitted television signal as narrow as possible, so that, at the receiving terminal, the signal components can be filtered out without the loss of too great a proportion of the random-noise spectrum.

An example of this class is due to Krivocheev (CCIR, 1958). The principle of operation is given in Fig. 3.9. A sinusoid of line frequency and appropriate amplitude has synchronising pulses added to the lowest point of each period (Fig. 3.9*a*) so as to simulate a video signal. At the measuring point, the received signal is passed first

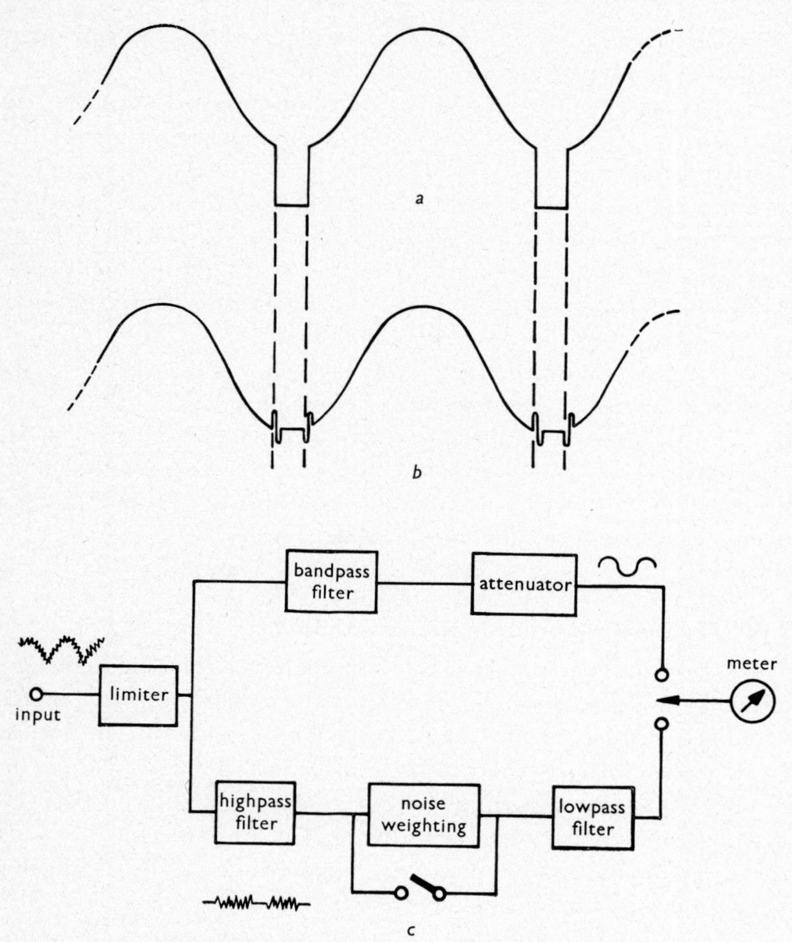

Fig. 3.9 Noise measurement with sinusoidal signal

a Sinusoid with inserted synchronising pulses
b Synchronising pulses removed by clipping
c Simplified circuit of sinusoidal method

through a limiting device, which removes the synchronising pulses, after which it splits into two channels (Fig. 3.9*c*). The upper channel contains a bandpass or lowpass filter which makes the sinusoid substantially free from noise. In the lower channel, the sinusoid is removed and the random noise passes through a weighting network and band-defining lowpass filter to the measuring device, which must possess sufficient amplification. In the original version, a variable attenuator is shown in series with the meter, but it would seem to be more convenient to insert it in the upper path as shown.

The measurement is carried out by adjusting the attenuator until the same reading is obtained from each path, when its reading will be a measure of the signal/noise ratio. The calibration can take into account the fact that the noise has been removed from a proportion of each line by the insertion of the synchronising pulse, and that a small proportion of the lower end of the noise spectrum has been removed. In a variant of the method, the metering device is replaced by an oscilloscope, and the two signals are added again in a proportion which is adjusted until the noise amplitude appears to be equal to the sinusoidal amplitude.

The difficulty of the method lies principally in the limiter, which must be very carefully designed to reduce to a minimum the spurious products resulting from the elimination of the synchronising pulses, and which places a limit on the largest numerical signal/noise ratio which can be measured. Nonlinearity distortion in the system under measurement will give rise to harmonics of the line-frequency sinusoid, which must be removed during filtration; it is stated in the description that harmonics above the 15th are negligible in practice. A more fundamental drawback for some purposes is that the random noise cannot be measured at any desired level of grey in the signal, and, in particular, the instrument cannot easily be adapted to measure the noise in the output from a camera channel. It is not made clear in the original description whether field synchronisation is included, but this would seem to add to a considerable complication.

Other methods have been proposed where the signal consists of synchronising pulses with grey or white bars, the whole signal being band-limited to, say, 1 MHz. However, the loss of the lowest 1 MHz of the noise cannot be tolerated when weighted measurements are made.

A frequency discrimination method of a quite different kind was proposed by the author (Weaver, 1959*b*), chiefly for measurements on camera tubes where it is impossible to obtain a signal which does not contain some kind of synchronising information, if only blanking pulses, and where it is also desirable to be able to measure at the various levels of grey.

The technique used for separating the random noise from the signal depends on a property of the spectrum of the television signal which was first pointed out by Mertz and Gray (1934): namely, that the power in a television signal is not uniformly distributed but is concentrated in areas around the harmonics of line frequency.

With a signal which corresponds to a uniformly illuminated field, this is not difficult to envisage. The picture signal on each line consists of a flat-topped rectangular wave, so that, if it were not for the presence of the field-synchronising information, the spectrum would have components at line frequency only. However, owing to the presence of the field information, there is a superimposed modulation at 25 Hz and harmonics thereof which produces sidebands extending symmetrically over a small area around each line-frequency harmonic. Notwithstanding this, the greater part of the frequency interval between successive harmonics is free from signal energy. This is shown diagrammatically in Fig. 3.10, and photographs of the spectrum of a noisy sine-squared-pulse and bar signal are given in Fig. 3.11. The resolution of the spectrum analyser used was not sufficient to define the 25 Hz and 50 Hz sidebands, which are just visible as a jagged outline of the line-frequency harmonics.

The spectrum of the random noise is, of course, quite different, in that the noise power is uniformly distributed as a function of frequency, so that the central region between two successive line harmonics contains random noise but no contribution from the signal. It remains to find some way of exploiting this possibility of differentiating between the random noise and the signal.

The block diagram of the method adopted is given in Fig. 3.12. The signal to be measured is first connected to the upper path, where it passes through a fixed attenuator, and then is applied to the input of a selective amplifier-detector in the form of a communication-type receiver provided with an output meter having a fairly long time

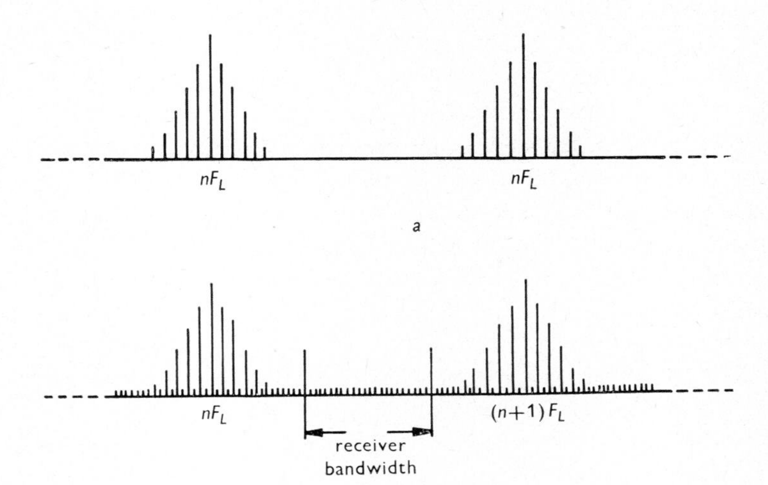

Fig. 3.10 Frequency-selective method of noise measurement

a Idealised noisefree spectrum of video signal
b As (*a*), but with random noise present
F_L = line-repetition frequency

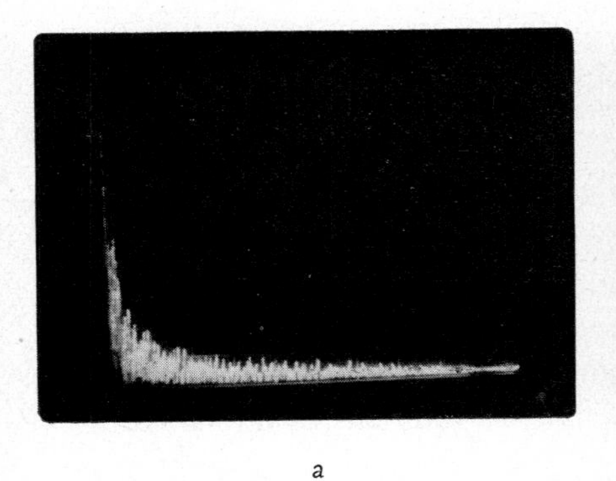

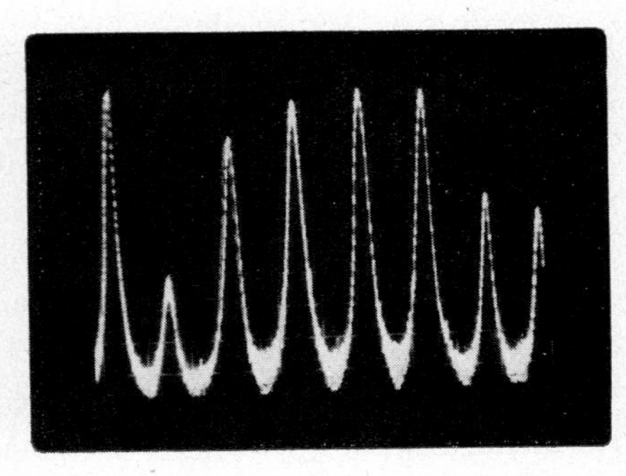

a b

Fig. 3.11 Video-signal spectrum

a 625-line pulse-and-bar signal (2*T*) with random noise present
b Detail of (*a*) in 2 MHz region

constant. The receiver most ususally employed because of its availability and suitability was an ex-Government service model with a total frequency coverage of 60–30 MHz and whose bandwidth could be altered in steps from 100 Hz to 6 kHz. It was modified by the addition of a buffer amplifier at the input aerial socket so designed that it always presented an accurate 75 Ω termination to the signal path. The purposes of the fixed attenuator is merely to prevent possible overloading of the receiver input stage. The other path contains a variable 75 Ω attenuator and a generator capable of providing 1 mW of random noise with a flat spectrum over a 5 MHz band. This

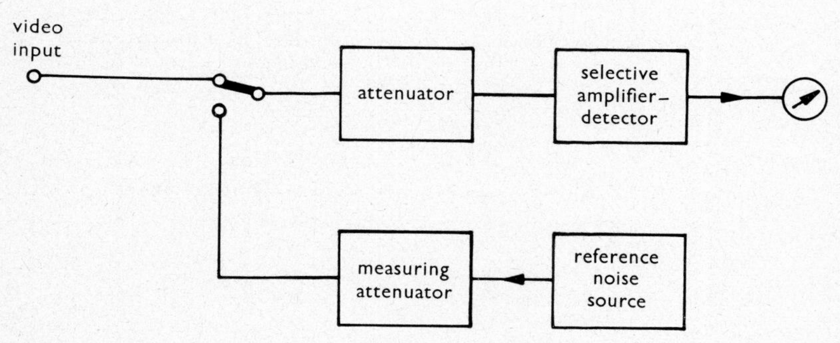

Fig. 3.12 Frequency-selective method

generator had previously been designed as a calibrating source of random noise. The noise is ultimately derived from a photomultiplier operating at a very low level of illumination which, owing to the gain contributed by the multiplier section, alleviates some of the difficulties associated with extremely high-gain, wide-bandwidth amplifiers.

To make a measurement, the receiver is set to the approximate frequency where a reading is required, and is tuned between two line harmonics, which are unmistakable, owing to the strong but very narrow signals which they furnish. It should then be found that a 'plateau' exists midway between these harmonics where the signal strength is constant. If no appreciably flat portion is found, the receiver bandwidth should be decreased; but since, for the sake of the sensitivity, the bandwidth should not be any narrower than is actually required, it is best to start with a fairly wide bandwidth and work downwards. In most instances, 2 kHz is adequate.

The receiver is then set to the middle of the range between two harmonics, and the gain is adjusted until a reference reading is obtained on the output meter. The switch is next thrown to connect the receiver to the reference noise path without touching the receiver in any way, and the attenuator is varied until the same reference meter reading is obtained as before.

The result of this procedure is not, strictly speaking, the measurement of signal/noise ratio, but the measurement of the random-noise power per unit of frequency, $\Delta w/\Delta f$ say, which for certain purposes can be even more useful since the noise spectrum is thereby obtained very accurately when a sufficient number of readings are taken over the video range.

Since the comparison random-noise generator furnishes 1 mW of random noise which is flat over a 5 MHz band, the corresponding value of $\Delta w/\Delta f$ is $10^{-3}/(5 \times 10^{3}) = 0{\cdot}2\ \mu\text{W/kHz}$, so that, if during a given measurement an attenuation of A decibels has to be inserted in the attenuator to provide the same reading in each switch position, the value of $\Delta w/\Delta f$ corresponding to the frequency of measurement is 0·2 W/kHz divided by the numerical factor corresponding to A decibels. The bandwidth of the receiver is the same for each measurement, and consequently disappears from the equation.

It should be noticed that the simple comparison technique employed removes three of the most important potential sources of error. The bandwidth of the receiver does not need to be measured, and it is of no consequence if, as always happens to some extent in practice, it varies with the frequency to which the receiver is tuned. The bandwidth is so small compared with the total width of the noise spectrum that no allowance need be made for the shape of the noise spectrum under measurement; over that very narrow band, the noise may be considered as flat whatever in fact its spectrum may be. Finally, the metering circuit used with the receiver is always operated at the same point for each measurement and its precise characteristics are therefore unimportant.

The noise power per unit of bandwidth can be converted into an equivalent signal/noise ratio by assuming that the noise power is uniformly distributed over the video band which is, say, F megahertz wide. Then, if the measuring attenuator is set to 0 dB, the total noise power in the signal is $0{\cdot}2 \times F \times 10^{-3}$ watts, to which there corresponds an r.m.s. voltage, assuming the power is developed in 75 Ω, of $\sqrt{(75 \times 0{\cdot}2\ F \times 10^{-3})}$ volts. The signal/noise ratio corresponding to the zero setting of the measuring attenuator would be, in decibels,

$$20 \log \frac{0{\cdot}7}{\sqrt{(75 \times 0{\cdot}2\, F \times 10^{-3})}}$$

or

$$10 \log \frac{0{\cdot}49 \times 10^{3}}{75 \times 0{\cdot}2\, F}$$

$$= 10 \log_{10} \frac{32{\cdot}67}{F}$$

However, there will be A decibels of attenuation in the measuring attenuator; so the signal/noise ratio in general is, in decibels,

$$A + 10 \log_{10} \frac{32{\cdot}67}{F}$$

It should be mentioned that this method has been in use both in the BBC and in industry (Turk, 1966; Knight, 1968) for a number of years, chiefly for the checking of image-orthicon camera tubes, and has been found very reliable and accurate within its limitations. The latter are not capable of being expressed concisely, since the possible accuracy is a function of the type of signal and the frequency of measurement. More detailed information is given in a BBC monograph (Weaver, 1959*b*).

Finally, the method can evidently be used as a means of measuring the power spectrum of the noise, whether or not signal components are present.

3.11 Time-discrimination methods

This group comprises methods in which time selection is the basic means for separating the noise from the signal. A typical signal will be the output from a camera which is viewing a uniformly lit field, i.e.

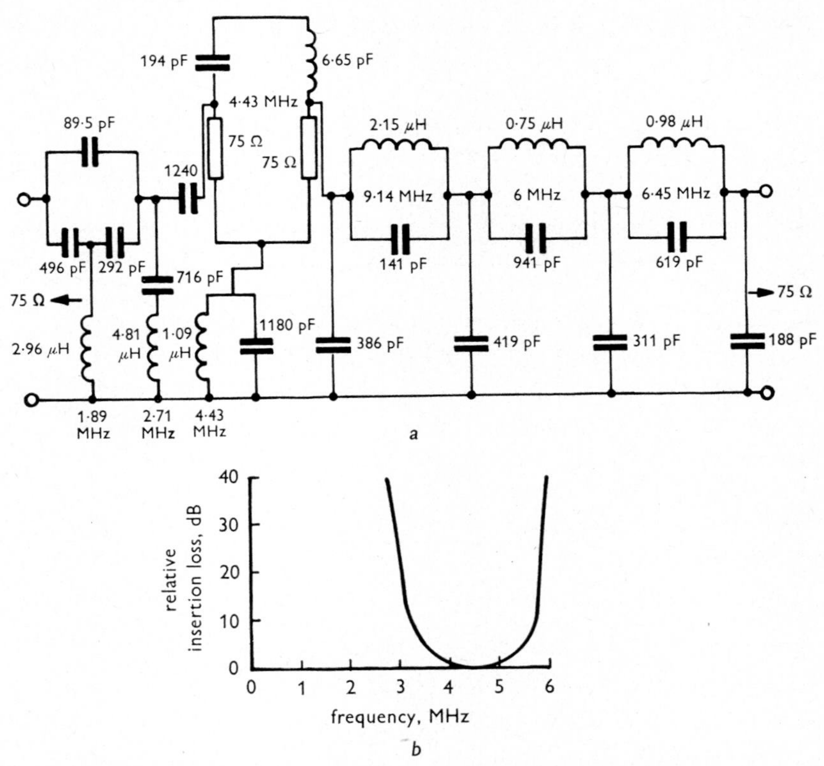

Fig. 3.13 CCIR weighting network for chrominance channel (system *I*)
a Circuit of weighting network, *b* Insertion loss of weighting network

Inductance, capacitance and resistance tolerances within ±1%
Inductances to be adjusted to give maximum insertion loss at indicated frequencies
Q factor ⩾ 100 for 3–6 MHz
Insertion loss ⩾ 35 dB above 6 MHz

it consists of synchronising pulses with a flat-topped bar of length equal to the active line period, and of any amplitude between zero and white level.

If a noisy signal of this type is passed through an ideal gate which is open for a duration somewhat less than the active line period and centred on the picture period, the output waveform will consist of rectangular bursts of noise, the total power of which is proportional to the noise power in the signal. A measurement of the gated signal is therefore equivalent to a measurement of the noise of the signal.

The advantage of this principle is its versatility, since it can be utilised for the testing of transmission systems or other apparatus which require a video signal to be present, for the measurement of picture sources where a uniformly illuminated field can be obtained but where blanking must be introduced, or even for the measurement of random noise in the absence of a video signal. Other types of interfering signal can also be measured. In each instance, completely objective measurements are possible with a satisfactory degree of accuracy over a wide range of signal/noise ratios.

A fairly early member of this family was designed by Edwardson (1961) for use in laboratory work on camera tubes. It utilises a straightforward gating technique with subsequent removal of the accompanying pedestal by a balancing process. It has high accuracy over a wide range of signal/noise ratios, but it is not suitable for operational use.

Another instrument was designed by Holder (1963) for general use in television broadcasting where stability, reliability and a minimum of controls are highly desirable. It utilises a simple, but very effective, device for ensuring that the separated noise signal contains a minimum of components of the original video signal.

The operation can be understood by reference to Figs. 3.14 and 3.15. In Fig. 3.14, the incoming signal takes two paths, one of which is devoted to the generation of modulating pulses. In the main signal path, it is suitably amplified, and passed through a highpass filter, which serves two purposes: it removes the greater part of the energy of the signal, thus leaving less for the gate to do, and at the same time it effectively clamps the signal by removing those components which give rise to a change of d.c. position with a change of signal content, and so stabilises the working point of the gate which follows. This is shown in Fig. 3.15 for a signal containing a rectangular pulse.

The gate operates as a linear modulator, and is so termed in the block diagram. The modulating signal consists of a rectangular pulse whose transitions are shaped by passing it through a Thomson network (Section 6.2) such as is used for the generation of sine-

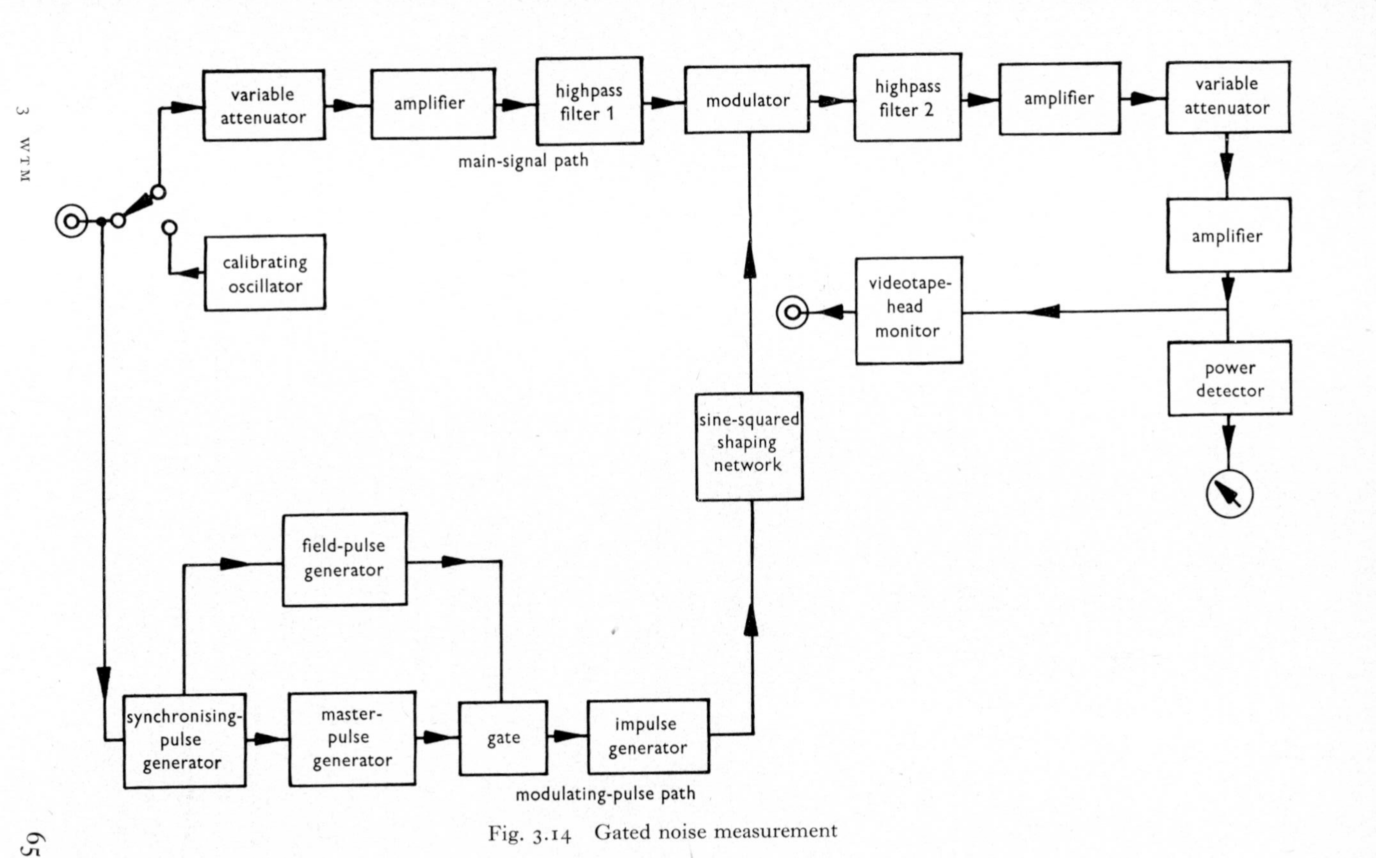

Fig. 3.14 Gated noise measurement

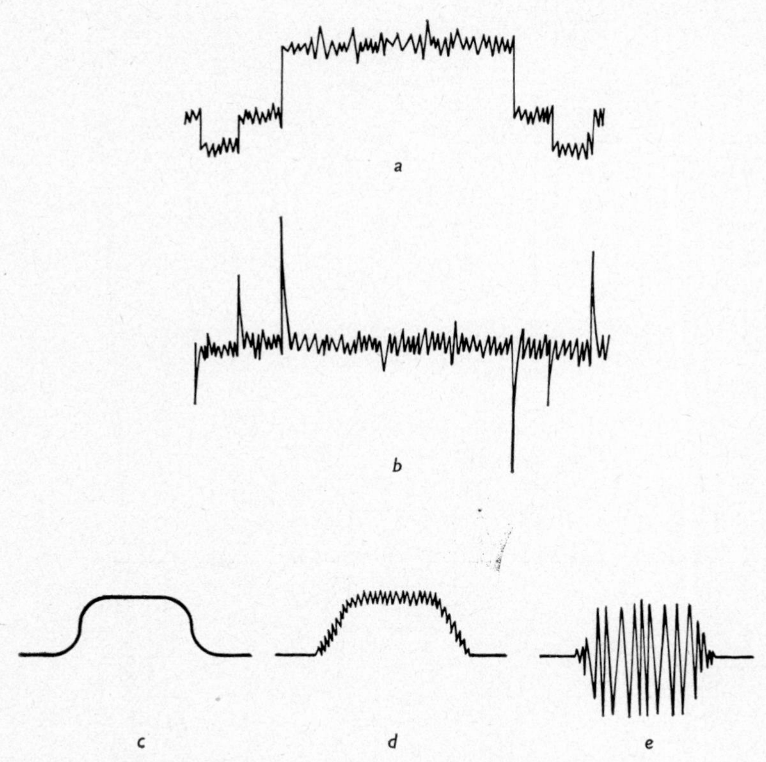

Fig. 3.15 Waveforms of gated noise-measuring set

a Noisy video signal
b As (*a*), but differentiated
c Gating pulse
d Gated output
e Amplified gated output after removal of gating pulse

squared-pulse and bar signals, with the time constant of the network chosen so that very little energy above 80 kHz is present in the modulating pulse.

This pulse is timed to occur in the centre of the active portion of the line, so that the output of the modulator has the appearance of the last waveform but one of Fig. 3.15, i.e. a pulse waveform of similar shape to the modulating pulse but containing the noise components which were present over the greater fraction of the active portion of the incoming signal. When this output signal is then passed through a second highpass filter with a cutoff frequency of 180 kHz, all components deriving from the signal and the modulating pulse are removed and the random noise is left in the form of line-frequency bursts.

The separated random noise is subsequently further amplified, an attenuator being provided to set the level, and finally measured in an approximately r.m.s. detector, the actual measurement being performed by operating the attenuator so as to set the output meter on a fixed reference value. As has already been explained (Section 3.8), this device minimises the errors due to the metering system. The gain of the overall equipment is standardised by the provision of a calibrating sinusoidal oscillator whose output level is set by means of a voltage-reference diode. The instrument indicates the signal/noise ratio directly at a point where the composite signal level is 1 V pk–pk.

The above description refers primarily to measurements on a waveform which contains no field information. If field-synchronising pulses are present, precautions must be taken to avoid errors due to the half-line pulses, and also the equalising pulses in 625- and 525-line measurements. The device adopted to avoid errors due to them is very simple: the modulator is blocked for the duration of the field-synchronising interval, so that the sampled noise pulses are not passed on to the output meter. The error that could be introduced is 4% at the most, according to the line standard, but in the event is removed by gating the calibrating oscillator in an identical fashion. If no synchronising information of any sort is present, the modulating-pulse generator can be switched to a 'free-run' position, whereby gating takes place as though picture information were present, and so the calibration of the instrument is retained.

An auxiliary function of interest concerns the measurement of signal/noise ratio with 4-head videotape machines. It is of importance to know whether the noise outputs of the heads are approximately equal or whether appreciable differences are present. The amplified gated-pulse sequence is rectified by means of a diode, and the envelope waveform is provided at a separate output terminal where it may be examined by an oscilloscope. Any difference in level between the successive groups of approximately ten lines is immediately visible, and may be measured. Although the actual noise power from each head is not provided, the differences, which are most significant, can be obtained.

The prime advantage of this method of obtaining gated bursts of the random noise essentially free from signal components is its very great stability. The separation of the noise from the signal depends on passive networks, and no balancing processes are involved. The price paid for this is a progressive loss of random-noise components below 180 kHz. While this is of negligible importance with unweighted noise, an appreciable error is introduced when a weighting network is used, owing to the much greater relative amplitudes at the lower end of the weighted-noise spectrum. This is overcome by an internal emphasis circuit, which gives a compensating boost to noise components in the range 200–500 kHz. When this is suitably carried out, the greatest error with any noise spectrum likely to be encountered is less than 0·25 dB.

When the input signal is a 1 V pk–pk composite signal, the maximum error is 0·5 dB for signal/noise ratios of 20–50 dB and 1 dB for 50–60 dB. With synchronising pulses only or random noise only, the accuracy is even better.

If it is important that the instrument should measure down to quite low frequencies with stability of operation, square-wave gating must be used in conjunction with some reliable means for ensuring a low proportion of residuals from the video signal carrying the random noise. An instrument of this type designed in the author's laboratory shows an advantage of at least 10 dB over that just described. The method used consists in introducing a second gating, by another rectangular pulse, whose width is chosen so as to remove the spurious signals left by the first gating operation. Naturally, the second gating operation also produces its own residuals; but, between the two gatings, the signal may be amplified to an extent which considerably improves the ratio of the separated random noise to the residuals from the video signal. In principle, yet more gating operations could effect further improvements, but, in practice, two seem to be amply

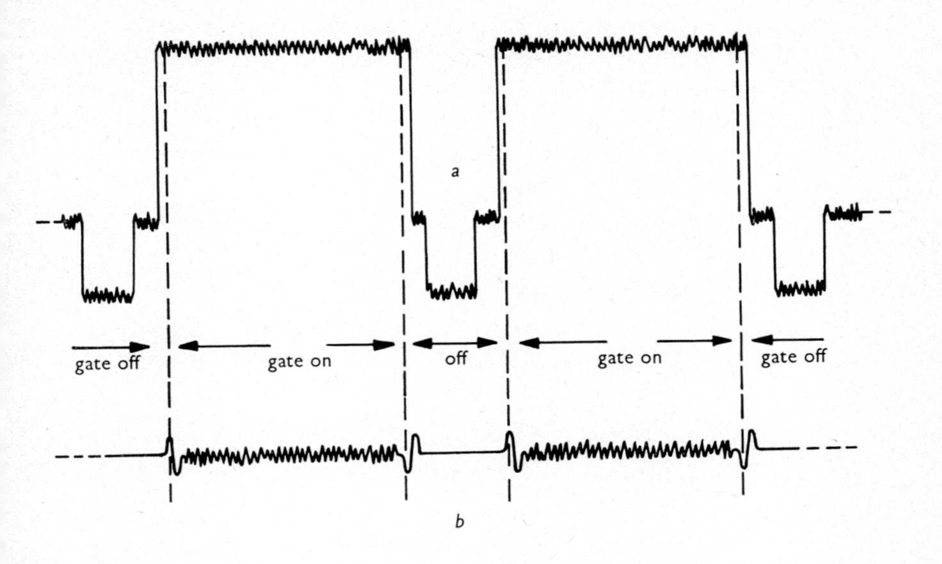

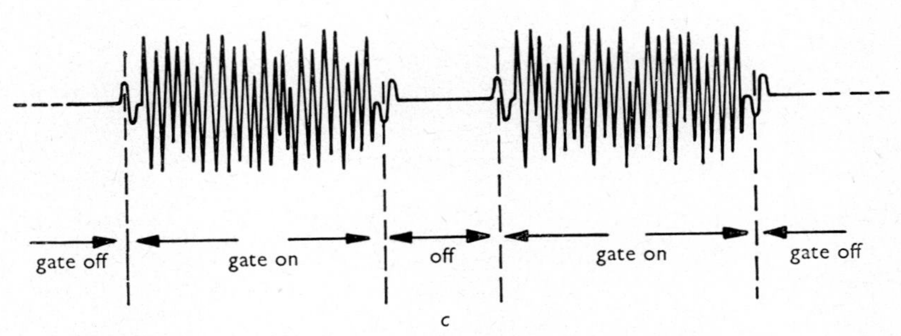

Fig. 3.16 Removal of signal components by double gating

a Noisy input signal
b Noise and signal residue after first gating
c Noise and reduced signal residue after second gating

sufficient. Fig. 3.16 shows the waveforms during successive gating operations.

The ability of instruments of this type to measure down to very low video frequencies can at times be embarrassing; e.g. even signal deformations such as line tilt may be interpreted as noise voltages, for which reason it is always advisable to make measurements through a suitable highpass filter unless very low-frequency noise is specifically to be included, when it may be necessary to introduce some waveform compensation.

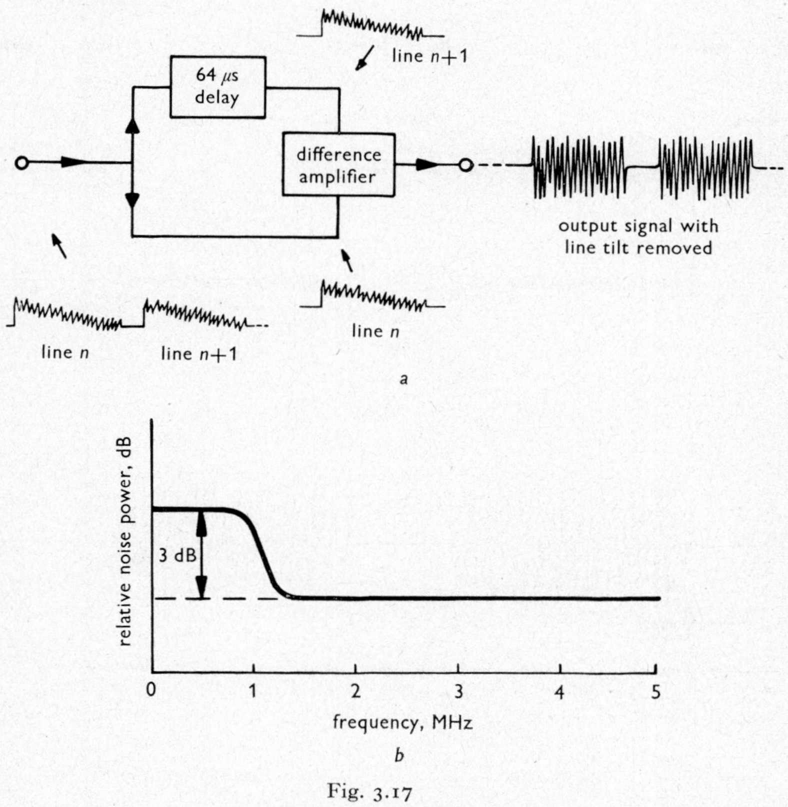

Fig. 3.17

a Drescher's method for removal of line-time distortion
b Modification of random-noise power spectrum

If a further degree of complication is no objection, the errors arising from waveform distortion, hum etc. can virtually be eliminated by a device described by Drescher (1963), and shown in Fig. 3.17. The principle of the method depends on the correlation existing between the unwanted components in each line and the next following line, which is unity for line-synchronous distortions, and which is still fairly high for other effects such as power-supply hum. Consequently, if, as in Fig. 3.17, the gated signal is divided into two identical paths, one of which is delayed by precisely one line period with respect to the other, and then these two signals are subtracted, all fully coherent distortion components will be completely removed, and other effects will be reduced in proportion to the degree of correlation.

On the other hand, the random-noise voltages from line to line are completely uncorrelated and will consequently add. Ideally, the improvement is very great, since all completely coherent line-time distortions are completely removed, other effects are reduced, and the noise power at the output is increased by 3 dB.

For this to be completely true, the bandwidth of the delay line must be at least equal to the nominal bandwidth of the video signal. Formerly, such a delay line using inductors and capacitors would hardly have been practicable, but ultrasonic delay lines are now readily available for delays of 64 μs with satisfactorily low distortion over bandwidths up to 5 MHz.

Particularly attractive for this use, in view of their low cost and small size, are the 64 μs ultrasonic delay lines now manufactured for Pal colour receivers. The bandwidth of the delay lines produced for colour receivers in the UK is somewhat greater than 1 MHz, which is sufficient to remove much of the power of the line-synchronous or near-synchronous distortion components. If a noise-weighting network is used between the gating stage and the subtracting device, the elimination should be virtually complete.

The 3 dB increase in the output-noise power will not be maintained over the whole of the videoband, since the bandwidth of the delay line is insufficient. Tuned amplifiers are required in association with the delay line, and can be arranged to cut off progressively around the band limit of the delay line, so that a transition occurs between subtraction when the random-noise power is increased by 3 dB and no subtraction where the noise power remains as before. The noise-power spectrum is accordingly modified by a curve such as is shown in Fig. 3.17*b*, which can easily be corrected before the noise is finally metered.

3.12 Sampling method

This interesting method, due to Wise and Brian (1970), has a number of very useful features. Basically, its operation depends on the sampling of a selected area of the picture by means of an extremely narrow pulse, the noise power being determined from an assembly of such readings. By choosing the type of picture signal, or by suitably locating the sampling points with respect to the picture material, it becomes possible to reject the signal information in favour of the random noise, and, to that extent, it resembles the methods of time discrimination described above. However, it also possesses some individual features of its own.

For simplicity it will initially be assumed that the sampling pulse is infinitely narrow, so that its spectrum has the form of an infinite array of equally spaced harmonics, all of equal amplitude. To form a stationary pattern of sampling points over the picture, the repetition frequency will be line-scanning frequency; the pulse harmonics for 625-line systems will consequently be spaced by intervals of 15·625 kHz. On the other hand, the spectral components of random noise possess a continuous distribution, even though the appearance of any given spectral component is determined statistically.

At any given instant, therefore, the frequency distribution in the neighbourhood of the *n*th line harmonic is as shown diagrammatically in Fig. 3.18*a*, where the line-frequency components f_L are indicated by arrow-headed lines and a noise component f_N by a dot-headed line.

The sampling is inherently equivalent to a multiplication of the spectral functions of the two waveforms, giving rise to sum and difference frequencies corresponding to all the spectral components taken in pairs. If the difference frequencies alone are considered, it will appear that every spectral component of the noise has a difference-component counterpart in the range $0–\frac{1}{2}f_L$. Moreover, since, by assumption, all the harmonics of f_L have equal amplitudes, all of these difference frequencies have the same relative amplitudes as their counterparts in the original noise signal. A large number of additional intermodulation products will also be produced (Fig.

3.18*b*); but, when these are eliminated, by a lowpass filter with a cutoff frequency at $\frac{1}{2}f_L$, a transformed version of the random noise is obtained, compressed in frequency range by a factor of more than 600.

This compressed signal is not a replica of the original, but it has similar statistical characteristics, and can be measured similarly. In particular, it is evident that the power in the two signals will always be linearly related. Of special interest are the components of the random noise which coincide with the harmonics of the line frequency.

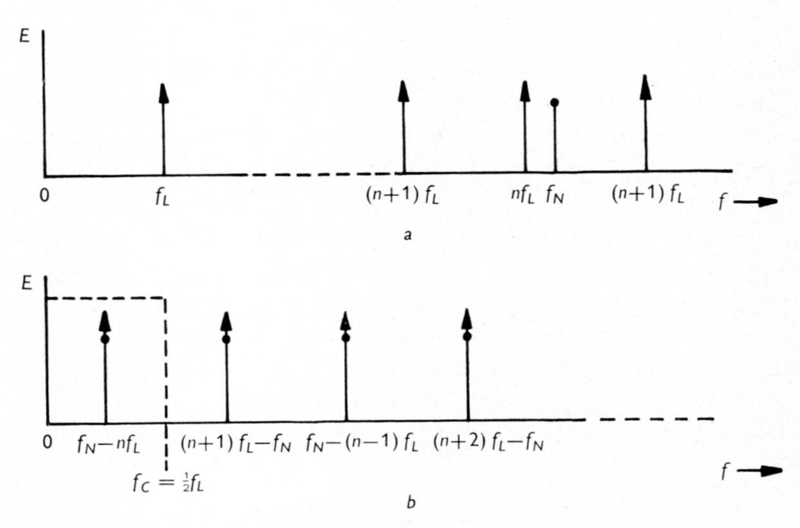

Fig. 3.18 Principle of sampling method of noise measurement
a Spectra of random noise and sampling pulse
b Spectrum of sampled signal

The difference frequency is then zero, which is tantamount to saying that picture information identical between sampling points appears only as a d.c. component and may be rejected. Since every line harmonic is suppressed in this instance, the sampling system behaves as a type of active comb filter as far as the input signal is concerned. On the other hand, this also means that any noise present in the signal to be measured which possesses some correlation with the location of the points in the picture where samples are taken will contribute too little to the output reading. To deal with such instances, e.g. the so-called 'curtain noise' in standards convertors, Wise and Brian arranged for sampling to take place when required diagonally across the picture.

As usual with techniques which involve time discrimination, the field-blanking interval represents a potential source of error, first because the noise in that region is usually different from that associated with the picture region, and secondly on account of the presence of the broad pulses and equalising pulses. This is dealt with in standard fashion by inhibiting the measurement rather longer than the duration of the field blanking, a correction being applied inside the instrument to compensate for this, e.g. by an increase in the internal amplifier gain; about 1 dB is required for the purpose. The transformed random noise may be metered by any suitable means, the use of a true-r.m.s. voltmeter being preferred for reasons given earlier.

A source of error hitherto kept in the background is the finite duration of the practical sampling pulse, which implies that the amplitudes of the line-frequency harmonics are now no longer everywhere equal. Some notion of the maximum tolerable duration can easily be obtained by noting that the envelope of the harmonics as a function of frequency in the spectrum of a rectangular pulse of duration is given by $A = (\sin \omega\tau)/\omega\tau$. The error in the measured noise power with flat noise is then given by

$$\text{error (dB)} = 10 \log_{10} \frac{1}{x} \int_0^x \frac{\sin \omega\tau}{\omega\tau} d(\omega\tau)$$

where x is the value of $\omega\tau$ at the upper limit of the frequency band. This definite integral, known as Si (x), cannot be evaluated analytically except in series form, but values are available from tables (Jahnke and Emde, 1945). In the instrument designed by Wise and Brian, the pulse duration was 20 ns, which, under the conditions just assumed, corresponds to an error of 0·1 dB, which is negligible for the purpose. In practice, the pulse is not rectangular and the noise is not necessarily flat, which will modify the error somewhat, but in no case seriously.

The reduction of the frequency band of the noise to the audio-frequency range has the advantage that high gain is easily obtained with stability; but, on the other hand, any supply-frequency ripple which finds its way into the amplifier is measured as part of the random noise, for which reason extensive electric and magnetic screening is required. In spite of this, the sampling instrument is claimed to measure signal/noise ratios up to 70 dB.

3.13 Measurement in the field-blanking interval

As will be explained in greater detail in Chapter 7, the measurement of signal impairments by test signals inserted into the field-blanking interval of the video signal is now acquiring a considerable importance since this provides a means of monitoring and controlling signal quality even during programme hours. The measurement of random noise can be included in this, and it has now been proposed internationally (CMTT, 1969) that a pair of lines in the field interval of 625-line signals, namely lines 22 and 335, should be reserved for this purpose. By this is meant that these lines must be clean at the point where the insertion test signal is originated, and at no subsequent point downstream must they be processed in any way which will modify the random-noise level acquired. No standard nomenclature seems yet to have been adopted for these; the most apt name would appear to be 'field-interval noise lines', but the rather illogical term 'quiet lines' is also heard.

One might imagine at first sight that it would merely be necessary to gate out the noise lines and measure the amplified noise signal, a suitable correction being made for the fact that only a fraction of the total signal noise has been measured. Such a course is not adopted, because of the very adverse peak-to-r.m.s. ratio of the resultant signal. For example, in the extreme case of a single noise line per picture, the quasi peak-to-peak noise voltage, taken over a sufficient number of fields, is the same as though the noise on all lines were being measured, but the noise power is reduced by a factor of 625. Thus, to drive the same power into the metering circuit as with all lines measured, the output transistors would have to deliver 25 times the peak-to-peak signal voltage.

Various devices have been used to deal with this situation. A method investigated in the author's laboratory makes use of a full-wave rectification of the noise power from the gated-out lines, which, at the low level concerned, is close to square-wave rectification. After suitable filtration to remove the higher-frequency components, an approximately rectangular pulse is obtained on each of the lines, which is then stretched by well known methods to obtain a long rectangular pulse whose duration is fixed but whose amplitude is proportional to the noise power in the line. Such a signal has a d.c component which can then be applied, after suitable amplification and integration, directly to a meter or digital voltmeter. It has been found that an overall accuracy within ± 1 dB is still obtainable at signal/noise ratios of 70 dB or greater, depending on the care taken in the instrumentation. This is perfectly adequate for practical purposes.

Another technique has been employed by D'Amato and Barbieri (1970). In this the separated noise signal is square-law-rectified by means of a halfwave rectifier, followed by an integrator which furnishes an output signal in the form of an approximately rectangular pulse with a duration just less than the active line period. The amplitude of this pulse is compared with a reference d.c. level, the difference signal being used to drive a decade attenuator until equality of levels is reached. The amplitude of the video signal is also maintained by an auxiliary control loop at a fixed reference level, so that the attenuator settings can be calibrated to indicate the signal/noise ratio directly. The circuitry is rather elaborate, but the attenuators can be made fairly simply to provide a data-coded signal for transmission to a distant point. It is claimed that the instrument is usable up to 70 dB, but the accuracy at this figure is not stated.

3.14 References

ALLNATT, J. W., and PROSSER, R. D. (1966): 'Subjective quality of colour television pictures impaired by random noise', *Proc. IEE*, **113**, (4), pp. 551–557

BARSTOW, J. M., and CHRISTOPHER, H. N. (1954): 'Measurement of random monochrome video interference', *Trans. Amer. Inst. Elect. Engrs.*, **73**, Pt. I, pp. 735–741

CCIR (1958): 'Apparatus for the measurement of signal/noise ratios in television signals', Document XI/25, Moscow

CCIR (1959): XIth Plenary Assembly, Document 11E, Annex D (XI)
CCIR (1966*a*): Documents of the XIth Plenary Assembly, Oslo, Recommendation 421–1, p. 83
CCIR (1966*b*): Documents of the XIth Plenary Assembly, Oslo, Recommendation 451
CMTT (1958*a*): 'The relationship between effective value and the quasi peak-to-peak value in noise measurements', Document 9E of the 2nd CMTT meeting, Monte Carlo
CMTT (1958*b*): 'Measurement of signal/noise ratio in video signals', Document 23E of the 2nd CMTT meeting, Monte Carlo
CMTT (1969): Document 1028E, section 2·8
D'AMATO, P., and BARBIERI, G. (1970): 'Signal to noise ratio automatic measurement in the blanking interval of a video signal', *IERE Conf. Proc.* 18, pp. 179–197
DRESCHER, F. (1963): 'Objektive Rauschmessverfahren im Videokanal', *Tech. Mitt. RFZ*, **7**, (2), pp. 49–58
EDWARDSON, S. N. (1961): 'An instrument for measuring signal/noise ratios', BBC Engineering Monograph 37
FIX, H., and KAUFMANN, A. (1960): 'Die spektrale Zusammensetzung der statischen Schwankungen bei der zur Zeit üblichen Fernsehkameraanlagen', *Rundfunktech. Mitt.*, **4**, pp. 60–65
GARUTS, V., and SAMUEL, C. (1969): 'Measuring conventional oscilloscope noise', *Tekscope* (Tektronix Inc.), **1**, (2), pp. 8–11
GILBERT, M. (1954): 'Visibility of random noise in a television picture', BBC Research Department Report TO48
GOUSSOT, L. (1959): 'Le brouillage des images de télévision par les signaux parasites', *L'Onde Élect.*, **39**, pp. 690–700
HOLDER, J. E. (1963): 'An instrument for the measurement of random noise in the presence of a television signal', *IEE Conf. Rep. Ser.* **5**, pp. 20–26
JAHNKE, E., and EMDE, E. (1945) 'Tables of functions' (Dover)
KILVINGTON, T., JUDD, D. L., and MEATYARD, L. R. (1954): 'An investigation of the visibility of noise on television pictures', Post Office Engineering Department Research Report 2289
KNIGHT, R. E. (1968): 'A broadcaster re-tests his image orthicons', *Roy. Televis. Soc. J.*, **12**, (3), pp. 58–61
MAARLEVELD, F. (1957): 'Measurements on the visibility of random noise in a 625-line monochrome television system', The Netherlands PTT Report 107RL
MERTZ, P. (1951): 'Data on random noise requirements for theatre television', *J. Soc. Motion Picture Televis. Engrs.*, **57**, pp. 89–107
MERTZ, P., and GRAY, F. (1934): 'A theory of scanning', *Bell Syst. Tech. J.*, **13**, pp. 464–515
MÜLLER, J., and DEMUS, E. (1959): 'Ermittlung eines Rauschbewertungfilters für das Fernsehen', *Nachrichtentech. Z.*, **12**, (4), pp. 181–186
PROSSER, R. D., and ALLNATT, J. W. (1965): 'Subjective quality of television pictures impaired by random noise', *Proc. IEE*, **112**, (6), p. 1099
PUTMAN, R. E. (1966): 'Measurement of signal/noise ratios', *J. Soc. Motion Picture Televis. Engrs.*, **75**, (3), p. 221
RICE, S. O. (1945): 'The mathematical analysis of random noise', *Bell Syst. Tech. J.*, **24**, (1), p. 55
TURK, W. E. (1966): 'The practical testing of camera tubes', *J. Soc. Motion Picture Televis. Engrs.*, **75**, (9), pp. 841–845

WEAVER, L. E. (1959*a*): 'The subjective impairment of television pictures', *Radio Electron. Engr.*, **36**, p. 170
WEAVER, L. E. (1959*b*): 'The measurement of noise in the presence of a television signal', BBC Engineering Monograph 24
WISE, F. H., and BRIAN, D. R. (1970): 'A new equipment for the measurement of video noise', *IERE Conf. Proc.* 18, pp. 49–58
YAMAGUCHI, Y. (1967*a*): 'Weighting function for evaluation of random television interferences with different standards', *J. Soc. Motion Picture Televis. Engrs.*, **76**, pp. 176–179
YAMAGUCHI, Y. (1967*b*): 'A new random noise measuring instrument for television signals', *ibid.*, **76**, pp. 180–182

4 INTERFERENCE

4.1 General

The term 'noise' or 'interference' is applied to a large variety of parasitic effects which afflict a television signal. In the present instance, it is proposed to discuss the rather untidy group of disparate phenomena which must be classified as noise, but which do not possess the statistical properties characterising random noise. For this reason, the group is sometimes also called 'nonrandom noise'.

Some justification for this distinction can be derived from the fact that it is ideally possible, if one takes sufficient precautions, to generate a video signal free, to any desired extent, from hum, ignition interference etc. On the other hand, random noise is a consequence of the granular structure of matter and statistical effects in a wide variety of active devices, and a given circuit under fixed conditions will have a limiting signal/noise ratio which cannot be bettered.

A factor common to most of these types of noise is that the picture impairment they cause more often depends on the peak-to-peak value of the waveform than on its r.m.s. value, and consequently, the signal/noise ratio is now usually expressed in terms of the ratio of the peak-to-peak signal to the peak-to-peak noise. The exceptions are noted below.

One particular difficulty with the measurement of interference phenomena is the simultaneous existence in most instances of several different types, and some means have to be found for distinguishing between the interference to be measured and the other kinds also present. Only occasionally can this be completely achieved by filtration, but it is a good rule to separate the obviously low-frequency and the obviously high-frequency effects by the use of the junction filter recommended for the measurement of random noise (Section 3.4.2). This consists of a pair of complementary highpass and lowpass filters with a crossover at 10 kHz, which is a convenient choice for most practical work. Apart from fairly simple precautions of this kind, the best one can do in most instances is to make use of the remarkable power of the eye to recognise patterns when the waveform is examined on an oscilloscope. When the sweep rate is carefully varied, it is possible to distinguish, and even to measure, very small amounts of interference, even when the frequency is far from stable, in the presence of a video signal and also other types of interference. This applies especially to impulsive and pattern interference.

In the discussion below of the various kinds of interference, a classification has been adopted which is to some extent arbitrary, but which has been found very convenient in practice. An even simpler broad classification made possible by the use of the junction filter, is a division into 'above-line-frequency noise' and 'below-line-frequency noise'. The names are derived from the original use of the junction filter with 405-line signals, but since we are not concerned with a precise demarcation, it is convenient to retain them.

4.2 Very low-frequency noise

This group of effects lies on the borderland between noise and what would otherwise be termed instability of the signal, because they take the form of a relatively rapid variation of the position of the black level of the video signal. This is sometimes also called 'd.c. wander'. Since a closer definition is desirable, very low-frequency noise will be arbitrarily classified as parasitic waveforms with a frequency, in so far as they can be said to possess a frequency, between 0·5–10 Hz.

The only really feasible method of measurement is by direct observation of the position of black level on the screen of a waveform monitor, either by using a model fitted with a very long-persistence phosphor, or by taking a photograph of the screen with a sufficiently long exposure.

This type of noise can be eliminated by a black-level clamp with a suitable time constant, unless nonlinear effects over the range of shift of the video signal are appreciable.

It is usual not to make a measurement but to rely on observation of a static pattern on a waveform monitor, and action is taken if any perceptible amount of d.c. wander is present, unless circumstances dictate otherwise.

4.3 Low-frequency noise

This will be defined as noise lying between about 10 Hz and 1 kHz. Low-frequency noise therefore principally consists of power-supply hum and hum harmonics with other miscellaneous effects, such as teleprinter clicks.

No standardised measurement method exists. The usual technique is to measure, whenever possible, in the absence of a signal. The apparatus or circuit concerned is terminated at each end of its correct impedance, and the output waveform is examined by a waveform monitor with a sweep speed low enough for two or three cycles of the supply frequency to be seen across the screen. This enables a visual distinction to be made between higher-frequency parasitic waveforms and power-supply hum, so that the peak-to-peak value of the latter can be estimated.

Usually, the magnitude of the video signal which would normally be present at the point of measurement is known sufficiently accurately, and the signal/low-frequency-noise ratio can consequently be calculated. If the circuit or the waveform monitor needs calibrating, a known waveform can be applied either to the sending end of the circuit or to the input terminals of the equipment under test. A similar estimate can also be carried out during programme transmission by making use of black-level or synchronising-pulse bottom.

However, the measurement of peak-to-peak value is only satisfactory for a limited set of circumstances and does not give an adequate idea of the corresponding picture impairment where the waveform differs appreciably from the sinusoidal or where the frequency is not 50 Hz.

The observer reaction to low-frequency noise has been investigated by Fowler (1951) for the U.S. 525-line system using a 60 Hz field frequency, and also by Savage* for the UK 625-line system with

* Savage, D. C.: BBC unpublished document

50 Hz field frequency. These measurements were used by Savage as the basis for the design of a network intended to weight any low-frequency interference in a manner analogous to that employed with random noise (Section 3.5). This has proved to be a good first attempt rather than a final solution, but the network represents a serious attack on the problem, and, for that reason, it is given in Fig. 4.1. It should be noted that it requires to be utilised in conjunction with a 1·5 kHz lowpass filter, and additional gain is needed to make up for the loss through the combined network.

Although the work has so far largely been concerned with monochrome systems, there seems to be no reason to believe that the addition of chrominance information will make any significant difference to the degree of picture impairment, provided, of course, that the same field frequency is employed.

It should be noted in passing that low-frequency noise effects of a periodic nature have a visibility which is critically dependent on the frequency, so that the tolerances to be applied to the signal/noise ratio must be chosen on the basis of the condition of worst visibility in each instance, since there is no guarantee that the frequency will not at some time take up a worst-visible position. However, this point will not be discussed further because we are here concerned with the measurement of the interference rather than with the applicable tolerances.

4.4 Periodic noise

Periodic noise will be defined as one or more parasitic waveforms having a strongly periodic character whose frequency lies above 1 kHz, and includes not only sinusoidal or quasisinusoidal waveforms, such as are caused by common-channel interference or direct pickup of broadcast radio stations in video circuits, but also periodic interference of a 'spiky' nature due to rotating machinery or power-supply systems such as transistor invertors.

The signal/noise ratio is again expressed in terms of peak-to-peak picture signal to peak-to-peak interference, since the noise, although periodic, may be far from sinusoidal in shape.

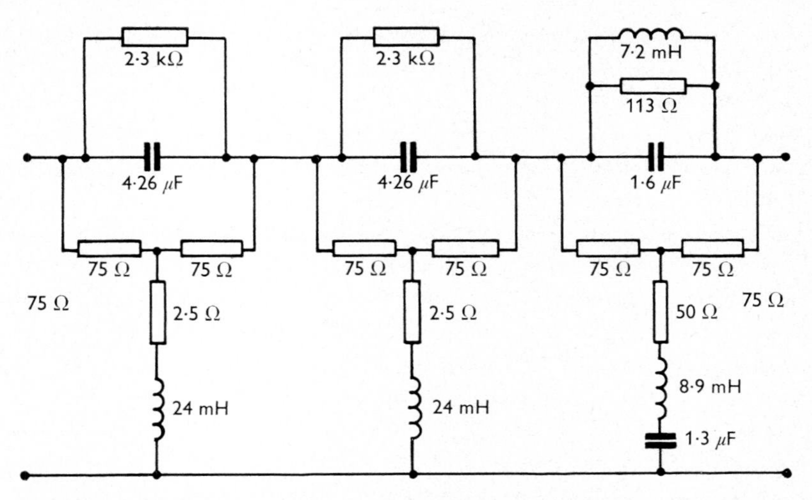

Fig. 4.1 Experimental weighting network for measuring low-frequency interference

Inductance, capacitance and resistance tolerances within ±2%

Frequency	Total loss of network and 1·5 kHz lowpass filter
Hz	dB
Sine wave	
0	60
50	40
100	28·5
150	22
200	17·5
300	12
400	9
600	6·5
1000	6·5
Square wave	
50	6
Differentiated square wave	
50	15

In common with low-frequency periodic noise, the degree of picture impairment produced is a function, not only of the position of the noise in the video spectrum, but also of the precise relationship between its frequency and the field and line frequencies. In general the frequency is not constant, and the changes with time of the pattern produced by the interference on the screen of the receiver add to the magnitude of picture impairment, so that it is necessary to use some arbitrary convention, such as measuring only at the precise frequencies at which the picture impairment is greatest.

Curves showing the level of sinusoidal interference as a function of frequency for the same degree of picture impairment are given by Goussot (1959) for monochrome 819-line signals. There is a clear worst point around 800 or 900 kHz (say 600 or 700 kHz for a 625-line system), although the variation below this frequency is only around 6 dB. Above this region of greatest impairment, the decrease in impairment with increasing frequency is quite considerable: in fact, in one instance, equivalent to a change in the level of interference of 34 dB. Colour signals will behave similarly, except for another region of increased impairment in the region of the sub-carrier. This behaviour has led to divergence of view on specifying limits on periodic noise of a sinusoidal or quasisinusoidal character, some administrations preferring to set limits according to a highly simplified version of the impairment/frequency curve and others preferring to lay down a value of signal/noise ratio so that the impairment is likely to be acceptable whatever the frequency.

The latter method makes it possible to simplify the measurement if a suitable waveform monitor is available with a high-gain pre-amplifier. Even when the random-noise level is fairly high it is usually possible to trigger the oscilloscope, so that an unstable, but recognisable, display of the periodic noise is obtained from which the peak-to-peak amplitude can be estimated. This method is crude in some respects, but has the virtue that the peak-to-peak value can be estimated even when the noise is not sinusoidal, and it avoids the necessity for drawing a distinction between sinusoidal and other types of periodic noise.

If the tolerance is laid down as a function of frequency, it becomes necessary to use either a range of bandpass filters followed by a high-gain amplifier, or a wide range receiver provided with a calibrating oscillator. If the bandpass filters are employed, it is still advantageous to measure the periodic noise with a suitable oscilloscope since the measurement may be then made with less error in the presence of high random-noise levels. The use of a bandpass filter and oscilloscope is recommended particularly for the measurement of nonsinusoidal noise which is known to be likely to occur within a certain range of frequencies, e.g. invertor noise (see also below).

4.5 Impulsive noise

This type is characterised by a peak-to-peak value which is high compared with the duration of the transient. It may, or may not, be periodic. The most notable type is automobile ignition interference, but atmospheric and manmade electrical discharges of various kinds can also give rise to noise of this kind.

It is usual to measure the peak-to-peak value only by means of an oscilloscope, although it sometimes appears that the peak-to-peak value alone does not sufficiently define the picture-impairment value of this noise, which also depends on the duration of the impulse and whether it is followed by 'ringing' due to shock excitation of some resonant system. The signal/impulsive-noise ratio is defined as the ratio of the peak-to-peak signal amplitude to the peak-to-peak noise amplitude expressed in decibels.

4.6 Invertor noise

It is not uncommon for the amplifiers used for video transmission over coaxial cable to be powered by d.c. invertors which have a switching frequency of a few kilohertz. It then becomes difficult to avoid some low level of interference from breakthrough of the switching waveform, which has a spectrum consisting of the fundamental component and harmonics decaying very rapidly with increasing order. The problem is complicated by the fact that it is not possible to control the switching frequency very closely; so these frequencies are randomly distributed about the nominal value. It is required to measure this interference in such a way that the result has some connection with the resulting picture impairment.

This was investigated by Allnatt and Bragg (1958) who came to the following conclusions. In view of the large preponderance of the fundamental component in the switching component, the interfering effect of a single invertor can be equated to that of a sinusoid of the same frequency. Furthermore, the combined effect of ten or more invertors on the same circuit is closely equivalent in picture impairment to random noise which is band-limited to the region corresponding to the spread of the switching frequencies. On this basis, a series of subjective tests was carried out and the relationship between signal/noise ratio and picture impairment determined. The results can be found in the original paper (loc. cit.).

The converse is also true, and the preferred method of measuring such invertor noise in cases where ten or more are present on the circuit is to measure with an r.m.s. meter via an amplifier and band-pass filter of bandwidth determined by the design of the invertors used. This filter must have very steep sides to the attenuation curve to prevent the inclusion of irrelevant noise. The case when the number of invertors lies between one and ten was not investigated by Allnatt and Bragg; however, in instances which have come to the author's attention, the interference could then be regarded as negligible.

4.7 Crosstalk

Crosstalk is most usually defined as the injection of an unwanted signal into a circuit from a similar circuit via a mutual impedance which may be resistive or reactive. An example is the crosstalk which may occur between sources in a mixer or switching matrix. The restriction to similar circuits avoids the inclusion of phenomena such as the injection of supply hum from a power cable into the sheath of a nearby coaxial cable.

The possibilities of crosstalk occurring are numerous and the likely sites of trouble are quite varied, so it is not possible to give more than the basic rules of procedure.

The definition of the amount of crosstalk will depend on circumstances, and a certain amount of common sense may have to be used. In most instances, the crosstalk will be of a video signal into another video-signal channel, and the crosstalk in decibels is then defined as 20 $\log_{10}$ of the ratio of the peak-to-peak picture component of the wanted signal to the peak-to-peak picture component of the crosstalk signal, although, in instances where the crosstalk signal is very distorted, the ratio of the two overall peak-to-peak amplitudes may have to be taken. For other types of crosstalk signal, e.g. subcarrier crosstalk, the ratio of the picture component of the wanted signal to the peak-to-peak crosstalk signal will normally be used, since this is the quantity of interest.

Since the crosstalk takes place via a mutual impedance, it is important to maintain the normal operating impedances of the circuits. A typical measurement procedure would be as follows. With all circuits operating normally, the peak-to-peak amplitude of the wanted signal is measured on a waveform monitor. The wanted signal is now removed and the input of the circuits is terminated correctly. Then the peak-to-peak amplitude of the crosstalk signal is measured, if necessary with the introduction of additional amplification. The crosstalk in decibels is then given by 20 $\log_{10}$ of the ratio of these two amplitudes. A discussion of crosstalk as a picture impairment is given by Fowler (1951).

There is one phenomenon, chrominance–luminance crosstalk, which is not dealt with here because of its anomalous nature. It takes the form in a colour video signal of a variation of luminance amplitudes in accordance with the chrominance amplitudes, and consequently partakes more of the nature of a nonlinearity effect than of simple crosstalk. However, in view of its close association with chrominance–luminance gain inequality, it is dealt with in Section 6.3.6.

4.8 References

ALLNATT, J. W., and BRAGG, E. J. W. (1968): 'Subjective quality of television pictures impaired by sinewave noise and low-frequency random noise', *Proc. IEE*, **115**, (3), pp. 371–375

FOWLER, A. D. (1951*a*): Observer reaction to low-frequency interference in television', *Proc. Inst. Radio Engrs.*, **39**, (10), pp. 1332–1336

FOWLER, A. D. (1951*b*): 'Reaction to video crosstalk', *J. Soc. Motion Picture Televis. Engrs.*, **57**, pp. 416–424

GOUSSOT, L. (1959): 'Le brouillage des images de télévision par les signaux parasites', *L'Onde Élect.*, (386), pp. 352–361

5 NONLINEARITY DISTORTION

5.1 General

Apart from one or two very particular exceptions, all television apparatus is required to be amplitude-linear; i.e. at all points within the working range, a given increment in the input voltage must always give rise to the same increment in the output voltage. The practical behaviour may be expressed mathematically as

$$\Delta e_0 = \frac{\partial e_0}{\partial e_i} \Delta e_i \quad (e_{min} \leqslant e_i \leqslant e_{max})$$

$$= g\Delta e_i, \quad \text{say}$$

where the partial derivative has been used because g is a function of frequency. It is also, in general, a complex quantity, so that the phase angle also may change as the input level is varied. The modulus of g is known as the 'incremental gain' or 'slope gain', which is not to be confused with 'differential gain' (Section 5.4.2).

Obviously, g will have an infinite range of both real and complex values, since it is a function of both amplitude and frequency, and some considerable degree of organisation is required if the nonlinearity is to be expressed in terms simple enough for use. With colour signals, nonlinearity implies interaction between the chrominance and luminance channels, and the consequent gain and phase changes must also be taken into account.

5.2 Line-time nonlinearity

Fortunately, the change of phase angle with input level can be disregarded as far as monochrome or luminance measurements are concerned, first because at the lower video frequencies, which form the bulk of the energy in the luminance signal, the phase change is normally negligible, and secondly because such changes, even if they became appreciable, would not affect the picture to any important extent.

Even so, the range of g is still infinite, so some still further simplification must be introduced. The method which is becoming universal is to define the nonlinearity in terms of the minimum and maximum values of g over the range of amplitudes assumed by the picture component of the video signal The synchronising-pulse region is treated quite separately

For practical reasons, g is not itself measured; but the minimum and maximum values of output voltage are found from a set of equal increments in input voltage The nonlinearity distortion is then defined (CCIR, 1966) as

$$D_{NL} = \frac{\Delta e_{max} - \Delta e_{min}}{\Delta e_{max}} = \frac{100(\Delta e_{max} - \Delta e_{min})}{\Delta e_{max}} \text{ per cent.}$$

Since $\Delta e_{max} = g_{max}\,\Delta e$ and $\Delta e_{min} = g_{min}\,\Delta e$, where Δe is the common increment in input voltage,

$$D_{NL} = \frac{(g_{max} - g_{min})}{g_{max}} = 1 - \frac{g_{min}}{g_{max}}$$

normally expressed as a percentage.

The problem of measurement in this instance now reduces to the provision of a suitable test signal which provides a range of equal increments of input voltage over the picture-signal range, followed by the accurate measurement of the minimum and maximum increments in the corresponding output signal. Two methods are recommended by the CCIR (1966), each of which has some merits and some demerits. Since each measurement makes use of a test signal whose duration is a complete line, the distortion measured is said to be the line-time nonlinearity. As a result of its application to the measurement of colour signals, the term 'luminance nonlinearity' is also encountered.

5.2.1 Sinewave on sawtooth

This variant is not in common use in the UK, but is standard in quite a large number of other countries. The basic test signal consists, as can be seen from Figs. 5.1*a* and *b*, of a linear line-time sawtooth waveform on which is superimposed a 100 mV pk–pk sinewave with a frequency equal to $0{\cdot}2 f_u$, where f_u is the nominal upper limit of the videoband; i.e. 600 kHz for the 405-line system and 1 MHz for 625-line systems. For this purpose, the bandwidth of all 625-line systems is assumed, by international agreement, to be 5 MHz.

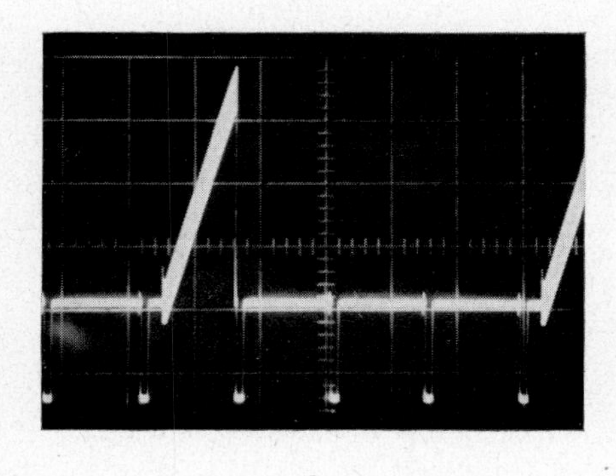

a

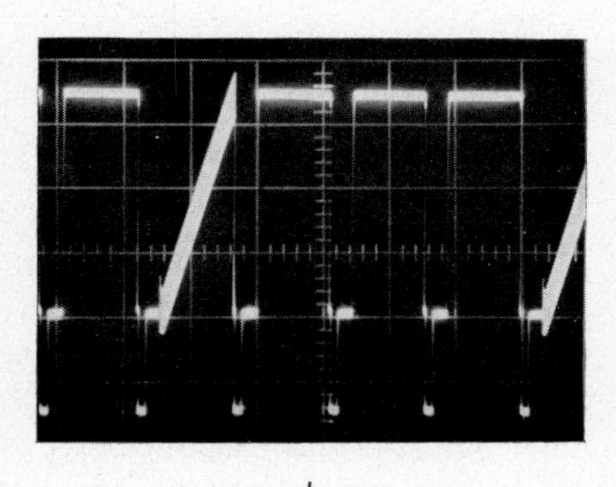

b

Fig. 5.1 Sawtooth waveform with superimposed 1 MHz sinewave
a CCIR waveform in '3 blacks' mode
b CCIR waveform in '3 whites' mode

This waveform is equivalent, in round numbers, to 50 equal increments of input level equally spaced along the active portion of the line, and, in an ideal case, would give rise to 50 equal steps of output level. The sawtooth may be inserted on every line; or, to explore the dynamic range of the video signal more satisfactorily, it may be combined with lines containing an optional white bar (Fig. 5.1*b*). The reasons for this are discussed in Section 5.2.4.

The measurement is carried out at the output of the system under test by passing the output signal through a highpass, or preferably a

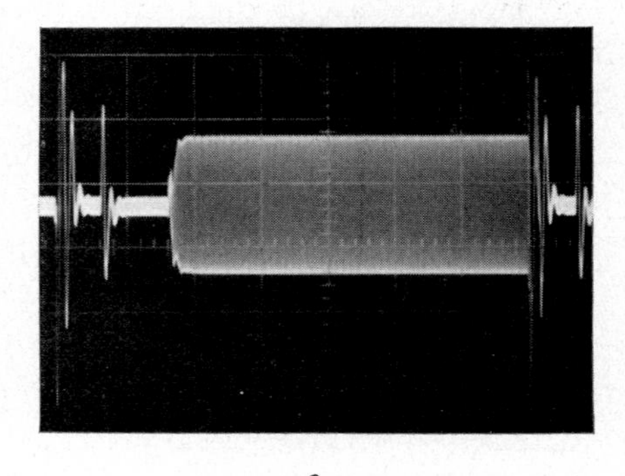

a

b

Fig. 5.2 Filtered 1 MHz waveforms
a No distortion
b With severe distortion

bandpass, filter, so as to pass the measuring frequency $0{\cdot}2f_u$ together with sufficient of the sidebands to avoid marked distortion at each end of the separated burst of sinewave frequency. The resultant signal, when displayed on a waveform monitor having a high enough gain, gives the appearance of Figs. 5.2*a* and *b*. The maximum and minimum amplitudes are measured by a graticule and the distortion calculated as described above.

The procedure has to be modified by the exclusion from the measurement of a small area at each end of the separated sinewave envelope, to avoid the effect of possible transient distortion arising from the filtration being treated as nonlinearity distortion. If the passband of the filter is made very wide, some frequency components of the sawtooth itself may be included, as well as an undue amount of any random noise which is present in the output signal of the equipment under test. On the other hand, if the bandwidth is kept rather

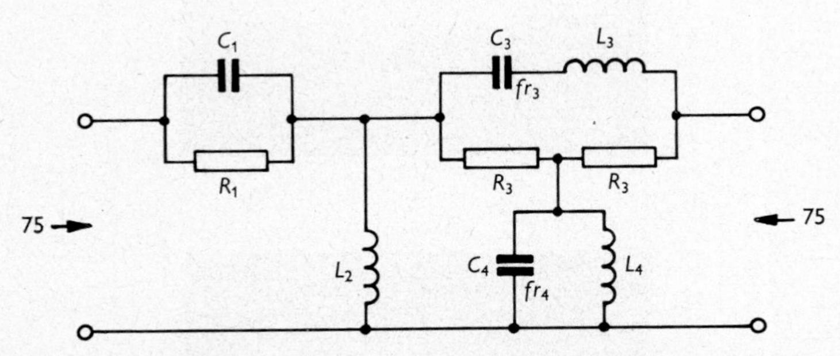

Fig. 5.3 Circuit of separating filter for sinewave-on-sawtooth method

$R_1 = R_3 = 75\ \Omega$	$C_1 = 10000$ pF	$L_2 = 50\ \mu$H
$fr_3 = fr_4 = 1$ MHz	$C_3 = 1250$ pF	$L_3 = 20{\cdot}3\ \mu$H
	$C_4 = 3600$ pF	$L_4 = 7\ \mu$H

small the noise and spurious frequencies will be kept at a low level, but some transient distortion is to be expected at the beginning and end of the display. A trace of such distortion may be noticed in the waveforms of Figs. 5.2*a* and *b*. A reasonable compromise is a total 3 dB bandwidth of about 600 kHz.

The circuit of a constant-resistance bandpass filter with element values for the 1 MHz used for the measurement of 625-line systems is given in Fig. 5.3. The highpass section improves the loss at low video frequencies.

5.2.2 Staircase method

An alternative method, which is preferred in the UK and some other countries, employs a staircase waveform consisting of a number of precisely equal increments in amplitude separated by flat 'treads'. The number of steps recommended for system *I* (UK) is five, although, for checking the linearity of transmitters and some studio equipment, seven or even ten steps may be used to obtain better resolution. In the USA, ten steps are almost universally standard. Fig. 5.4 shows the system-*I* waveform.

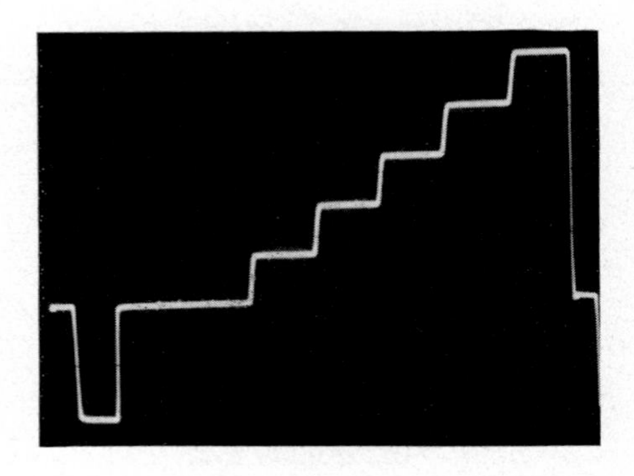

Fig. 5.4 Basic staircase waveform for nonlinearity distortion measurement

The most obvious method of measurement is the estimation of the heights of the successive steps against the graticule of a waveform monitor, but this is far from satisfactory, since the nonlinearity distortion depends on the difference of quantities which cannot be measured with any degree of accuracy.

A much more satisfactory procedure was introduced by the UK Post Office; it is now the recommended method for the UK system *I* (CCIR, 1966). It utilises a special network of bandpass form which effectively differentiates the staircase and converts the transients resulting from the transitions into well-shaped pulses of approximately sine-squared form. The network is given in Fig. 5.5, and the derived train of pulses in Fig. 5.6*a*. Provided certain conditions discussed below are observed, these pulses will have amplitudes linearly proportional to the corresponding step amplitudes.

Under practical conditions, the distortion is quite small, so that the inequality between the pulses is also small, which means that these

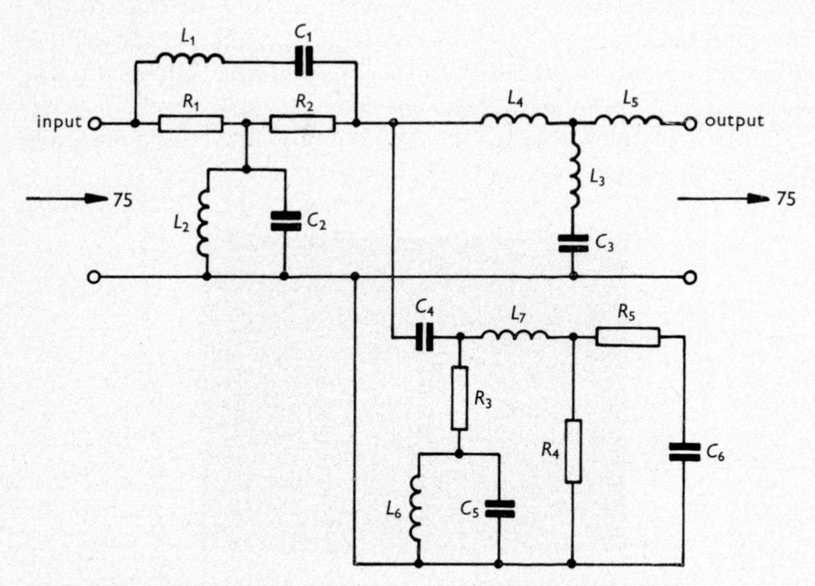

Fig. 5.5 Circuit of constant-resistance differentiating network

$L_1 = 11{\cdot}2\ \mu H$	$C_1 = 6260$ pF	$R_1 = 75\ \Omega$
$L_2 = 35{\cdot}25\ \mu H$	$C_2 = 1989$ pF	$R_2 = 75\ \Omega$
$L_3 = 3{\cdot}08\ \mu H$	$C_3 = 3407$ pF	$R_3 = 75\ \Omega$
$L_4 = 33{\cdot}55\ \mu H$	$C_4 = 3649$ pF	$R_4 = 166{\cdot}6\ \Omega$
$L_5 = 6{\cdot}25\ \mu H$	$C_5 = 2871$ pF	$R_5 = 62\ \Omega$
$L_6 = 27{\cdot}85\ \mu H$	$C_6 = 4810$ pF	
$L_7 = 12{\cdot}1\ \mu H$		

Half-amplitude duration = 1 μs
All tolerances are ± 1%

small differences may be measured with greater precision by increasing the gain of the waveform monitor to a sufficient extent and 'windowing' the display, since any nonlinear effects in the monitor will evidently have only a 2nd-order effect on the accuracy of the measurement. In this manner, amplitude differences as low as 0·1% can be estimated with relative ease, although it is necessary to have available a waveform monitor having a sufficiently high gain, say at least 10 mV/cm. An example is given in Fig. 5.6*b*. A device which has proved to be extremely useful is to combine the network with a small

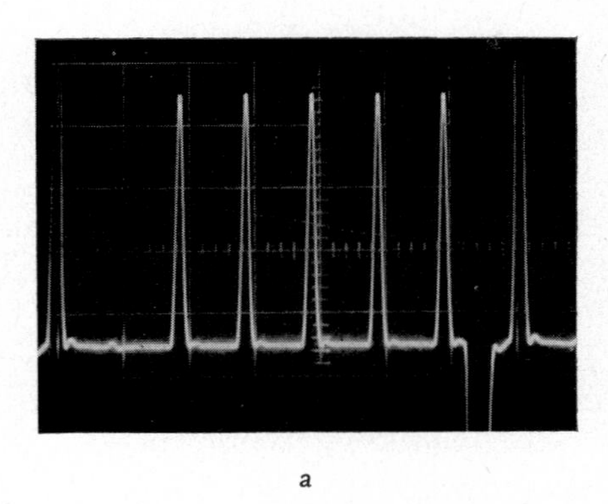

a

b

Fig. 5.6 Differentiated staircase waveforms
a Complete waveform
b Windowed waveform (error on first step = 0·25 %)

battery-driven transistor amplifier, preferably equipped with a means for adjusting the d.c. level of the output signal in case the waveform monitor has a limited range of vertical shift.

This filtration technique has shown itself to be extremely useful and practical under operational as well as under laboratory conditions. An incidental advantage is the quantised nature of the resulting information. Not only is it very simple to pick out the tallest and the shortest pulse from which the nonlinearity distortion can be calculated by the method of Section 5.2, but it becomes extremely convenient to record for future reference the relative amplitudes of all the pulses, thus providing the amplitude–amplitude transfer function in a skeletonised form. From such information, valuable deductions about the stability of the amplitude–amplitude transfer may be drawn.

However, if the highest possible measurement accuracy is desired, there is one special precaution which must be taken in constructing the network when it is in the form given in Fig. 5.5: namely, that L_2 must be designed to have the highest possible Q factor; ideally the Q factor should be infinite. The error does not arise from any change in the transfer characteristic of this portion of the network; if it were so, the change could be compensated, at least to a fair degree of approximation, by adding a resistor R_x of suitable value across the series arm, as is shown in Fig. 5.7*a*. The real reason can be seen by a consideration of Fig. 5.7*b*, which is the simplified equivalent circuit of Fig. 5.7*a* at very low frequencies, where it becomes clear that the input of the network is coupled to the output via the resistor r, which is the effective series resistance of the inductor L_2 at low frequencies. As a result, the output waveform has added to it an approximate sawtooth component whose amplitude is directly proportional to r. The resulting tilt on the train of pulses can form a serious source of error when the distortion measured is small.

The error can be entirely eliminated by a classical device which has been used to improve supply-line decoupling in amplifiers as well as for uses similar to the present. The principle is explained in Figs. 5.7*a* and *b*. L_2 is replaced by a bifilar winding having the same inductance L_2 on each side; the windings are connected in 'series opposing'; i.e. the two corresponding earthy ends are joined. Then, since the mutual coupling factor is close to unity as a result of the bifilar construction, it follows that the two upper ends of the transformer are at the same potential. Hence they can be considered to be connected together, except as regards the equivalent series resistances of the windings, which are now separated and no longer provide mutual resistive coupling between the two halves of the network, so the possibility of error from this source is removed. The Q factor of L_2 is now no longer particularly important, although, if it should be rather low, it may be necessary to add the shunt resistor R_y to restore the network performance at the expense of a slight increase in the basic loss. The capacitor C_2 is shown split into two halves, which are connected

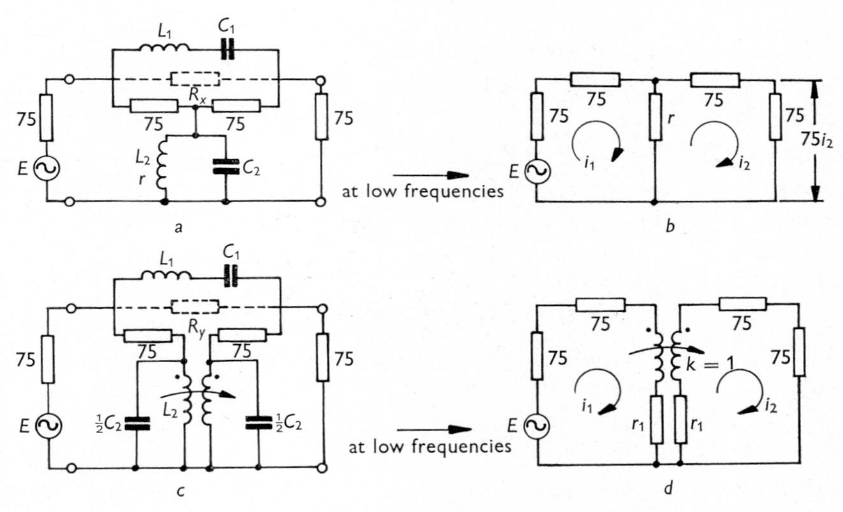

Fig. 5.7 Removal of input–output crosstalk in constant-resistance networks

a Network in standard form
b L.F. equivalent circuit of (*a*)
r = l.f. series resistance of L_2
c Modified network
d L.F. equivalent circuit of (*c*)
r_1 = l.f. series resistance of each bifilar winding

symmetrically to the two windings of L_2. This is not strictly necessary, but, in practice, it is to be preferred, since it tends to reduce spurious effects at the higher frequencies from one winding acting as a section of open-circuited line.

A much simpler network, also originating from the UK Post Office (1961), is shown in Fig. 5.8. As illustrated, it has been modified from the original form by the introduction of the improvement of

Fig. 5.7. The output waveform furnished by this network when fed with a square wave having a short risetime is given in Fig. 5.9*a*, with Fig. 5.9*b* providing more detail of the positive-going output pulse. It is clear that its shape is very similar to that of a sine-squared pulse. However, this network should be used with a little caution, since it is not constant-resistance, and if reflection effects resulting from a poor impedance match are able to influence the behaviour of the apparatus under test, misleading answers may be obtained.

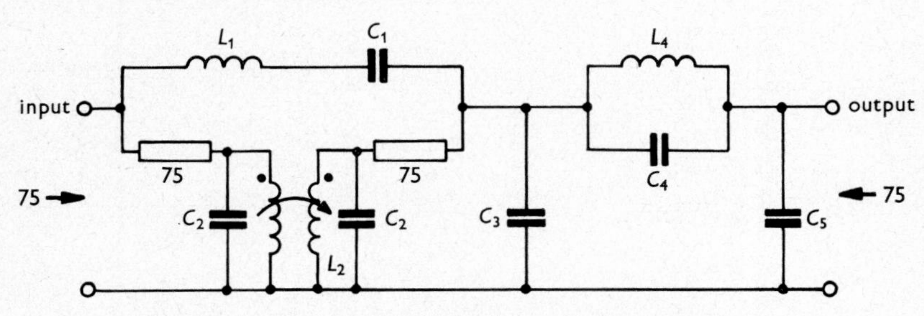

Fig. 5.8 Circuit of simplified differentiating network

$L_1 = 11{\cdot}2\ \mu H$	$C_1 = 6265$ pF	$fr_1 = fr_2 = 0{\cdot}601$ MHz
$L_2 = 35{\cdot}2\ \mu H$ (bifilar)	$C_2 = 995$ pF	$fr_4 = 1{\cdot}552$ MHz
$L_4 = 24{\cdot}6\ \mu H$	$C_3 = 1471$ pF	
	$C_4 = 428{\cdot}3$ pF	
	$C_5 = 8300$ pF	

Half-amplitude duration = $1\ \mu s$
All resistances and capacitances are nominal $\pm 1\%$
Resistances in ohms
Inductances adjusted to resonance at frequencies given

Another fairly simple alternative to the original network has been proposed by Thiele (1967), and has the merit of presenting a constant resistance at each end. The modification of Fig. 5.7 may equally well be applied to L_3 and, in fact, was advocated by Thiele, but has been omitted from the circuit of Fig. 5.10 to make the basic circuit a little clearer. The square-wave response of this network is given in Figs. 5.9*c* and *d*, which should be compared with Figs. 5.9*a* and *b*. It is clear that the pulse delivered by the Thiele network is less symmetrical than that from the UK Post Office networks, and, consequently, the former is somewhat less efficient in reducing random noise.

The effect of the finite Q factor of the shunt inductor of the input section of each of these networks may also be eliminated by a very simple device due to White and Heinzel (1970). It consists merely in removing the right-hand-side member of the pair of 75 Ω resistors

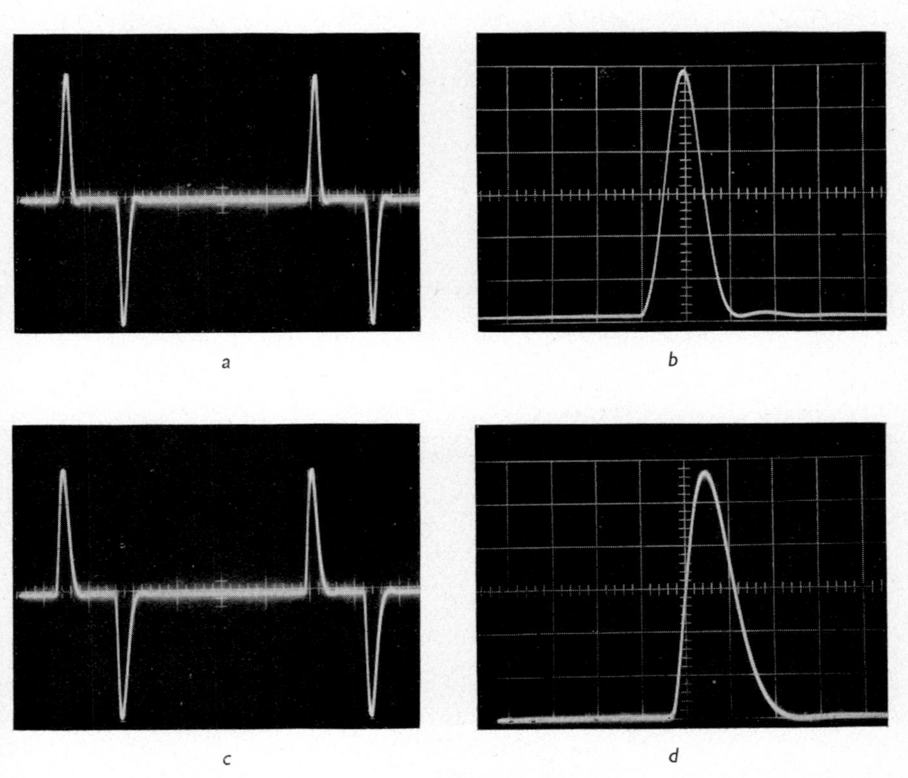

Fig. 5.9 Square-wave responses
a Network of Fig. 5.5
b Detail of pulse in *a*
c Network of Fig. 5.10
d Detail of pulse in (*c*)

forming the bridge, e.g. R_2 in Fig. 5.5, which thereby makes it impossible for the effective series resistance of the shunt inductor to form a common coupling element between input and output of the first section. This bridge resistor is not required in principle to make the input section have a constant resistance, provided that it is presented by the second section with an impeccable 75 Ω termination. This rules out the use of the simpler network of Fig. 5.8 for the majority of purposes, and places a greater onus on the output termination of the network if reflection effects at the input are to be avoided. On the other hand, the bifilar winding has no effect on the network behaviour, and is consequently to be preferred in spite of the slightly greater complication.

It may be surprising, on consideration, that this technique should prove so satisfactory in practice when the quantity actually measured

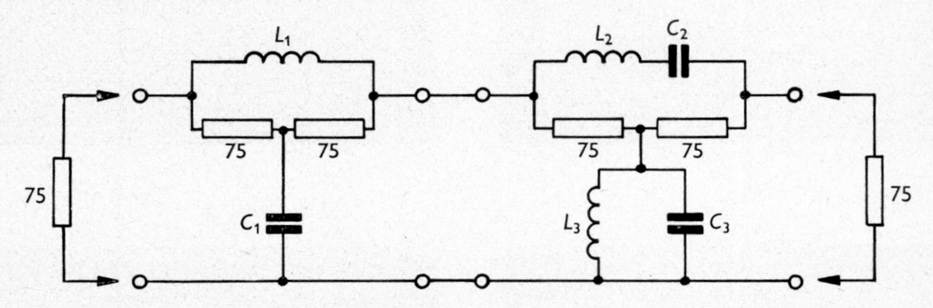

Fig. 5.10 Circuit of proposed constant-resistance differentiating network

$L_1 = 32{\cdot}3\ \mu H \pm 1\%$ $C_1 = 5750\ pF \pm 1\%$

$L_2 = 20{\cdot}4\ \mu H$ } resonate to 404·6 kHz $C_2 = 7600\ pF \pm 1\%$

$L_3 = 42{\cdot}7\ \mu H$ } with C_2 and C_3, respectively $C_3 = 3620\ pF \pm 1\%$

Half-amplitude duration = 1 μs

Resistances in ohms

is not the step height but the maximum rate of change of amplitude over the transition. If all the transitions were identical in shape, the step height and the rate of change would evidently be linearly related; but, under practical conditions, it is unlikely that this would be completely true.

The saving factor is the low upper cutoff frequency of the differentiating network, say, a maximum of 1 MHz for 625-line systems. The process can most simply be understood by considering the edge shaping performed by the network to take place in two distinct stages: namely, shaping of the transition, performed by the Thomson filter section, followed by differentiation. Since no nonlinear process comes between the two filtrations, they are commutative and can be connected in either order. Then, provided the risetime of the step transition is sufficiently short, the shape of the transition at the output of the first section of the network is determined entirely by the transfer function of this section, and the differentiating section is presented with a series of transitions of identical shape, although, in general, of different amplitudes. The amplitudes of the output pulses from the differentiating network are thus in practice strictly proportional to the amplitudes of the corresponding steps.

It follows that the inherent error in the measurement is reduced by making the upper cutoff frequency as low as possible, and this is also of assistance where measurements have to be made under high-random-noise conditions. The lower limit is set by mutual interference between successive pulses in the differentiated train as the pulse duration increases with decreasing bandwidth; a practical figure is about 500 kHz for the networks of Figs. 5.5 and 5.8 depending somewhat on circumstances. The Thiele network of Fig. 5.10 is rather less favourable in this respect owing to the increased trail of the back edge of the pulse, as can be clearly seen in Figs. 5.9*c* and *d*; the upper cutoff frequency must accordingly be raised, compared with the other two networks.

The inherent error is also reduced by decreasing the risetimes of the steps, although there is a limit here if the staircase is also to be used for chrominance-nonlinearity measurements (Section 5.4), since it is inadvisable for the bandwidth of the steps to be so wide that spectral components of appreciable amplitude fall into the chrominance channel. A simple way of ensuring that all conditions are simultaneously met is to generate a staircase waveform with fast transitions and subsequently to pass it through a Thomson network whose first rejection frequency is situated at colour-subcarrier frequency, although it has not proved difficult to design generators in which this precaution is superfluous.

The principal source of error in this type of measurement arises from the difficulty of constructing generators in which the step amplitudes are all precisely equal, although it has been found possible with care to reduce the differences to no more than 0·1%. Where necessary, the errors are noted and used as corrections for the measured results. Another source is waveform distortion of the signal, for example, line tilt, and it may be necessary, under some circumstances, to consider equalising this before nonlinearity measurements are made. In particular, the generator should have no trace of any such distortion. In this respect the sinewave-on-sawtooth test signal is less susceptible to error, as a result of the rejection of the low frequency components by the bandpass filter.

5.2.3 Small nonlinearities

The measurement of very small amounts of nonlinearity distortion such as may be found, for example, in a single video amplifier, presents a special problem, whether the staircase or the sinewave-on-sawtooth waveform is chosen. A standard device is the tandem connection of a number of units, say, ten, which are then measured overall, and the distortion of each is assumed to be the overall value divided by the number in use. This assumes, of course, that the error is systematic and that the individual distortions are roughly equal, which has to be proved in a given instance.

One of the great obstacles to accurate measurement with small amounts of distortion is the need to amplify the output signal considerably to display the small differences involved. A considerable improvement can be effected by the use of a high-grade difference amplifier with the signal applied to one of the inputs after differentiation, and a flat pedestal applied to the other. Thus the difference amplifier acts effectively as a form of clipper of very low inherent distortion, with the added advantage that the output windowed signal can be calibrated by changing the amplitude of the pedestal by known amounts. Further information on a technique of this type can be found in the paper by White (1970).

An elaboration which automatically avoids a number of the inherent errors consists in directing the output test signal, as it is received, into two paths by a resistive pad. One of these paths contains the apparatus under test and the other a variable delay line adjusted so as to compensate for the delay of the test object. These paths terminate in the two inputs of a high-grade difference amplifier.

The gains of the two paths are adjusted to give zero output from the difference amplifier for one selected step, and the delay is adjusted to minimise edge effects on the steps. The differences between the amplitudes of the output steps can now be measured and then converted into percentages of the original step amplitudes simply by removing the input from the test path and terminating it, after which attenuation is added in the other path until the observed step amplitude has the same value as was obtained with the difference measurement. This process is carried out for each of the steps in turn.

This is a convenient and accurate technique, and can be used quite generally whenever both input and output of the apparatus under test are simultaneously available. The need for an adjustable delay line is a certain drawback; but, on the other hand, it does not need to be of superb quality, and the bandwidth required is not large. Although it has been described in terms of the staircase waveform, it should be evident that the technique can also well be applied to the sawtooth waveform.

5.2.4 Change of average picture level

The picture information contained in the video signal can vary from virtually zero to the other extreme of a completely white field in each line. Hence, whenever the d.c. component of the signal is lost as a result of an a.c. coupling, the signal will sweep over different portions of the amplitude–amplitude transfer characteristic, depending on the nature of the picture carried by the signal. This means that, in general, the nonlinearity distortion is a function of the average picture level; and consequently, any single measurement may give an entirely erroneous impression of the behaviour of the equipment under test. The CCIR has accordingly recommended that the measurement of nonlinearity distortion should be carried out at two values of average picture level (CCIR, 1966).

It would be pertinent to consider what exactly is meant by 'average picture level'. Two definitions are possible, depending on whether or not one includes the blanking intervals in the calculation. The latter seems by far the more reasonable, since it makes the average picture level 0% for an all-black picture and 100% for an all-white picture. This also agrees with the definition given in the IRE 'Standards on television measurement of differential gain and differential phase' (1960), and has the further advantage of being independent of the television standards used.

The average picture level (a.p.l.) will therefore be defined as the signal level with respect to blanking level during the active picture periods only, averaged over a field and expressed as a percentage of the difference between blanking and white levels. The sawtooth and

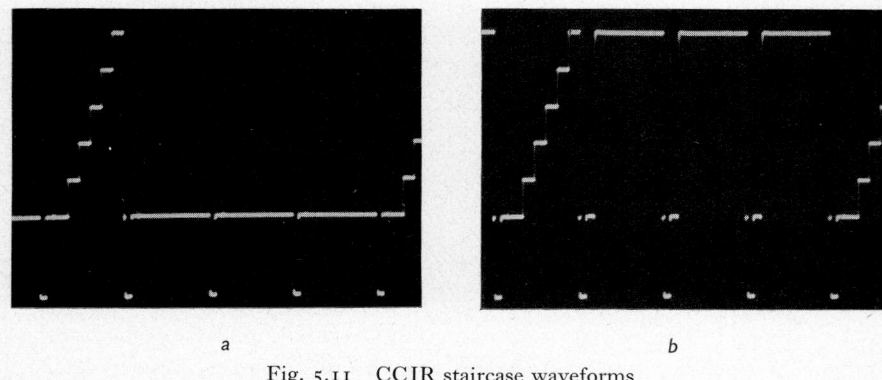

Fig. 5.11 CCIR staircase waveforms

a '3 blacks' mode
b '3 whites' mode

staircase signals described above then have average picture levels close to 50%, which appears to be only slightly higher than the value for representative programme material (Quinn and Newman, 1965).

The simplest way of achieving a variation in average picture level in practice is to insert the actual test signal, either staircase or sawtooth, on one line only of a sequence, the other lines being at either white or black level. The CCIR standard is one test line followed by three lines bearing either an all-black or an all-white signal, as is illustrated in Figs. 5.1 and 5.11, corresponding to average picture levels of about 12.5% and 87·5%, respectively. The nonlinearity is then measured by one of the methods described above at each of these values of average picture level, and the worse of the two readings is taken as the significant figure. It is usually worth while to record both readings in each instance, since this information is often of value as a guide to the stability of the equipment under test. The test-signal generator should be designed so that either of these alternative signals can be obtained by throwing a switch; and it has been found very useful to arrange for the changeover to be affected by a relay, so that it can be actuated either manually or automatically at a chosen rate when the generator and waveform monitor are separated.

An alternative form preferred by some administrations, particularly those using 525-line standards, employs a sequence of one test line followed by a group of four lines either entirely white or entirely black, corresponding to average picture levels of 90% and 10%, respectively. This alternative to the CCIR-preferred signal has the minor advantage that, in a fullfield signal, i.e. one containing the field information, the sequence automatically starts correctly at the beginning of each field, with the result that even the simplest triggering circuit on the waveform monitor can be used without interference on the display from the field synchronisation. This is because the number of lines in all standards, except the now obsolescent 819 lines, is divisible by five. However, no difficulty is found in practice with the CCIR signal in designing the generator to provide the correct sequence, and the added complication is small. The slightly more extreme values of average picture level given by this signal are perhaps less advantageous than those afforded by the CCIR signal, but this is very much a matter of opinion.

It is worth noting that the change of average picture level is apt to disturb the synchronisation with some waveform monitors, and it is preferable to employ an external trigger, either derived from the generator, for local measurements, or from a simple synchronising separator when the generator is remote from the monitor. However, television waveform monitors are likely to cope with such signals without any special precautions.

5.2.5 Comparison of waveforms

The principal claim made for the sawtooth is that it is very much simpler than the staircase to generate to a high degree of linearity, which is evidently desirable when small amounts of nonlinearity are to be measured. This, it must be admitted, is true, but, on the other hand, the difficulty of designing a linear-staircase generator are sometimes exaggerated, and the extra expense of the circuitry is not forbidding.

Another claim made for the sawtooth is that its resolution is inherently much greater than that of the staircase, but this is not a very convincing argument, since the curvature of the transfer characteristic is so slight that the shape of the curve is quite adequately defined by five ordinates. There are, however, certain rare, but nevertheless important, occasions when a 10-step staircase is preferred. These arise especially during alignment in situations where it is possible for a maladjustment to produce a 'kink' in the curve which is more easily detected with a greater number of steps than five. An instance is the alignment of certain transmitters for optimum linearity. Nevertheless, it must be emphasised that a 5-step staircase is still recommended for the normal nonlinearity measurement.

A simple modification to the standard 5-step staircase has been found useful for development work and acceptance testing. It was built into a generator designed under the author's direction, but it may also be fitted to an existing generator. The device consists in the introduction of a known attenuation into the staircase component of the waveform, leaving the synchronisation component intact, together with a variable pedestal sufficient to shift the pedestal over at least the complete amplitude range of the video signal. Then, for example, a 6 dB attenuation of the staircase component is equivalent to doubling the number of steps as far as resolution is concerned, and the variable pedestal makes it possible to check for the sudden onset of distortion in the synchronising and superwhite regions, as well as to explore the normal-picture region in greater detail. Typical waveforms are shown in Fig. 5.12.

The resolution of the staircase waveform is naturally less than that of the sawtooth, but this is offset by the quantising of the information in such a way that the shape of the transfer characteristic can be unambiguously recorded with little effort and in standardised form.

As regards the networks used for evaluating these two waveforms, the preferred differentiating network of Fig. 5.5. is more complicated than the 1 MHz separating filter of Fig. 5.3, but not to the extent that this should be taken as a deciding factor. In favour of the staircase is the rather better noise rejection which may be obtained without unduly influencing the measurement accuracy, at least for operational

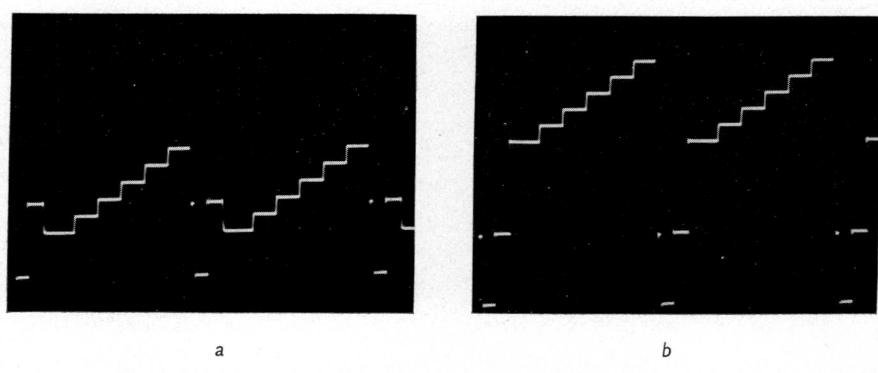

Fig. 5.12 Use of pedestal for increase resolution
a Below black level
b Above white level

purposes, and the protection given against the interpretation of filter distortion as nonlinearity distortion. It is, however, more vulnerable to low-frequency waveform distortion.

To sum up, the arguments are by no means conclusive in either direction, and careful comparative tests show no significant differences when the two waveforms are used to measure the same circuit or equipment. However, apart from the significant operational advantages of the staircase, there is the pragmatic reason for preferring it: namely, that it has been internationally accepted for the measurement of chrominance nonlinearity, and also as a component of insertion test signals (Chapter 7). Consequently, it does not seem reasonable to insist on the employment of the sawtooth for one area of measurement alone, where in any case the staircase would function at least equally well.

5.2.6 Overload test

With long distribution links, it is possible for the gain to rise by a small, but significant, amount above unity; and it is as well to ensure that the undue amount of distortion cannot occur even under near-fault conditions. It has accordingly become customary to ensure, usually under acceptance or postmaintenance conditions only, that the equipment has a sufficient margin of safety. This does not apply when the system contains some inherent limitation, such as an amplitude limiter. The standard amplitude increase hitherto has been 3 dB, but under present conditions, when limits are held more rigidly than previously, this seems a little generous.

Such a measurement would be meaningless if the accuracy of the test signal were to be impaired by the increase in its level, and the generator must be designed so that this cannot occur. The simplest and safest method is to make the output level, say, always 3 dB above 1 V pk–pk; but under normal conditions of use, a 3 dB pad is inserted in series with the output. For overload tests, the pad is switched out, preferably by a switch which has deliberately to be held in the overload-test position so that it cannot be actuated inadvertently during normal maintenance testing. This pad has the slight incidental advantage that the return loss of the output of the generator is improved by 6 dB.

5.3 Dynamic-gain distortion

One particular aspect of nonlinearity as it affects distribution networks is sometimes distinguished from nonlinearity distortion as defined above, especially in the USA, since it appears to give rise to a variation in the gain of the system, rather than a modification of the shape of the transfer characteristic as a consequence of a change in the average picture level. It has been called, not very appropriately, ‘dynamic gain’ (Rhodes, 1970), and is considered to apply either to the luminance or to the chrominance signals; in the former case, the picture and synchronising regions may be treated separately. The distinction between the two forms of distortion is not recognised in international standards except as regards the synchronising region; this is discussed below.

Another type of dynamic-gain distortion occurs transitorily as a result of the d.c. step applied to the system by a quasi-instantaneous change in the average picture level. It seems to have no standardised name, but is often referred to, for obvious reasons, as 'signal bumping' (Section 5.3.2.)

5.3.1 Synchronising-pulse nonlinearity

The requirements for the linearity of the transfer characteristic over the synchronising-pulse region are far less severe than over the picture region, since no question arises of any change in gradation. The principal need is to ensure that the picture/synchronising-pulse ratio does not depart too far from the nominal value and that no significant displacement of the leading edge of the synchronising pulse with changes in picture content can occur.

The measurement is preferably carried out with the two CCIR waveforms described above, and the amplitude of the synchronising pulse is found for each of the two conditions. The differences between the measured values and the nominal value should not be greater than the allotted tolerance for the circuit in question.

Since the bottom of the synchronising pulse may very well not be flat under the conditions of test, the amplitude is defined as that between the midpoint of the base of the pulse and blanking level. The reference value is calculated from the amplitude of the staircase or bar waveform in the test signal, which implies that the signal from the generator must be initially very carefully adjusted to the correct picture/synchronising-pulse ratio. The distortion, as a number, is expressed as the percentage ratio of the greater of the two errors in the synchronising-pulse amplitude to the staircase or bar amplitude.

5.3.2 Bump test

A sudden cut during a programme from bright to very dark scenes, or vice versa, is equivalent to the addition of a step waveform to the d.c. component of the video signal. Since the signal path, in the case of a long distribution network, will contain *CR* time constants, feedback amplifiers and regulated power supplies, the signal will not necessarily recover from this transient in a well behaved manner. A considerable overshoot may follow the transition, and, in bad cases, particularly with feedback amplifiers having an insufficient margin of stability at very low frequencies, a damped oscillation of considerable magnitude may ensue.

Such amplitude changes are not a cause for alarm in themselves, since they will be removed by subsequent clamps, but a strong possibility exists that severe transient signal distortion may occur owing to the signal being driven into a nonlinear portion of the transfer characteristic as a result of the unusually wide amplitude excursion. Such distortion affecting the picture component of the video signal will not be important, provided that it has a duration of a few fields only, since it will not be perceived by the eye. However, if the synchronising pulses are badly crushed, there may be a momentary loss of synchronisation, or 'picture roll', which is very disturbing to the viewer. It is consequently essential to ensure that this condition can never occur.

The test is usually carried out with a waveform which can change between all black and all white bars within a line or so, and no loss of d.c. can be allowed in the output coupling circuit. The standard CCIR waveform with fast switching between the two states is sometimes also employed, but it is a less severe test.

As the changeover takes place, the synchronising region of the video-test waveform is examined on a waveform monitor having a very slow sweep. A long persistence phosphor is essential, and a storage oscilloscope, if available, is invaluable. It is very advantageous if the waveform generator is so designed that the mark/space ratio of the bump-test waveform is precisely 50:50, since this ensures that the transition always occurs at the same point on the screen. Even so, it is useful to take a photograph of the transition, since the measurement of the amount of synchronising-pulse crushing is thereby greatly facilitated. Such a photograph is given in Fig. 5.13, where, for the sake of clarity, a gross fault condition is shown.

The test waveform started in the all-black condition, i.e. synchronising pulses only, and then abruptly changed to all white bars. At the instant of transition, the picture signal momentarily underwent a violent positive excursion, indicated by the fine white line, and very quickly afterwards was driven downwards into a region of severe non-

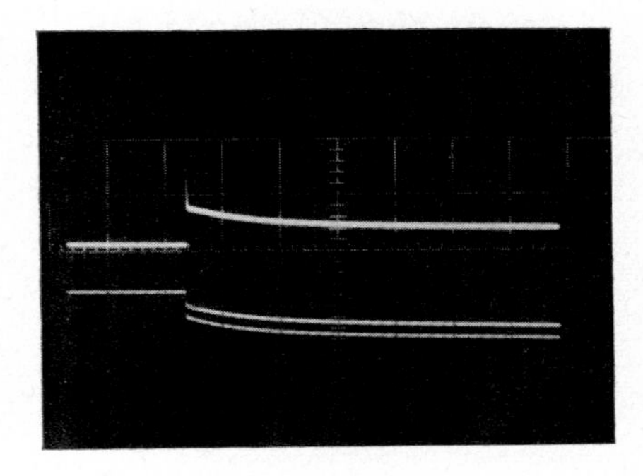

Fig. 5.13 Response of poor circuit to bump test

linearity, indicated by the compression of the synchronising pulses to about one-quarter of their normal amplitude. On the other hand, the bar amplitude was reduced by only 8%, which, provided it lasts for not more than a few fields, is not of much significance. It is important that the system should also be tested with a transient of the opposite polarity, i.e. by a change from all white lines to all black lines.

It is advisable in practice to ensure that all equipment is capable of handling a highly oscillatory signal of this type generated at some other point in the system. There is a need for some standardised test waveform for this purpose. An empirical modification of a feedback video amplifier to give an oscillatory signal when driven by the bump waveform has been found suitable. An overshoot in the region of 30% is suggested as a reasonable compromise.

5.4 Chrominance nonlinearity distortions

With a purely luminance signal, the effect of nonlinearity takes the form of a change in the shape of the transfer characteristic corresponding to the range of video-signal amplitudes. However, when the signal contains separate and independent luminance and chrominance information, intermodulation effects between the two channels occur in addition to the individual channel distortions. We are concerned here with three effects: luminance-conscious changes in the amplitude and phase shift of the subcarrier and a chrominance-conscious change in the amplitudes of the luminance components. The first two are universally known as differential-gain and -phase distortions, respectively; not particularly appropriate names, but too well established to be replaced. The third is known as chrominance–luminance crosstalk (in the USA, axis shift); and since it has a particular importance in connection with the measurement of the linear chrominance distortions as well as in its own right, it has been thought proper to discuss it in that context (Section 6.3.6).

5.4.1 Test signals

The two test signals recommended by the CCIR for measuring differential phase and gain (CCIR, 1966*b*) are based on the monochrome (luminance) signals discussed above. However, in this instance, the sawtooth and staircase are utilised to examine the behaviour of a sinusoidal subcarrier test signal over the range of amplitudes assumed by the luminance component of the video waveform. They are subsequently discarded and the actual measurement is carried out on the subcarrier component only.

In the form in which they are actually used, these waveforms become a linear line sawtooth to which 100 mV pk–pk of subcarrier has been added (Fig. 5.14*a*), and a 5-step staircase to which has been added 140 mV pk–pk of subcarrier (Fig. 5.14*b*). It is highly desirable that both forms should be used at the two values of average picture level recommended by the CCIR, i.e. a single test line followed by either three black lines or three white lines, so that the dynamic range of the transfer characteristic is more adequately explored. Both waveforms normally have the colour burst added. The staircase in the two CCIR modes with colour burst is illustrated in Figs. 5.15*b* and *c*, and the detail of the test line is shown in Fig. 5.15*a*.

It will be noticed that the subcarrier is not added to the whole of the test line but only to the active portion, attention being paid to the necessity of providing an adequate duration of subcarrier at black level, since, as will be explained below, this is required as a reference

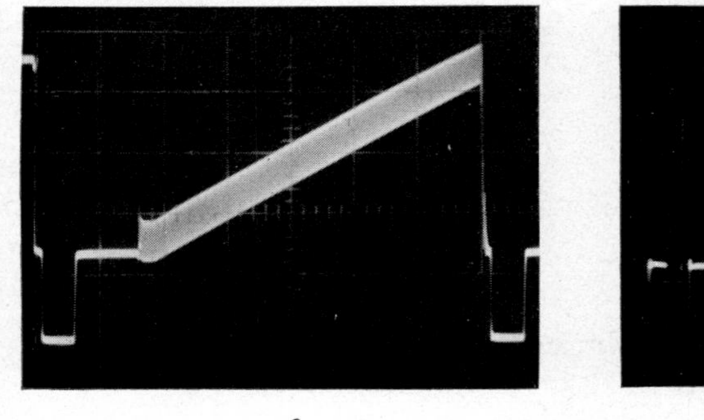

a

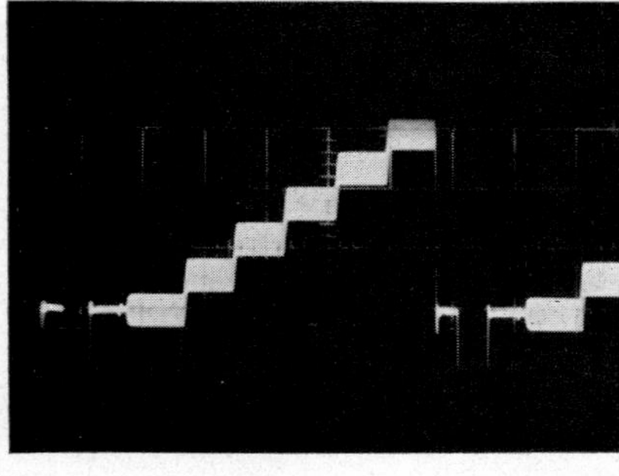

b

Fig. 5.14 Test waveforms for chrominance nonlinearity measurement
a 1 MHz on sawtooth with pedestal
b Subcarrier on staircase

for the measurement. In passing, it might be remarked that, in Fig. 5.14*a*, the waveform includes a pedestal, so that the reference level is actually above black level. The need to blank the subcarrier except during the active line period introduces some difficulty in the design of the generating equipment, and special precautions must be taken to avoid introducing into the test waveform itself distortion of the type it is planned to measure.

The question whether or not the all-black and all-white lines in the CCIR waveform should also contain the superimposed subcarrier can only be answered with reference to the particular test equipment to be employed. This is considered below in the description of the measurement techniques available.

The superimposed-subcarrier amplitude should in principle be as small as possible so that only a differentially small fraction of the transfer characteristic is explored at each of the luminance levels, but if the amplitude were reduced too far, the signal/noise ratio would severely restrict the accuracy of measurement. On the other hand, if the subcarrier were made large, the result would be in error as a

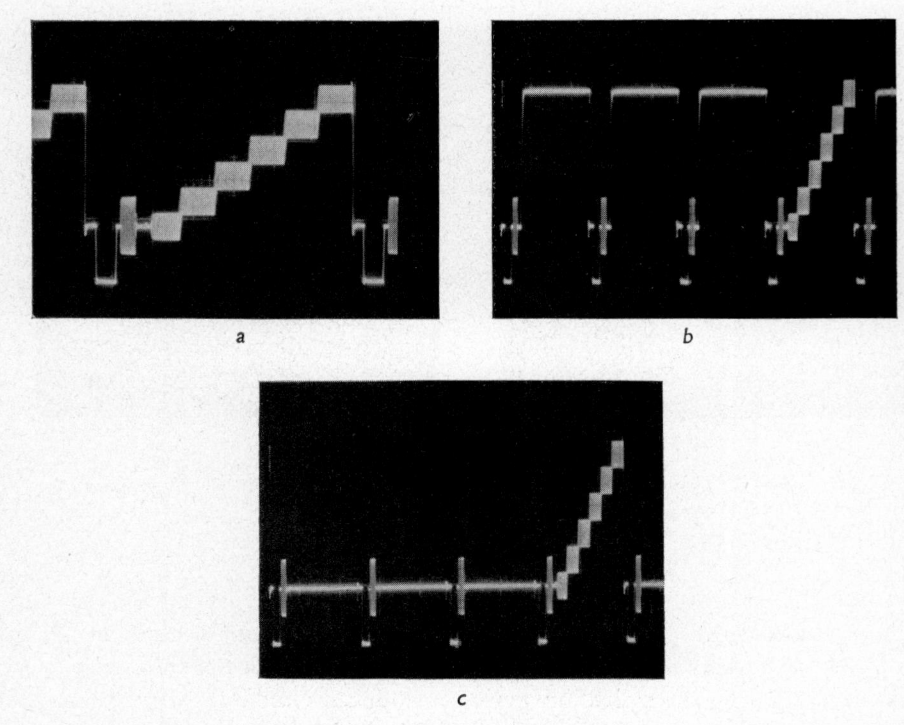

Fig. 5.15 Staircase waveforms for chrominance nonlinearity measurement

a Detail of test waveform
b CCIR '3 whites' mode
c CCIR '3 blacks' mode

result of the measurement being taken along a chord of the transfer characteristic instead of effectively along the tangent at the point corresponding to the desired luminance level. About 100 mV pk–pk has been found to be a good working compromise, and this figure is employed with the sawtooth since it also agrees with the standard level of overlaid sinewave used for the measurement of line-time nonlinearity. 140 mV is preferred with the staircase, since it is equal to the height of each step, which makes it very simple to standardise the subcarrier level by setting the top of one subcarrier block to the same level as the bottom of the immediately higher block. This condition is shown in Fig. 5.15*a*, although, as so often occurs in waveform photographs, the overexposure in the regions of low writing speed which is forced on one by the need for revealing some detail in the regions of high writing speed, i.e. during the transitions, brings about a spread of the image which appears to show that the subcarrier amplitude was actually greater than 140 mV pk–pk, even though visually the adjustment was exact.

It is sometimes argued in favour of raising the subcarrier level superimposed on the luminance test waveform, that it increases the sensitivity and improves the signal/noise ratio, while providing a more severe test of the chrominance nonlinearity. This is fallacious; what is really required is an accurate picture of the shape of the differential-phase and -gain curves, and all that a large subcarrier amplitude achieves is an averaged value, which very frequently is considerably less than the maximum value registered when a small level is employed. Attempts to compare the readings found with a colour-bar signal, considered, for this purpose as an extreme form of chrominance nonlinearity test signal, have shown that it is quite possible to find that the differential phase, for example, appears to be near zero, when small-signal measurements have disclosed that the distortion is actually quite severe. These considerations do not necessarily apply to the testing of transmitters where anomalous types of nonlinearity distortion are possible.

The requirements of the luminance waveform which carries the subcarrier are in most respects less severe than in the case of monochrome- or luminance-channel measurements, since its function is to situate the subcarrier at a number of points spaced over the transfer characteristic of the apparatus to be measured, and consequently, small deviations from strict linearity only give rise to 2nd-order errors in the measurement. However, one important precaution must be taken, particularly with the staircase waveform. This is to ensure that no possibility exists of spectral components being present in the vicinity of subcarrier frequency which would pass into the chrominance-measuring circuits and be interpreted as part of the superimposed subcarrier.

The most satisfactory method of dealing with this has been found to follow the generated luminance waveform by a filter having a reasonable loss at subcarrier frequency. Normally, for the sake of the luminance of monochrome nonlinearity measurements, it is desirable for the staircase waveform to have well shaped, symmetrical transitions. This may economically be achieved at the same time if the filter concerned is a Thomson filter (Chapter 6) whose cutoff frequency is chosen so as to locate a frequency of infinite loss at subcarrier frequency.

So far, no guide has been given as to which of these two waveforms is to be preferred. For some years, the sawtooth has been preferred on the Continent of Europe and the staircase in the UK; more recently, the staircase has been gaining ground in the USA, Canada and some other countries.

The CCIR and CMTT have furthermore recommended the staircase for use in insertion test signals both for the measurement of differential phase and gain and for the measurement of line-time nonlinearity, and such signals will continue to be used on an ever increasing scale (Chapter 7). Moreover, compact test signals having the same form as the insertion test signals will also be very widely employed (Chapter 6). In view of this, it seems highly probable that the sawtooth will eventually be universally replaced by the staircase, except for certain special purposes, both for chrominance and luminance measurements.

A quite different test waveform is employed in the so-called 'Bell–Kelly' method (Kelly, 1954), which is still widely used in the USA, largely owing to some excellent equipment designed for this method. The waveform is a 15·750 kHz sinewave on which is superimposed a small amplitude of colour subcarrier, usually 140 mV. For 50% average picture level, the amplitude of the sinewave is the normal picture amplitude, in US terms 100 IEEE units. Line synchronising pulses are inserted in the troughs of the sinusoid.

At the receiving terminal, these synchronising pulses are removed by a clipper, when the subcarrier may be separated in a simple high-pass or bandpass filter. The remainder of the measurement follows very much on the lines described below, for both differential gain and phase. To avoid a potential source of error, the synchronising pulses should be filtered to remove spectral components from the subcarrier region.

The disadvantage of this signal is its only very approximate resemblance to a video signal, and in particular the impracticability of inserting field blanking. Changes in average picture level can be accommodated, but only by the expedient of increasing the amplitude to 194 IEEE units and then measuring between the appropriate ranges of the signal.

5.4.2 Differential gain

This is a variation in the amplitude of the subcarrier component of the video signal with variations in the amplitude of the accompanying luminance component. Its magnitude (Fig. 5.16) is defined specifically (CCIR, 1966*b*) as the largest change in the amplitude of a small super-imposed subcarrier component as the luminance varies between black level and white level, expressed as a percentage of the subcarrier amplitude at black level. For two values of average picture level, the worse of the two readings is taken as the significant figure.

The measurement technique is very simple. The superimposed subcarrier is separated by means of a highpass or bandpass filter, preferably the latter to minimise the random-noise level. The amplitude of the envelope of the subcarrier is then measured on a waveform monitor, which must be capable of considerable degree of amplification without distortion, so that one side of the envelope may be 'windowed' when very small amounts of differential gain have to be measured. Although this method is not altogether elegant, it nevertheless makes possible the measurement of differential gains down to, say, 0·1%.

Just as with the sawtooth method of measuring the line-time non-linearity, the filter used must have an adequate bandwidth to avoid falsifying the differential-gain reading (Fig. 5.17). Fig. 5.17*a* shows the distorted test waveform before separation of the subcarrier component; the distortion has deliberately been made rather large for purposes of illustration. Fig. 5.17*b* is the subcarrier component separated by means of a bandpass filter with a total 3 dB bandwidth of 2 MHz, and the transition on the first step, which is the principal

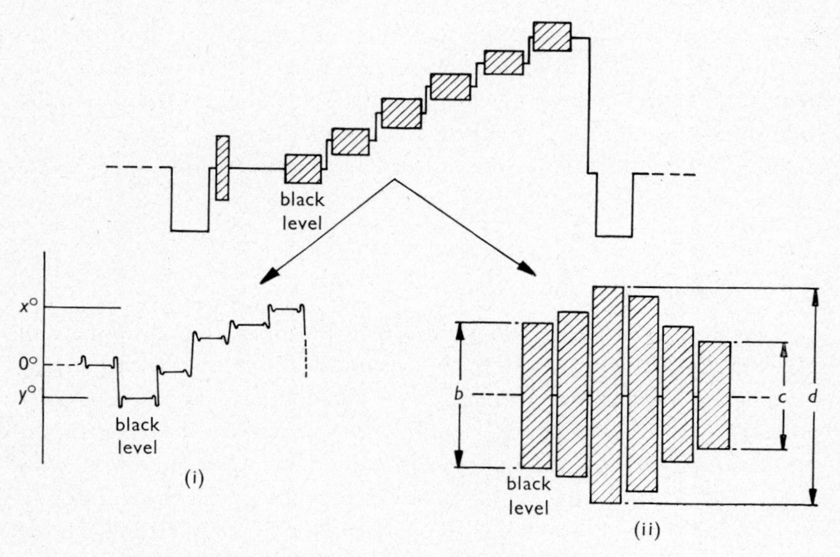

Fig. 5.16 Definitions of differential phase and gain

(i) Differential phase, expressed as $+x$ or $-y$ degrees, whichever is numerically greater

(ii) Differential gain, expressed as $-100\ (1-c/b)$ or $+100\ (d/b-1)$ per cent, whichever is numerically greater

distortion, is very well defined. The amplitude of this first step is 28·5% of the half-envelope amplitude at black level, so that, if this were a practical measurement, the differential gain would be given as 28·5%.

When the bandwidth of the filter is reduced to 300 kHz (Fig. 5.17*c*), the transitions are appreciably blurred, but there is still no difficulty in measuring the amount of the differential gain, although, if a sawtooth had been employed instead of a staircase, the first and last 6% of the effective line time would have to be ignored. This emphasises the importance of providing a sufficiently long duration of subcarrier at black level in the test waveform.

Finally, Fig. 5.17*d* shows the effect of reducing the 3 dB bandwidth of the filter to 100 kHz. The first step is now completely missing, and without further information one would have to call the

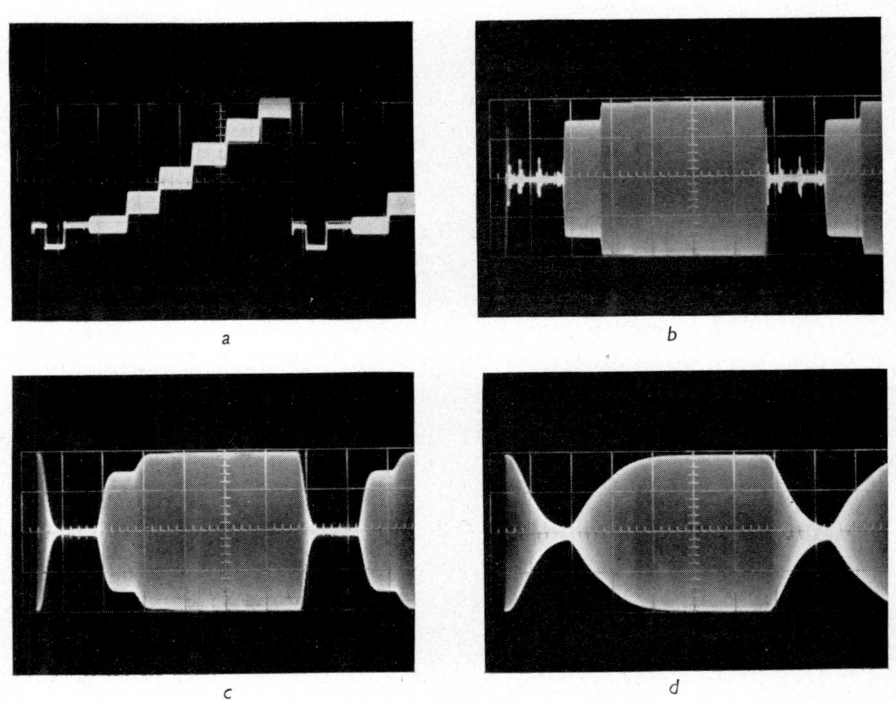

Fig. 5.17 Differential-phase measurement

a Distorted staircase waveform
b Subcarrier separated by 2 MHz bandpass filter
c Subcarrier separated by 300 kHz bandpass filter
d Subcarrier separated by 100 kHz bandpass filter

differential gain zero. Very obviously, a bandwidth of 300 kHz should be regarded as an absolute minimum. A reasonable working compromise between low-noise bandwidth and accuracy is a bandwidth of 1 MHz. The filter of Fig. 5.3 may be used with the element values suitably scaled for the appropriate subcarrier frequency if rather better noise protection is required and a bandwidth of 600 kHz is acceptable.

The direct measurement of differential gain from the subcarrier envelope has the merit of great simplicity, but is not without disadvantage when small values have to be measured, since a considerable burden is then placed on the vertical amplifier of the waveform monitor as a result of the overloading required to reveal the small differences, and errors become possible.

It is consequently preferable to obtain the subcarrier envelope wherever possible by demodulation, which results in a signal of lower frequency, Even so, one is not completely out of difficulty, since, either the full rectified envelope signal must be applied to the waveform monitor, still with some risk, if somewhat lower than before, of distortion, or the envelope can be employed as an a.c. signal, in which case calibration is less straightforward. A useful device in this situation is that recommended for the measurement of small nonlinearity distortions (Section 5.2.3), in which the demodulated envelope is applied directly to one input of a high-grade difference amplifier and a flat pedestal is applied to the other, by means of which the majority of the envelope amplitude may be removed. At the same time, calibrated changes in the pedestal amplitude may be used as a means of measuring the irregularities in the envelope which represent the differential gain.

Although, by convention, only a single figure is required to define the differential gain, it is nevertheless often useful to note the values given by the various steps of the staircase waveform, in view of the information which this can afford on the stability of the distortion characteristic. It is also useful to allocate a sign to the differential gain, and, for this purpose, a subcarrier amplitude greater than that at black level is called positive.

The use of a single figure to define the differential gain implies that a given magnitude of distortion corresponds to the same degree of picture impairment irrespective of the luminance level at which it occurs. For the relatively low values measured with modern equipment, this assumption can be taken to be true, although it is doubtful whether it would still be valid for large amounts.

5.4.3 Differential phase

Differential phase is the change in the phase of the colour-subcarrier frequency arising from a change in the amplitude of the accompanying luminance signal. It follows from the fact pointed out at the beginning of the Chapter that the derivative of the output voltage with respect to the input voltage is in general a complex quantity. The significance of differential phase arises from the impairment it produces in the colour picture, more severe with the NTSC system than with Pal and Secam, although with Pal it is also a function of the decoding system employed.

The accepted definition of differential phase (CCIR, 1966*b*) is as follows: When a small amplitude of subcarrier, ideally differentially small, is superimposed on a luminance signal which varies between black and white levels, the differential phase is the numerically greatest change in the subcarrier phase with respect to the phase angle at black level, for the complete range of luminance amplitude. The measurement is preferably carried out at the two values of average picture level afforded by the CCIR recommended signal, and the larger of the figures is taken as the differential phase. Although the sign of the phase angle is often ignored, it may well be useful to record it, a phase angle greater than that of the black level being positive. With large values of differential phase, there are reasons for believing that the shape of the curve has some effect on the degree of picture impairment, but, with modern equipment, the values obtained should in general be small enough for this to be ignored. The definition of differential phase when the luminance waveform is a staircase is shown in Fig. 5.16.

The measurement process with differential phase is decidedly more complex than with differential gain, owing to the need to derive from the separated subcarrier a signal measuring its phase shift with respect to the phase shift at black level. The basic method most often used is best illustrated by a reference to Fig. 5.18, which is the highly simplified block diagram of a test set for the measurement of differential phase and gain designed for operational and investigational purposes. As will be seen, the differential gain is obtained for a negligible extra expenditure.

For phase angles to be measured, two signals are needed, one to be measured and another as a reference. The first is derived in the upper path of Fig. 5.18 by passing the distorted staircase or sawtooth through a bandpass filter to remove the luminance component, which is now redundant; amplifiers are provided on each side of the filter as buffers. The output of this path is identical with the separated subcarrier used for measuring differential gain, so that an output can be made available, as shown, for this purpose.

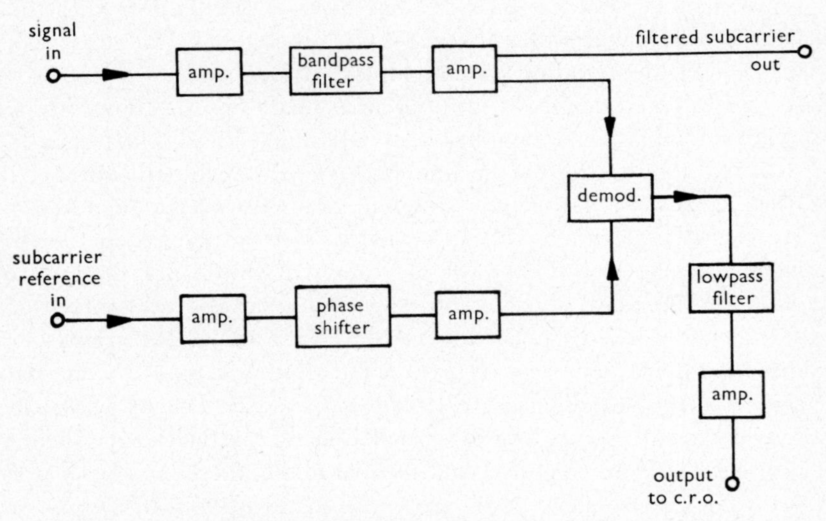

Fig. 5.18 Differential-phase-measurement equipment (simplified)

The phase reference in the lower path of the block diagram may be supplied from the generator or a local coherent source if these are available at the point of measurement; otherwise, it may be derived from the colour burst on the test signal by means of a standard burst-locked oscillator. An alternative which has been employed is the extraction of the subcarrier from the test signal by a very narrowband crystal filter. Since there is an increasing tendency, however, to include the burst in colour test signals, and since burst-locked oscillators are standard component items in a great deal of colour equipment, the former method seems preferable.

Yet another means of obtaining the phase reference is possible if one is prepared to restrict measurements to the two CCIR waveforms only, and provided that the generator superimposes the subcarrier on the three black lines and three white lines as well as on the

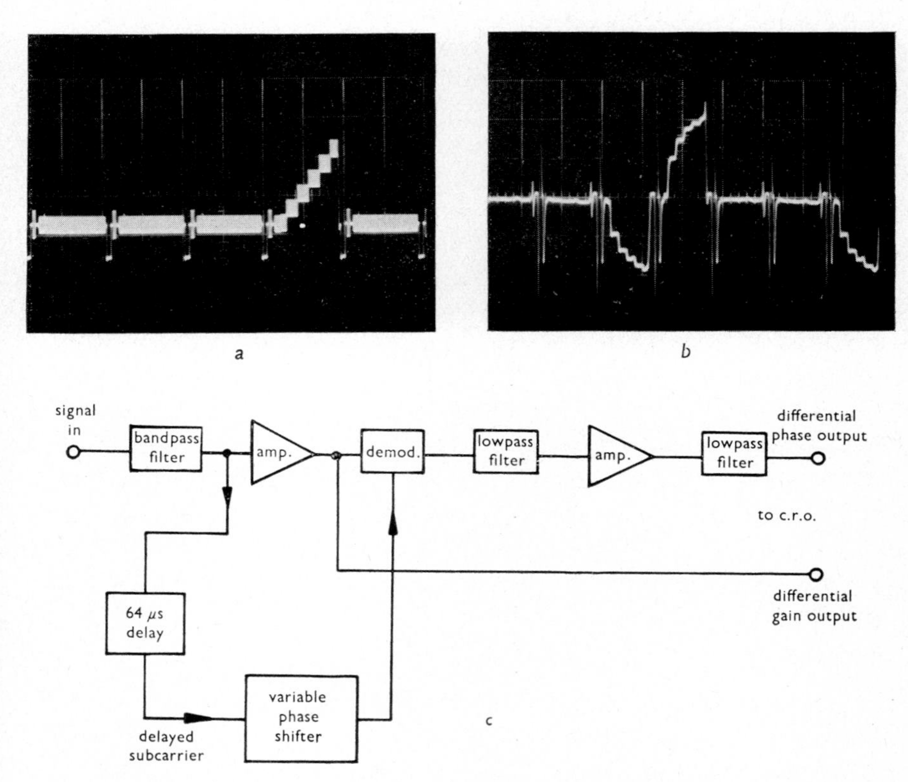

Fig. 5.19 Delay-line method of differential-phase measurement
a Input waveform
b Output waveform
c Measuring equipment (simplified)

staircase or sawtooth (Fig. 5.19). It utilises a standard 64 μs ultrasonic delay line of the type used as a component of delay-line Pal receivers to delay a feed of the distorted test signal. Clearly, this device offers a line of subcarrier coincident with the test line, which fulfils the conditions for a phase reference, since the latter only needs to be available for the duration of the staircase or sawtooth. Otherwise, there is no essential difference between the two methods of measurement.

The two signals may now be utilised as the inputs to a phase demodulator. If a true phase demodulator of the discriminator type is employed in this position, the output is a linear function of the phase shift between the signal to be measured and the reference signal, and should be independent of any amplitude variations of the measured signal due to differential-gain errors. The variable phase shifter in the reference path may then serve a double purpose: to set the phase of the reference signal to that of a suitable reference point on the measured signal, and, if it is calibrated, to measure the differential phase by introducing a known phase shift into the reference path so as to shift the display on the waveform monitor by a known amount. A very effective variant has a small phase shifter in the reference path which is gated in and out at a halfline rate, with the result that a line-repetitive display shows two traces displaced vertically by the amount of the phase shift introduced (Fig. 5.23*a*). In this instance, the calibrating phase shift was 1°, corresponding to half a graticule division; hence the differential phase was slightly more than 3°.

A somewhat simpler form of phase demodulator can be formed from a standard 4-diode bridge modulator by introducing a fixed phase shift of 90° into the reference path. The output of the bridge for constant amplitudes is then a d.c. level proportional to the sine of the angular deviation from the exact quadrature condition, i.e. sufficiently nearly proportional to the angle itself for angles up to, say, 10°. It suffers from the disadvantage that an error is introduced by any amplitude variations of the measured signal such as would be introduced by simultaneous differential gain distortion, which may be obviated by the inclusion of a limiter in series with this signal path. Alternatively, the need for a limiter can be avoided if the phase shifter is composed of two portions, one widerange and uncalibrated if desired so that the quadrature phase condition can be established for a selected point on the waveform to be measured, and the other of more restricted range but accurately calibrated.

To make a measurement of differential phase, the widerange phase shifter is initially adjusted until the angle between the reference subcarrier and the subcarrier at black level is precisely 90°, when the output from the modulator becomes zero. This condition can be identified by comparison with other points on the output waveform at which the modulator output is known to be also zero, such as any areas which do not originally contain subcarrier. Relatively wide, flat areas of the test waveform are the most suitable, and the three black and three white lines of the CCIR waveforms lend themselves particularly well to this purpose, as can be observed on the right-hand side of Fig. 5.23*a*. By double-triggering the waveform monitor, such zero reference lines can very conveniently be superimposed on the rest of the waveform. For this use, these intervening lines must be free from subcarrier in the test waveform. If this is not possible, the bottom of the synchronising pulse will serve.

The reason for the choice of the zero modulator output condition is its complete independence of the amplitudes of the two subcarrier inputs. It follows that measurements free from differential-gain error, and also free from all approximation, can be obtained once the widerange phase shifter has been set to give zero output at black level by employing the calibrated phase shifter to bring to the same zero line any desired region of the demodulated output waveform. The reading then provides the differential-phase error at the point of measurement. The bridge modulator is therefore used less as a phase demodulator than as a null detector. The measurement of the differential phase as defined above consists in carrying out this procedure for the step having the greatest deviation in the waveform.

The method of measuring the differential phase when the phase reference is derived from the original signal delayed by a whole line period is basically identical with that described above (Fig. 5.19). There is, however, a difference between the two phase-demodulated waveforms which, although not significant from the point of view of the measurement, ought nevertheless to be mentioned. This is the appearance in the output waveform, with the delay-line method, of a pair of identical but mutually inverted signals in successive line periods (Fig. 5.19*b*). This is because the two inputs to the phase demodulator are now identical waveforms delayed by a complete line period, and therefore contain two sets of test waveforms and their corresponding phase references, except that the sequence of test waveform and reference is inverted in successive lines. Hence two demodulated waveforms of opposite polarity are obtained. The measurement can, of course, be carried out equally well with either.

Another technique with some special advantages has been used in a commercial instrument (Fig. 5.24). The distorted test waveform is sampled by two gates whose widths are slightly less than the tread of the staircase waveform. Gate 1 takes a sample from the black level of each staircase line, and gate 2 a corresponding sample from a selected step of the staircase. This results in two trains of subcarrier bursts which may then be fed into phase-locked oscillators to regenerate continuous trains of subcarrier, whose relative phase angle is equal to the angle between the two subcarriers at the points of sampling when the circuits are correctly adjusted.

The two regenerated subcarriers are then applied to a phase comparator of a suitable type, and a d.c. output is obtained whose magnitude is proportional to the angle between the subcarriers at the points of sampling, i.e. to the differential-phase error at the chosen staircase step. Stray errors of phase in the two paths are taken into account by a preliminary zeroing procedure in which the movable gate 2 is also positioned at black level and a zero-setting phase shifter operated until zero output is obtained from the phase comparator, i.e. until the phase shifts through the two paths are equal.

The great advantage of this technique is that the demodulated output is now no longer a repetitive waveform which must be displayed on a monitor, but a current which can, either directly or after suitable amplification, operate a meter, preferably calibrated to read the phase angle directly with respect to black level. Suitable signal delays for the gates are provided on a switch so that the readings on the successive steps of the staircase can be obtained immediately, and, of course, the largest of these readings is the differential phase. The same principle could evidently be used to drive a digital counter which would then show the phase angle directly, with the consequent benefits of automatic selection of the largest reading, printout of results etc. Although the description has been given in terms of the staircase waveform, for which this method is evidently particularly adapted, it can also be employed with the sawtooth.

Another very considerable advantage stems from the phase-locked oscillators, which produce a very significant improvement in the random-noise level associated with the regenerated subcarrier by virtue of their action as very narrowband filters; this is supplemented by the smoothing in the output d.c. path and the metering circuit. However, these benefits are not fully realised unless a very high standard of stability can be maintained throughout, particularly in the triggering circuits and the phase-locked oscillators. The latter in particular have a rather difficult task to perform, and must be carefully designed.

5.4.4 Differential phase: insertion test signals

The measurement of differential phase using insertion test signals poses a special problem with the reference required for the demodulation of the phase errors. First, measurements are not infrequently needed when no burst is available, e.g. during the distribution of monochrome programme material, to prepare for a later colour programme. Secondly, even with the burst, the conditions are not optimum for the measurement of differential phase using a subcarrier reference regenerated from it as a result of the omission of the burst over part of the field-blanking region.

One solution makes use of a variant of the delay-line technique mentioned above, and, in fact, the present version of the insertion test signal has been framed with this in mind by the inclusion of a long subcarrier burst in the line following that containing the staircase (Section 7.5). This long burst is of such a duration, and is so located, that it can serve as the reference for the staircase subcarrier after a delay of a complete line period.

This method is simple and very effective, and its only real disadvantage concerns its use for continuous monitoring. During the measurement, the phase angle between the test subcarrier and its reference has to be brought to a standard value, usually zero, over the area corresponding to the initial black-level step of the staircase.

This is accomplished with a variable phase shifter. Now although the temperature coefficient of the delay line likely to be employed for this purpose is extremely small, it is not completely negligible, and some change will occur with time in the phase angle; although this is very unlikely to be of importance over normal measurement periods, it might be embarrassing over long periods of continuous operation unless special precautions are taken. In a less restricted sense, the reference subcarrier burst monopolises a considerable portion of an insertion test line at a time when demands on signal space in the field-blanking region are already fairly heavy. Indeed, it is very probable that the colour insertion test signal might have been framed quite differently if it had not been necessary to take the subcarrier burst into account.

All such disadvantages can be avoided by an ingenious proposal due to Voigt (1970), which accepts that it is perfectly feasible to replace the coherent subcarrier reference derived from the signal by a noncoherent reference locally generated. Phase measurement between two noncoherent waves is likely to disturb the purist, who can only conceive of a phase angle between two sinusoids of the same frequency, but the practical engineer should find the notion interesting, if unorthodox.

The basis of the proposal may perhaps best be visualised as follows. Let the difference between the reference frequency and subcarrier be written in the form $d\beta/dt$, where β is the phase angle, since, by definition, frequency is the time rate of change of phase. If one starts at a reference instant, say, when the two waves are simultaneously passing through zero with the same slope, their relative phase shift will increase linearly at the rate of $d\beta/dt$, until it again reaches coincidence after a time equal to the reciprocal of the frequency difference. Hence this period corresponds to 360° of phase difference. From this, one can simply deduce that a frequency difference between subcarrier and reference of 10 Hz will take, in round numbers, 28 μs to reach a phase difference of 0·1°, assuming that the two waves are in coincidence at the commencement of the period. In other words, since 28 μs is the useful duration of the subcarrier superimposed on the staircase in the insertion test signal, one may use a reference having a frequency differing by as much as 10 Hz from the subcarrier frequency and obtain an error in the phase angle of only 0·1°, and that solely at the end of the measuring period.

The basis of the technique may be understood from Fig. 5.20, which is highly simplified. The subcarrier separated from the insertion test signal by means of a bandpass filter is compared with the local oscillator in a phase demodulator at a point just preceding the actual measurement period, and the resulting output signal is used to drive a widerange voltage-controlled phase shifter operating on the

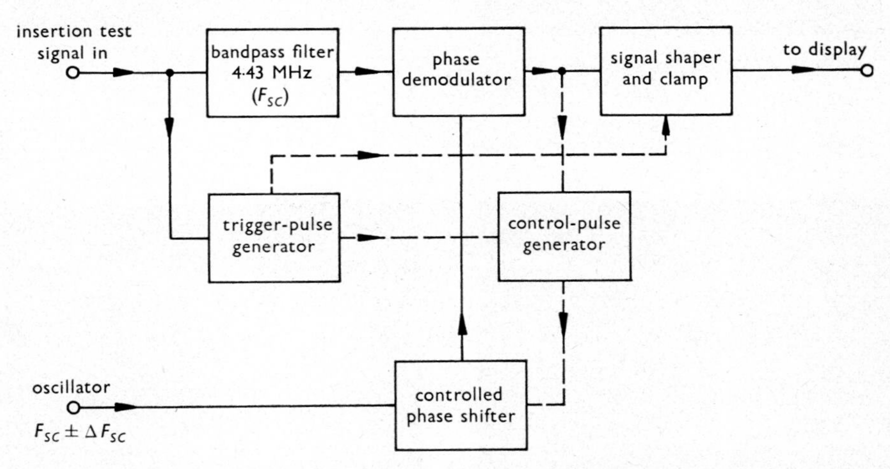

Fig. 5.20 Differential-phase measurement (simplified) with noncoherent reference

——— main path; – – – – control path

reference frequency. This part of the circuit forms a closed-loop phase-locking system which pulls the reference frequency into coincidence with the subcarrier within 2–3 μs, after which that condition is maintained unchanged for a period longer than is required for the measurement.

The number of cycles of the subcarrier required for locking is relatively small, so that the initial portion of a lengthened first staircase step provides an adequate sample, or, if the $20T$ pulse forms part of the waveform, its chrominance component may alternatively be pressed into service. The waveform in the control loop is shown in

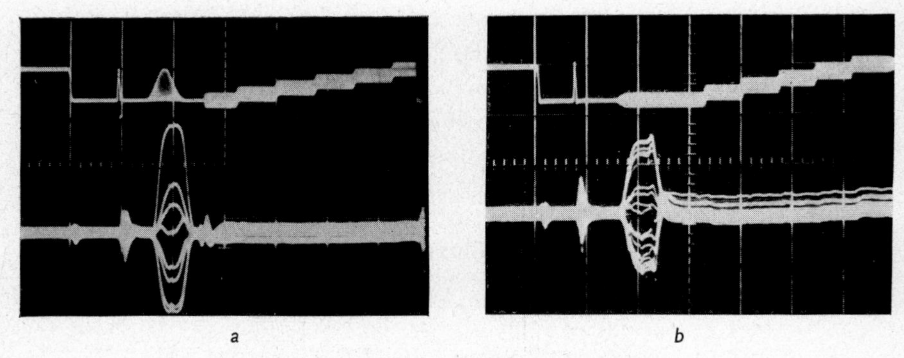

a b

Fig. 5.21 Operation of phase-locking circuit

a Locked by 20T chrominance pulse (40°/division)
b Locked by start of lengthened first staircase step (40°/division)

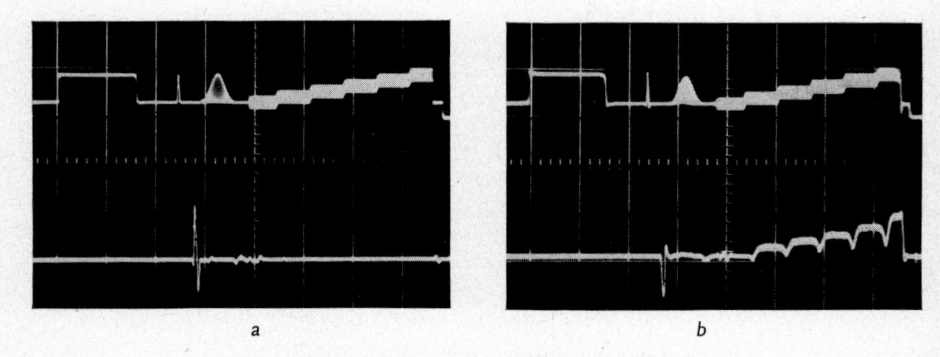

a b

Fig. 5.22 Output waveforms of circuit of Fig. 5.20

a Equipment in closed circuit (5°/division)
b Measurement of transmitter–receiver combination (5°/division)

Figs. 5.21 *a* and *b* for each of these. The large number of loops visible in the photographs is due to the fact that, with an exposure long enough to include a number of fields, the phase angle to be corrected will take on a number of values over the range $\pm 180°$, depending on the frequency difference between the subcarrier and the reference.

The incoming subcarrier is then demodulated with the reference, the result being as indicated in Figs. 5.22*a* and *b*. For convenience, the portion of the waveform corresponding to the phase angle at black level is clamped to a fixed d.c. potential; at the same time, the clamp may be arranged to remove some of the transient effects resulting from the operation of the phase-locking system.

The circuitry actually used is naturally somewhat more complex than that indicated in Fig. 5.20; e.g. the processes of phase demodulation are separated between the test signal and the phase-lock loop. Likewise, the phase-locking system is considerably more elaborate than is shown. On the other hand, the oscillator is relatively simple, since a tolerance of ± 10 Hz on the reference frequency is more than adequate, corresponding to a maximum error in the differential phase of $0{\cdot}1^\circ$, and a wider tolerance would be allowable in many instances.

Considering the few cycles of subcarrier frequency in the sample used, the phase-locking circuit can be made to operate in the presence of a surprisingly high amount of random noise. Measurements have shown that the lock will hold with signal/noise ratios at least as poor as 26 dB. Naturally, the display will be degraded by the superimposed random noise, but this can be dealt with by sampling equipment of the type described in Section 5.5.

Although the predominant advantage of this technique lies in the measurement of insertion test signals, it is evident that the equipment, with very little modification, may be made to operate on a line-repetitive signal, or on the CCIR signal.

5.5 Noise reduction by sampling

It is by no means uncommon to find, when differential phase is measured by one of the direct, nongating methods described above, that the accuracy is impaired by the random-noise level present on the waveform-monitor display of the demodulated signal. This is minimised by reducing the bandwidth of the chrominance channel in the measuring apparatus as far as is practicable; but the improvement achieved is insufficient.

This situation sometimes occurs on very long-distance international links, but the most frequent area of difficulty lies in the measurement of differential phase on videotape machines. These normally possess the combination of low differential phase with a random-noise level very satisfactory for programme purposes, but, owing largely to the effect of the tape noise in the chrominance band, the degree of interference with the phase-demodulated staircase waveform is quite severe. Fig. 5.23*c* shows a random-noise level corresponding to a rather poor,

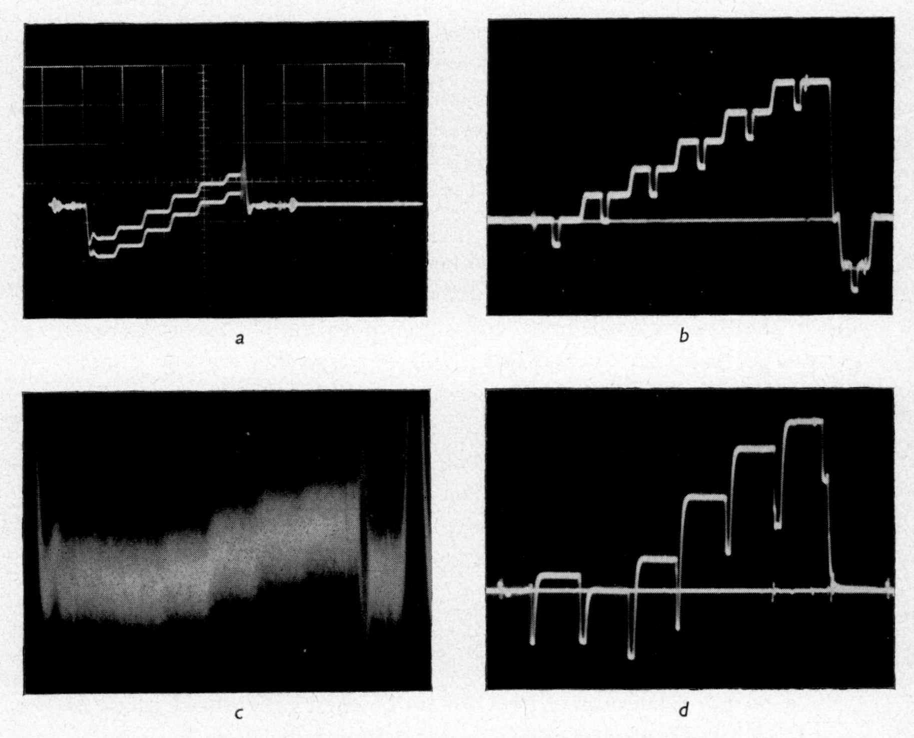

Fig. 5.23 Differential-phase-measurement waveforms
a Calibration by trace displacement (shift = 1°)
b Location of sampling instants
c Normal output waveform in presence of noise
d As (*c*) after sampling and staircase regeneration

but still usable, machine superimposed on a demodulated staircase waveform with a differential phase about three times greater than would normally be expected for this type of machine. The illustration consequently does not exaggerate, all the more since the combined effects of the γ of the cathode-ray tube and the characteristic of the photo-

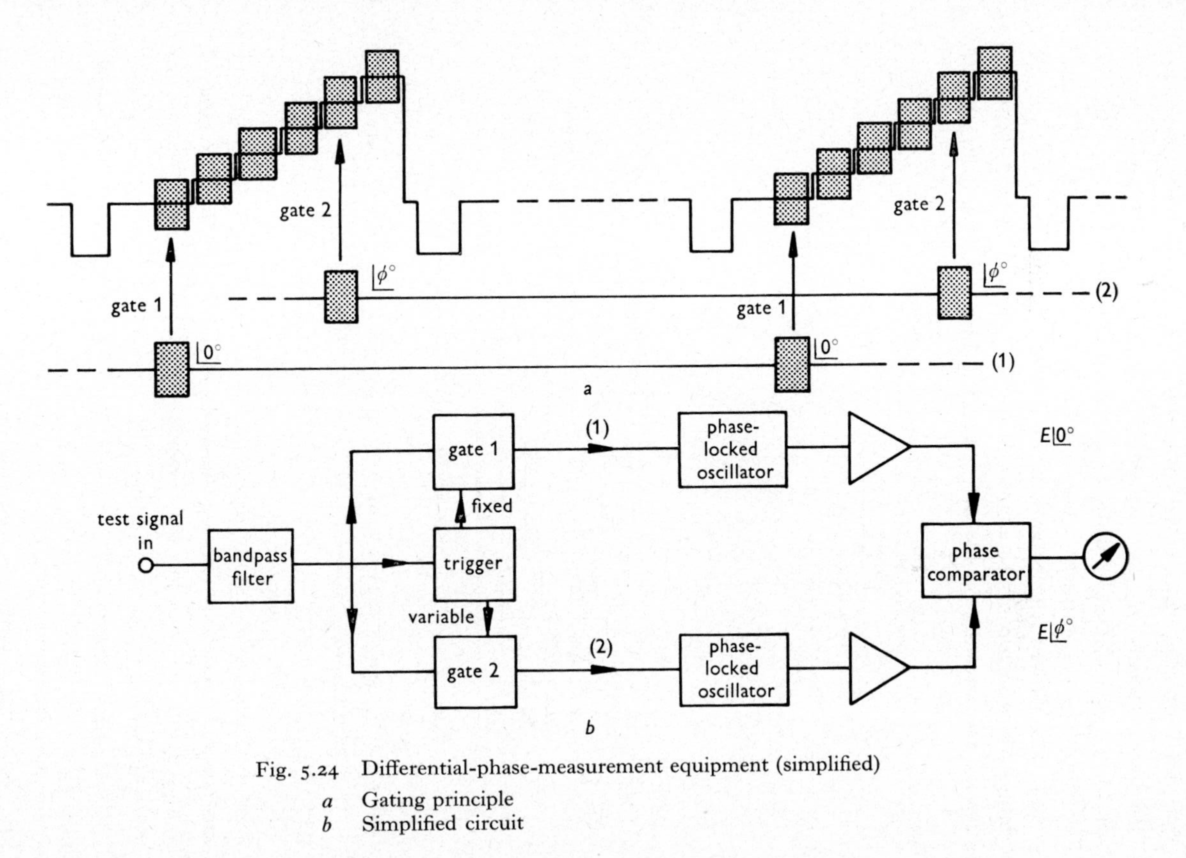

Fig. 5.24 Differential-phase-measurement equipment (simplified)

a Gating principle
b Simplified circuit

graphic emulsion more clearly define the outline of the noise than is the case visually.

This situation can be dealt with very effectively by the application of sampling techniques. Two variants are available, both of which may be explained with reference to Fig. 5.25. In the first, the demodulated staircase or sawtooth waveform is sampled at a number of equidistant points over the active period. This is shown in Fig. 5.23*b*,

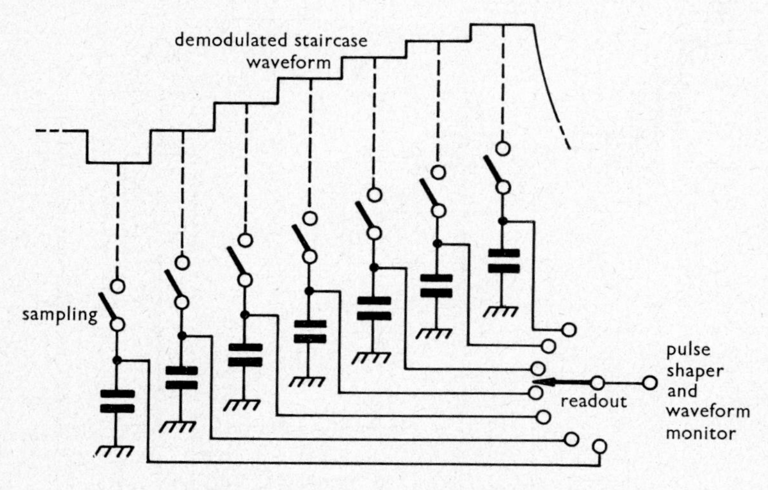

Fig. 5.25 Basic principle of sample-and-hold system

where the waveform is a staircase, and hence six samples are taken corresponding to the six luminance levels of subcarrier, together with an extra sample of a point which is not phase dependent, say in the synchronising-pulse region. The pulses shown are not the actual sampling pulses but markers, and this waveform is available from the instrument, to ensure that the sampling instants are correct when the test waveform is originated elsewhere and may consequently be nonstandard; to deal with this, the pulses may be moved over a range. The sampling-pulse duration is not critical; in the realisation described, it was about 1 μs.

Each sampler may be considered as a switch which at every cycle of the test signal connects a particular step to a capacitor whose leakage is kept at a low value. A readout switch examines the voltage across each of the capacitors sequentially, and converts this into a pulse of the duration of the original step and having an amplitude proportional to the capacitor voltage, so that the staircase is reconstructed if the original signal was a staircase, and converted into a staircase if it was a sawtooth.

The readout takes place at a lower rate than the sampling read-in; and, after each readout, the capacitor is discharged so that the cycle may recommence. In the example used here, the ratio of sampling to readout was 120:1; so the voltage across each capacitor at the instant of readout was proportional to the total charge drawn by the capacitor over 120 sampling periods, i.e. the sampled step amplitude together with the superimposed random noise was integrated over 120 sampling periods, thus achieving a substantial improvement in the signal/noise ratio of the reconstructed-staircase waveform. The process is equivalent to a filtration by means of a digital lowpass filter, and has the advantage that, although the noise power is reduced in the proportion of the effective noise bandwidths of the signal before and after filtration, and the uncertainty in the amplitude of each step is diminished in proportion, the general shape of the waveform can be preserved or even improved as a consequence of the reconstruction of the steps.

A waveform very similar to that just described is shown in Fig. 5.23*d*. The thickness of the horizontal portions is due to the residual low-frequency components of the random noise. In this method, which is now obsolete but which was an invaluable tool for many years, the apparent improvement in the random noise was slightly less than 40 dB, depending to a certain extent on the source of the noise.

At first sight it might seem one is getting something for nothing here, but of course this is not so. The price paid for the improvement in the signal/noise ratio is a reduced rate of presentation of the information, but the redundancy in the original display, in the present instance 15625 s^{-1} against a visual requirement of, say, 50 s^{-1}, is such that no hardship is felt. In the event, the repetition rate of the final display was around 65 Hz as a result of the introduction, every second line, of a calibrating trace. This was derived from the sample taken at the point in the demodulated waveform containing no phase information; and the resulting pulse, unlike the other six, was stretched to occupy a line period. When the waveform monitor was suitably double-triggered, a display similar to Fig. 5.23*d* was obtained. Any change in the calibrated phase shifter in the measuring apparatus then moved the regenerated staircase vertically with respect to the fixed calibrating trace, which was utilised as an index against which the step heights could be measured in terms of the calibration of the phase shifter. The absence of parallax and the freedom from errors due to drift of the waveform-monitor display make this device a great convenience, and improve the accuracy to a very useful extent.

The other variant of this method, which was used in a later version of this equipment, starts with a sampling process identical with that described above, but the storage and readout systems differ somewhat from the previous ones. Each of the storage capacitors is connected to the readout switch through a series resistor and shunt capacitor, and the readout is now nondestructive; in other words, no charge is taken from the final shunt capacitor. As a result, this capacitor acquires a potential equal to the mean potential across the charging capacitor.

The implication of this nondestructive readout is that it may be performed at any convenient rate; in the practical equipment, it was found most convenient to make this equal to the repetition rate of the original signal to be measured. While the advantage with line-repetitive or fullfield signals is not particularly large, a great benefit is obtainable with insertion test signals whose repetition rate is only 25 or 50 Hz. Provided that the charge lost by the capacitors between successive samples can be made sufficiently low, the readout may be carried out at a rate faster than the read-in, which overcomes the disadvantage of the former method where the final repetition rate is far too low for a comfortable visual display. The rate of presentation of information is naturally unaltered, but flicker is removed and the brightness is increased.

The effectiveness of the noise reduction can be estimated by a comparison of Figs. 5.23*c* and 5.23*d*, which are the same signal before and after processing; for convenience, the gain in Fig. 5.23*c* was reduced by a factor of three compared with Fig. 5.23*d*. The degree of discrimination achieved is revealed by the fact that the amplitude of the first step between the calibrating trace, i.e. the horizontal straight line and the flat top of the step, is equivalent to 0·3°. Tests made in the absence of noise have failed to show any perceptible degradation of the accuracy as a result of the sampling process.

In fairness, it should be pointed out that the visual display looks slightly less impressive than the photograph as a result of the integration of the residual very low-frequency noise on the steps, visible as a thickening of the trace. Since the residual noise is still random, the movements of the various steps is incoherent and gives rise to a slight 'jellyfishing' of the display, which, however, can easily be allowed for with a little practice. As can be seen, the total range of movement is small unless the noise level becomes exceptionally severe.

So far, the emphasis has been on the measurement of differential phase, because it is there that the difficulty of coping with the random-noise level is more severely felt, but it would nevertheless be advantageous to measure differential gain by the same type of sampling method. Unfortunately, this presents a certain difficulty, in that the signal obtained by filtering out the subcarrier component from its luminance waveform requires some processing before sampling.

The filtered subcarrier cannot be sampled directly, since any non-integral relationship between the subcarrier frequency and the sampling rate leads to a disturbing noise component, and it is not convenient to establish this integral relationship. It consequently becomes necessary to demodulate the subcarrier before sampling in such a manner that the envelope is undistorted, and in particular that even quite severe random-noise levels do not adversely affect the demodulated waveform. If the preferred method is synchronous demodulation, care has to be taken that a spurious component of differential gain is not introduced by the differential phase which is almost certain to be simultaneously present. However, fullwave rectification is much simpler, and the distortion is inherently low since the rate of change of subcarrier amplitude in the signal is normally very small.

These noise-reducing sampling methods, and in particular the second, offer the facility of an output reading on a meter or a digital display device, because an output analysis is obtained which is proportional in amplitude to the required reading. It may then be stretched in time to any desired amount and this stretched waveform utilised to drive the final display device. Such a facility is of great use in the measurement of insertion test signals (Voigt, 1970), and offers an immediate application to the field of automatic monitoring.

5.6 References

CCIR (1966*a*): Documents of the XIth Plenary Assembly, Oslo, **5**, Recommendation 421–1

CCIR (1966*b*): *ibid.*, Recommendation 451

IRE (1960): 'Standards on television measurement of differential gain and phase', *Proc. Inst. Radio Engrs.*, **48**, pp. 201–208

KELLY, H. P. (1954): 'Differential phase and gain measurement in colour television systems', *Trans. Amer. Inst. Elect. Engrs.*, **73**, Pt I, pp. 565–569

POST OFFICE Engineering Department Research Report 20661 (1961)

QUINN, S. F., and NEWMAN, P. M. (1965): 'Distribution of average levels in television programmes', *Electron. Lett.*, **1**, p. 261

RHODES, C. W. (1970): 'Measurement of non-linear distortions in colour television systems', *J. Soc. Motion Picture Televis. Engrs.*, **79**, (1), pp. 28–30

THIELE, A. N. (1967): 'A Thompson type bandpass filter for television linearity measurements', Australian Broadcasting Commission Report 49

VOIGT, K. (1970): 'A new method for measuring differential phase distortions with insertion test signals and without transmitting a reference signal', *IERE Conf. Proc.* 18, pp. 171–178

WHITE, N. W., and HEINZL, J. J. (1970): 'Techniques for measuring small-order non-linear distortions', *ibid.*, pp. 223–231

6 LINEAR WAVEFORM DISTORTION

6.1 General

A video channel should ideally resemble an ideal lowpass filter. If the upper limit of the videoband is f_u, the insertion gain function $G(\omega)\,e^{i\beta(\omega)}$ should ideally be given by

$$G(\omega) = \text{constant} \qquad (0 \leqslant \omega \leqslant f_u)$$

$$G(\omega) = 0 \qquad (f_u \leqslant \omega \leqslant \infty)$$

$$\frac{d\beta(\omega)}{d\omega} = t_g = \text{constant} \quad (0 \leqslant \omega \leqslant f_u)$$

$$\beta(0) = \pm n\pi \qquad (n = 0, 1, 2, \ldots)$$

Normally, the gain constant is unity, as for the lowpass filter. The phase response is defined, not in terms of the phase angle $\beta(\omega)$, but in terms of its derivative $d\beta(\omega)/d\omega = t_g$, the group delay or envelope delay. Since linear waveform distortion only is considered here, it will be assumed that both the gain and the delay are independent of the signal amplitude.

The second of the conditions imposed on the phase/frequency response of the channel, $\beta(0) = \pm n\pi$, is the requirement for zero phase-intercept distortion. It is included for the sake of completeness, but it is not likely that any distortion from this cause will be encountered in practical television.

However, the relationship between the departures from constancy of the gain-frequency and group delay-frequency characteristics and the resulting waveform distortion is far from simple, since it depends not merely on the magnitude of a given error and its position in the video spectrum, but on its shape. Also, unless the distortions are very small, they cannot be considered independently. Even more complex is the relationship between these distortions and the corresponding picture impairment which is the final criterion, since, after all, the ultimate aim is to provide the viewer with a picture of an acceptable standard.

If tolerance schemes for gain and delay errors are drawn up so that no combination of error curves can give rise to more than a permitted amount of picture impairment, the requirements are likely to be unduly severe in the majority of instances, which is uneconomical, since every standard of quality has its price; as the standard is raised, the cost tends to increase disproportionately. Conversely, if the tolerances are increased, it becomes possible at times for intolerably high picture impairments to appear.

Quite apart from any questions of economics, the traditional steady-state method of test and measurement is indirect, because it attempts to relate gain and delay errors to the corresponding waveforms, which in turn have to be related to picture impairments, whereas it is perfectly possible to eliminate the first and most dubious step by using a waveform directly for testing circuits and apparatus.

What is really required is a standard waveform which typifies the television signal and which can be reproduced with such high accuracy that even very small departures from the correct shape can be recognised, together with a means of estimating the magnitude of an error in terms of the subjective impairment of a picture.

The problem was understood even in the comparatively early days of television by a minority of farsighted engineers, and a number of solutions were proposed. By far the most elegant and effective is the 'sine-squared-pulse and bar' method devised by N. W. Lewis (1954) and his coworkers, notably I. F. Macdiarmid (1952 and 1959) and W. E. Thomson (1952). This method, which is explained below, has been in use in the UK since 1954, and has already been adopted by a number of foreign administrations, by some in a modified form. Nevertheless, old habits die hard, and a surprising resistance to the introduction of the newer methods is often encountered.

6.2 Monochrome sine-squared-pulse and bar methods

6.2.1 General considerations

The conditions which a waveform for video testing purposes must fulfil are severe:

(*a*) It must, as far as possible, be typical of a video signal.

(*b*) It must have a very simple and easily definable shape, so that errors can be recognised easily.

(*c*) It must have a spectrum essentially zero above some upper limiting frequency which can be made to correspond with the nominal upper limit of the videoband, so that the waveform to be measured is not confused by misleading and irrelevant information corresponding to the behaviour of the system under test to distortions at out-of-band frequencies.

(*d*) It must be capable of being generated in such a way that individual generators differ from each other to the least possible extent.

The waveform chosen by Lewis and Macdiarmid is based on the sine-squared pulse defined by

$$G(t) = \sin^2 \frac{\pi}{2} \frac{t}{\tau} \quad (0 \leqslant t \leqslant 2\tau)$$

where $\tau = 1/f_c$ and f_c is a nominal upper limiting frequency. This can be rewritten in the so-called 'raised cosine' form as

$$G(t) = \frac{1}{2}\left(1 - \cos \frac{\pi t}{\tau}\right) \quad (0 \leqslant t \leqslant 2\tau)$$

which is recognisable as a single period of a sinusoid with a frequency of $\frac{1}{2}f_c$.

The corresponding spectrum is

$$A(f) = \frac{\pi^2}{\pi^2 - \omega^2\tau^2} \frac{\sin \omega\tau}{\omega\tau}$$

The shape of the ideal sine-squared pulse, and its spectrum, are shown in Figs. 6.1*a* and *b*. It will be seen that the pulse has a half-amplitude duration τ, and the width at the base is 2τ.

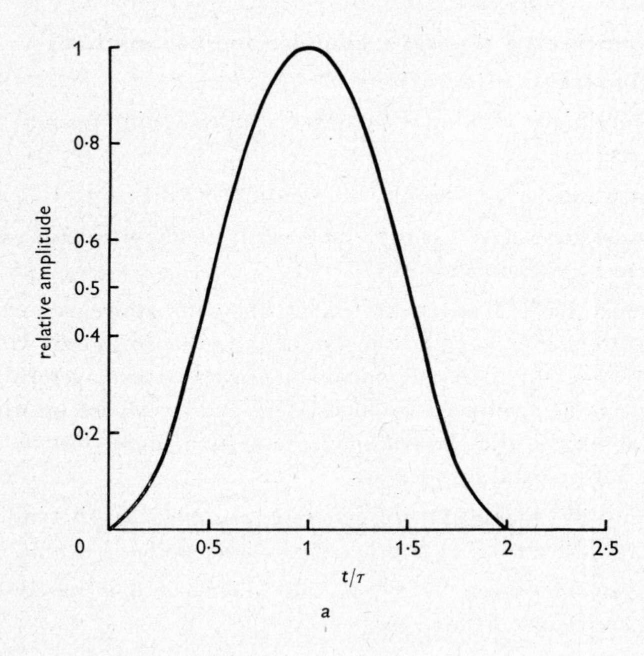

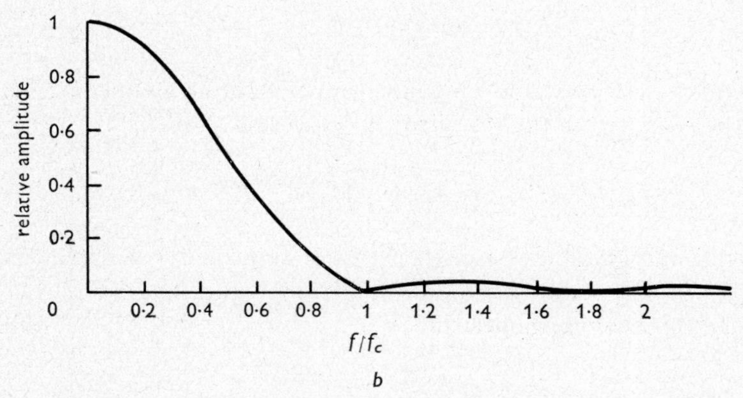

Fig. 6.1 Ideal sine-squared pulse and its spectrum

a Pulse
b Spectrum

However, the problem of reproducibility of the waveform remains. This was overcome by an ingenious principle based on a ladder network (Fig. 6.3) designed by Thomson (1952), which has an impulse response very closely approximating the ideal sine-squared pulse, as can be seen by a comparison of Figs. 6.1 and 6.2.

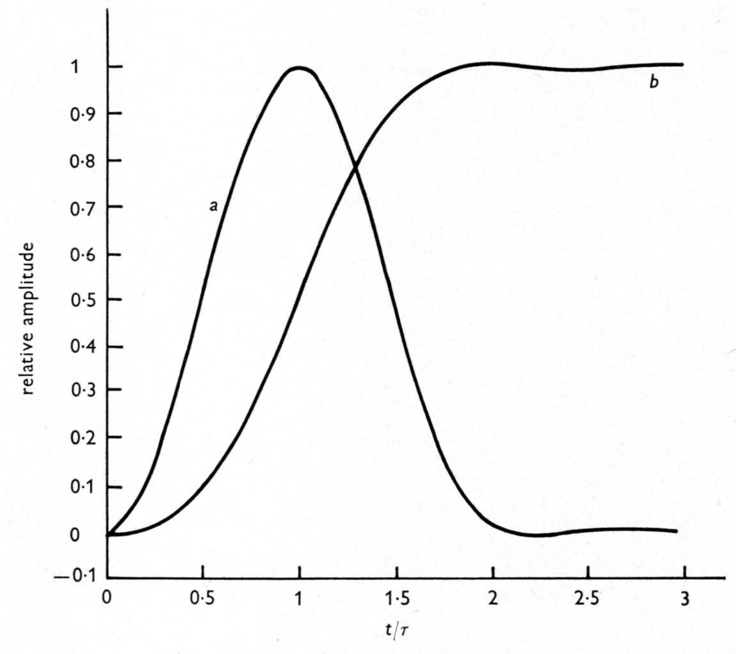

Fig. 6.2 Calculated response of sine-squared network

a To impulse
b To step function

This network is not unduly complicated, and the element values are reasonable, so that it can be manufactured with great accuracy. Provided its terminations are correct and free from stray capacitance, a sufficiently narrow pulse applied to the input will give rise at the output terminals to a sine-squared pulse whose shape, even if not perfectly sine-squared, is a very close approximation to the ideal and, at all events, is reproducible within very close limits.

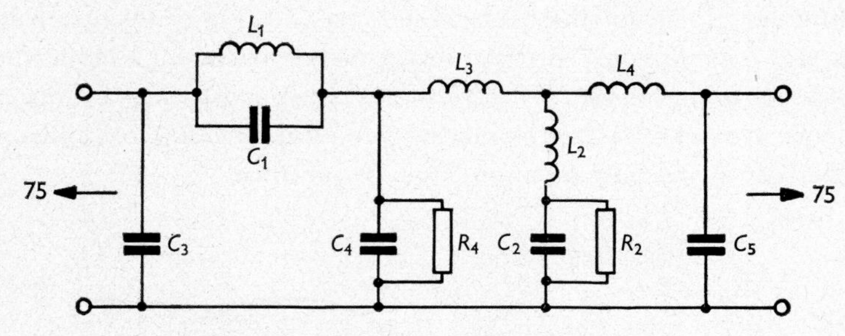

Fig. 6.3 Sine-squared-pulse-shaping network

Component	625-line Half-amplitude duration = 100 ns	= 200 ns	Tolerance	Q factor*
	μH	μH	%	
L_1	0·9477	1·895	±1	⩾ 70
L_2	0·185	0·37	†	⩾ 50
L_3	1·854	3·709	±1	⩾ 100
L_4	1·821	3·642	±1	⩾ 100
	pF	pF	%	
C_1	47·53	95·07	±2	
C_2	1301	2601	±0·5	
C_3	45·55	91·11	±2	
C_4	339·8	679·7	±0·5	
C_5	99·87	199·7	±2	
	Ω	Ω	%	
R_2	1300	1300	±5	
R_4	5100	5100	±5	

* L_2 should be adjusted to make insertion loss a maximum at 10·26 mHz for 100 ns network, and at 5·13 mHz for 200 ns network

† Q factor should be measured at 7 MHz for 100 ns network, and at 3·5 MHz for 200 ns network.

Network should, if necessary, be connected between masking pads designed to present impedances of 75 Ω ±1 % to network under operating conditions

Ideally, the driving pulse should be infinitely narrow, but a satisfactory approximation is not at all difficult to provide with present techniques. A good rule of thumb is to use a driving pulse whose half-amplitude duration is not greater than 0·2τ, for the following reasons.

Narrow pulses of good shape are normally roughly sine-squared, and the spectrum (Fig. 6.1*b*) only deviates from flatness at 0·2f_c by 10%, which has only a very small effect on the output-pulse shape.

Accordingly, the output pulse does not change significantly with the width of the driving pulse, provided that the latter is not greater than one-fifth of the half-amplitude duration of the desired sine-squared pulse since the spectrum of the driving pulse under these conditions only varies between complete flatness and a droop of 10% at f_c. Curves showing the output-pulse duration for a number of input-pulse widths and pulse types are given by Macdiarmid (1952).

It is important to realise that there is a small but significant difference between the practical sine-squared pulse produced by the Thomson network and the ideal sine-squared pulse. The sine-squared network has a very slightly different amplitude/frequency response from the spectrum of the ideal pulse, the most significant factor being the smaller amount of energy transmitted above f_c, which gives rise to a very small 'ring'. In addition, although the Thomson network has a 7th-order maximally flat group-delay response, the delay error at the highest transmitted frequency is not completely negligible, giving rise to a small skew-symmetrical distortion term.

The net result of these shortcomings is a small, highly damped 'ring' following the pulse, which must be kept within tolerances. In generators designed by the BBC, the first overshoot is maintained within the limits of $0{\cdot}9 \pm 0{\cdot}5\%$, the second within $0{\cdot}5 \pm 0{\cdot}4\%$, and no other irregularity before or after the pulse is allowed to be more than 0·2%, all of these being percentages of the peak amplitude of the pulse.

The slight disturbance introduced by this small 'ring' causes a small asymmetry of the flanks of the pulse with respect to the peak, which gives rise to no inconvenience but which is sometimes discovered with surprise by users who have not been forewarned.

It is sometimes asked why the network has not been improved since these shortcomings, small though they are, must necessarily reduce the accuracy to some extent. The reason is that the Thomson network represents a remarkably good compromise between performance and cost and ease of manufacture. The performance is perfectly adequate for all except measurements in instances where the distortion is exceptionally low, its reproducibility is very high, its cost is reasonable and its size can be made very small for use in miniaturised equipment. Thomson networks on printed boards measuring $2\frac{1}{2}$ in × $1\frac{1}{2}$ in × $\frac{7}{8}$ in have been made in large numbers to a high standard of accuracy for use in apparatus designed in the author's laboratory. Encapsulated versions of the network are also available commercially.

As so very often happens in engineering, when the performance reaches a given standard of excellence, a further fractional improvement has to be paid for at an exorbitant rate in complexity and expense. It is accordingly not possible to have a really significant improvement in the shape of the pulse without a fair degree of extra complication. Nevertheless, for certain specialised purposes, it may in the future be worth while to design generators in which improved phase linearity produces a pulse in which the usual slight asymmetry is absent. This would, at the same time, have the effect of approximately halving the present overshoot; of course, an identical overshoot would lead the pulse. To get rid of the overshoots altogether would increase the width of the pulse spectrum, which is undesirable. Such generators would be more expensive than the standard form, but the increased cost would be offset by the smaller number required.

Guinet (1969) points out that a simpler network is possible if one relies on a critical driving pulse duration to provide part of the spectrum shaping. This shifts part of the responsibility from the network to the pulse generator, and for the normal waveform the gain is debatable. However, this technique may well prove advantageous where a better shaped waveform is required.

6.2.2 Choice of parameters

At this point, it has been established that practical means are available for the generation in a highly reproducible manner of a test pulse whose shape is a very close approximation to the theoretical sine-squared pulse. This pulse can be defined by a single time parameter equal to the half-amplitude duration of the pulse which, to avoid confusion, has so far been designated τ. The practical form of the pulse has an advantage over the theoretical sine-squared pulse, in that its spectrum contains a smaller amount of energy above the nominal limiting frequency $f_c = 1/\tau$, and can thus be passed with only an extremely small amount of distortion through an ideal low-pass filter with cutoff f_c.

From now on, the practical sine-squared pulse will be called 'sine-squared pulse' without the qualification, unless that should be necessary for clarity.

It is standard practice to define the waveform in terms not of τ but of a parameter T which is one-half of the reciprocal of the nominal upper video frequency for the system concerned; i.e. $T = 1/(2f_u)$, where T is 333 ns for the 405-line system and 100 ns for the 625-line 5 MHz systems. Slightly anomalous are the UK 5·5 MHz 625-line system (system I), in which for the sake of uniformity, the value of 100 ns for T has been adopted, and the 525-line systems in which the bandwidth has been rounded off to 4 MHz to give a T value of 125 ns.

This convention now appears to clash with requirement (*c*) of Section 6.2.1 above, since the spectrum of the T sine-squared pulse has twice the nominal bandwidth of the video signal. In fact, two values of the parameter τ were specified by Lewis. The first has $\tau = 2T$, so that the spectrum of the pulse has the same width as the bandwidth of the video signal, and requirement (*c*) is satisfied. This was called by Lewis the 'maintenance test', since it is satisfactory for ensuring that the performance of equipment and links is up to standard.

However, the fact that the spectrum of the $2T$ pulse has a value of zero at f_u, the upper limit of the videoband, and is 6 dB down at $1/(2f_u)$ means that a progressively severe weighting is applied to gain and phase errors in the upper two-thirds of the band. In some respects, this is no bad thing, because a somewhat similar distribution applies also to the spectral components of a typical monochrome television signal, and moreover it is known that the tolerance of the eye to image defects originating in a particular area of the spectrum increases as this area approaches the upper band limit owing to psychophysical effects. It could therefore be argued with justification that the shape of the spectrum of the $2T$ sine-squared pulse is an advantage of the method. This also applies to the luminance channel of a colour signal.

Nevertheless, not all monochrome television pictures contain comparatively little fine detail of significance, and some indeed contain substantial and important amounts of spectral energy in the upper half of the videoband. Means must therefore be available for carrying out a more searching test at the higher frequencies than can be furnished by the $2T$ pulse. For this reason the T pulse with a spectrum only 6 dB down to the videoband limit is also provided. In many instances, this signal will contain a great deal of irrelevant information owing to out-of-band effects, and a mathematical analysis is needed to sift the relevant part of the test from the irrelevant.

This was termed by Lewis the 'acceptance test', since it is required more for the exploration of the full capabilities of the system under test than for the measurement of deviations from standard performance. The need for the acceptance-test method has diminished somewhat with the increasing use of the sine-squared-pulse and bar waveforms elaborated to test systems for their ability to transmit colour signals, as will be explained later.

6.2.3 Complete monochrome waveform

It has been shown that the $2T$ sine-squared pulse is the result of passing a very short impulse through a suitably shaped video channel having the same nominal upper band limit as the television system used, so that the sine-squared pulse can be considered as typifying the transmission of fine detail. Incidentally, Springer (1963) found that the waveform at the output of a good image-orthicon camera channel which is viewing a very fine vertical white bar on a black field bears a surprisingly close resemblance to the sine-squared pulse, although, since the bandwidth of the camera channel was not restricted, the resemblance was closer to a T than to a $2T$ pulse.

A television picture, however, rarely fails to contain significantly large areas of approximately uniform tone, and it is essential that the corresponding signals should also be transmitted with the smallest possible amount of distortion. Indeed, these larger areas usually contain the bulk of the significant information, and so may be claimed to be more important. Hence the sine-squared pulse needs to be supplemented by a bar which is made approximately equal in duration of one-half of the active line period, i.e. 40 μs for the 405-line system and 25 μs for the 625-line and 525-line systems. This is sufficient to reveal comparatively long-term effects, such as those which cause horizontal shading and streaking effects on the picture, as well as very long-term echoes. The chosen bar length makes the average picture level roughly 50%, which is similar to normal programme material, and at the same time any bar slope is accentuated.

The bar transitions are standardised as for the pulse, i.e. the pulse is initially generated with transition times much shorter than the pulse half-amplitude duration τ; in practice, not more than $0{\cdot}2\tau$ is adequate. This bar and the very narrow pulse are added together in the correct time relationship, and the combined waveform is passed through the appropriate Thomson filter, so that both waveforms have a shape which is determined by the network alone.

The bar thus generated is for convenience commonly called a 'sine-squared bar', although, of course, it is a misnomer. Each transition has approximately the slope of a halfcycle of the nominal upper frequency. The risetime between the 10% and the 90% points is $0{\cdot}98\tau$; i.e. equal to τ for practical purposes.

The presence of the bar serves another very useful purpose, in that it forms a reference for the amplitude of the picture portion of the signal. The area under the pulse represents the d.c. component of the pulse, which must remain constant in a lowpass system. Any change in bandwidth must alter the transition time of the leading and lagging edges of the pulse and thus increase its half-amplitude duration, with the result that the peak amplitude of the pulse must change to maintain the area constant. In other words, the pulse amplitude is a linear function of the bandwidth. On the other hand, the duration of the bar is so great compared with its risetime that only an insignificant change occurs in the width of the flat top. The amplitude of the bar, more specifically the amplitude at the centre of the flat top of the bar, is therefore a measure of the steady-state gain of the apparatus under test. The centre of the bar is adopted as a reference, since it is the point least likely to be affected by distortions of relatively long duration.

The bar/pulse ratio, defined as the ratio of the amplitude at the centre of the bar top to the amplitude at the peak of the pulse, is accordingly a measure of the bandwidth of the apparatus under test, provided that the ratio has previously been made unity at the output of the generator.

To test television apparatus, it is essential to have synchronising pulses available; so, standard synchronising pulses are inserted. These also serve as a means of triggering the waveform monitor used for the observation of the waveforms. Although many tests may be carried out using a purely line-repetitive waveform, it is essential to

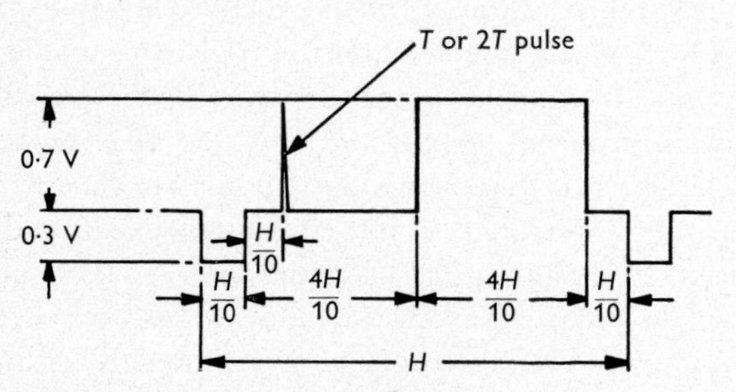

Fig. 6.4 Sine-squared-pulse and bar waveform

be able to provide the full field-synchronising waveform when required. To keep the cost and complexity of the generator to a minimum, it is most convenient to make use of external pulses from a synchronising-pulse generator, since, in almost all instances, they are available in the areas where tests using a fully composite waveform are required. A sketch of the line waveform is given in Fig. 6.4.

In the above, no provision has been made for tests at any frequency lower than line frequency. In this frequency range, the typical distortion is a 'tilt', e.g. from inadequate *RC* time constants.

In principle, such distortions are more tolerable, since appropriate information is available in the waveform to correct them by means of a clamp. Consequently, the information is of lesser importance than the information obtained during testing about distortion at the higher frequencies. Nevertheless, it may be desirable to investigate what low-frequency distortion exists.

This is done within the framework of the sine-squared-pulse and bar signal by doubling the length of the bar and removing it during every alternate period of 10 ms. The result of this (Fig. 6.5) is seen to be the equivalent of a 50 Hz square wave with added line synchronising pulses. It is convenient to arrange for this signal to be synchronised either with externally provided field-synchronising pulses or with the 50 Hz supply, at will.

In some administrations, it is customary to use a signal pedestal or 'set-up' i.e. a value of black level differing from blanking level by a few millivolts. The sine-squared-pulse and bar test signals may like-

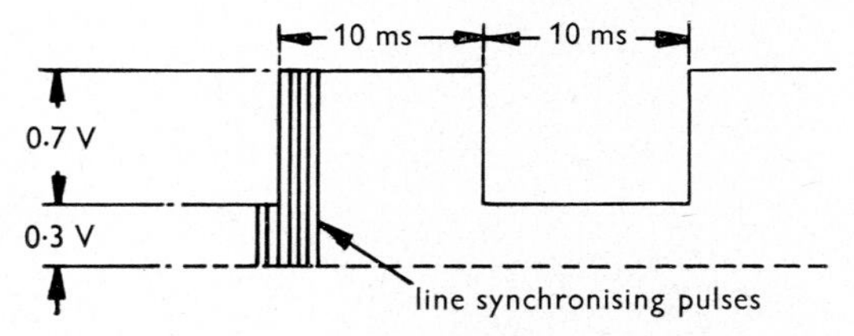

Fig. 6.5 50 Hz square-wave test signal

wise be mounted on a pedestal if preferred. However, apparatus handling colour signals must cope with negative transitions which enter the subblanking region, and consequently, as the trend for making replacement equipment colour-capable continues, the use of a pedestal increasingly loses its point.

6.2.4 *K*-rating system

One of the very great advantages of the sine-squared-pulse and bar method is the way in which a serious attempt has been made to relate the distortion of the waveform to the subjective reaction of a viewer to the corresponding impairment. This is done by a systematic process which allocates to a waveform with a given distortion a unique factor, known as the *K* factor or *K* rating, which is a measure of the subjective picture impairment.

Two distinct methods of *K* rating the apparatus or circuit under test are prescribed according to whether the maintenance-test method using the $2T$ pulse-and-bar signal, or the acceptance-test using the

T pulse-and-bar signal is employed. The former is very simple, and an almost immediate answer can be obtained, whereas the latter method provides a more searching test for a considerably increased degree of complication. The fact that two K ratings may apply to the same circuit need not cause confusion if it is remembered that the acceptance-test rating measures the ability of a system to transmit any monochrome television signal corresponding to the line standard in use, whereas the maintenance method measures the day-to-day performance of a system previously found satisfactory by the acceptance test. The two methods are indeed basically the same, but the acceptance test measures additional parameters.

The picture impairment corresponding to a given distortion of the sine-squared pulse was originally determined as a function of the subjective viewer reaction to pictures distorted by single echoes of various magnitudes and delays (Lewis, 1954). It is known that any band-limited signal can be considered as the sum of a series of component signals, each having the same upper limit of the frequency spectrum f_u and spaced at maximum intervals of $1/(2f_u)$ (Shannon, 1948). This applies in particular to the sine-squared pulse (Ville, 1951).

It is therefore possible to consider the distortion term of the sine-squared-pulse signal under examination, i.e. the difference between the original signal and the distorted signal, as the sum of a number of echoes each of sine-squared shape. Consequently, the problem of dealing with distortions of arbitrary shape can be reduced to the simpler problem of finding the subjective impairment due to echoes of the original sine-squared pulse. In fact, in the maintenance method, a further simplification was introduced by taking the distortion to be equivalent to a single echo term having the amplitude and delay of the most significant deviation of the distorted sine-squared pulse from the undistorted. While this is not strictly accurate, the practical simplification introduced by this assumption is considerable, and experience has shown that any errors which may be introduced are not important in practice particularly when, as should be the case, the greatest amplitude of the maximum echo term is, at the most, only a very small percentage of the pulse amplitude.

The subjective impairment due to a single echo term with various amplitudes and delays was determined by displaying standard pictures distorted in this manner to a panel of observers, who awarded each picture a rating on an impairment scale. From the results, the tolerance diagram was prepared (Fig. 6.6*b*). It can be seen that the closer its spacing to the pulse, the greater the amplitude the echo may have for the same subjective impairment; e.g. an echo at a spacing of $2T$ from the peak of the pulse can have four times the amplitude as an echo at a spacing of $8T$ for the same degree of impairment.

Certain other types of distortion must be dealt with, in particular bandwidth limitation; also relatively long-term distortions, which are best revealed by the bar waveform. While such distortions can also be considered as patterns of echoes of the sine-squared pulse, they cannot be simplified to a single significant echo, and must therefore be treated somewhat differently. Accordingly, tests were made on the subjective impairments due to bandwidth limitation and long-term distortions using the same panel and test conditions, to establish their relationship with the single echo spaced $8T$ from the pulse.

The final result was that a long-term distortion which gives a maximum change of $\pm K$ per cent. relative to the centre-point of the bar and a pulse/bar ratio of $100 \pm 4K$ per cent. are both approximately equivalent to an $8T$ echo, which is K per cent. of the pulse amplitude. All these distortions may therefore be expressed in terms of the $8T$-echo amplitude.

All these distortions are likely to occur simultaneously, whereas it is highly desirable to be able to assign a single rating factor to the circuit under test. It was therefore decided to take the largest equivalent $8T$-echo magnitude as the rating factor for the routine-test method. For example, if the distortion of the bar top corresponds to an echo magnitude of 2%, the pulse/bar ratio corresponds to a magnitude of 1%, and the largest significant echo as measured on the pulse corresponds to a 3% $8T$ echo, the apparatus under test is said to have a routine-test (or maintenance-test) K rating of 3%. This rating was found by Lewis (1954) to correspond to a picture impairment which is perceptible but not disturbing, although it must be remembered that this refers only to the particular test conditions used for the experiments and should not be taken as an absolute judgment.

The acceptance-test method makes use of the relationships described above and additional conditions which impose more severe requirements on the transmission of the upper portion of the video spectrum than can be achieved by the routine test.

The significant difference between the two tests lies in the use of the T pulse-and-bar signal for the acceptance-test method. As has already been explained, the spectrum of the T pulse is not zero at the nominal upper limit of the videoband as with the $2T$ pulse, but has a relative amplitude of 0·5 at that point, with the result that the distorted test signal may contain irrelevant and misleading distortions from effects occurring between f_u and $2f_u$.

This irrelevant information is removed, and the test is standardised by making use of the fact that the T pulse-and-bar test signal at the output of the apparatus under test implicitly contains within its shape the information from which can be reconstructed the waveform which would have been produced at the output of the test object if the input waveform had possessed a spectrum flat to f_u and zero from f_u upwards, in other words a pulse of the well known $(\sin x)/x$ type. This pulse can be derived either from the sine-squared pulse or the bar; but the pulse is used since its spectrum over the nominal bandwidth of the system more closely corresponds to that of the $(\sin x)/x$ pulse and the accuracy is accordingly likely to be better. This topic will be discussed again in greater detail below.

From mathematical operations on the T-pulse response, four conditions can be laid down, two of which are equivalent to the conditions already imposed on the $2T$-pulse shape and pulse/bar ratio for the routine test. From these, an equivalent K rating is derived, which in effect imposes a considerably more stringent condition on the apparatus or circuit tested than does the routine test.

However, the acceptance test is considerably more difficult and time-consuming than the routine test, even when a computer is used to perform the calulations, to the extent that it is only comparatively rarely attempted. Fortunately, on all except the now obsolescent 405-line system, the new sine-squared-pulse and bar colour-test waveforms described below make the acceptance test less necessary.

6.2.5 Routine test: practical

Three separate graticules are required for the measurement (Figs. 6.6*a*, *b* and *c*); these measure

(*a*) pulse/bar ratio and distortion of the top of the bar

(*b*) pulse shape

(*c*) 50 Hz square-wave distortion.

Although only one set of limit lines is given in each instance, it is found most suitable in practice to engrave the graticule with two sets, namely those corresponding to $K = 2\%$ and $K = 4\%$. Other values are found by interpolation. If it is not unusual to have to measure appreciably poorer K ratings, other graticules with different engravings may have to be provided.

The procedure is then as follows:

(i) The oscilloscope controls are adjusted so that the waveform coincides with the reference points B, W, M1, M2. The rating corresponding to the maximum deviation (positive or negative) of the top of the bar is measured as K_{bar}, say. Note that readings are not taken for points on the waveform within 640 ns of the bar edges to avoid confusing short-term high-frequency distortions with the long-term distortion to be measured.

(ii) The K rating corresponding to the peak of the pulse is also measured as, say, K_{pb}. For this, the gain is kept as in (i).

(iii) The time-base speed is then increased to correspond to the horizontal scale of the 'submarine' graticule; i.e. if the $2T$ intervals are 1 cm apart on the graticule and T = 100 ns, the time-base rate will need to be precisely 200 ns/cm. The amplitude and shift controls are then adjusted so that the baseline of the pulse coincides with the horizontal line through the point 0 on the 'submarine' graticule, the peak of the pulse lies on the upper horizontal line, and the pulse is symmetrically disposed within the vertical lines. Note that the peak of the pulse will not be precisely on the central vertical line even when the pulse is undistorted, owing to the slight residual asymmetry described above.

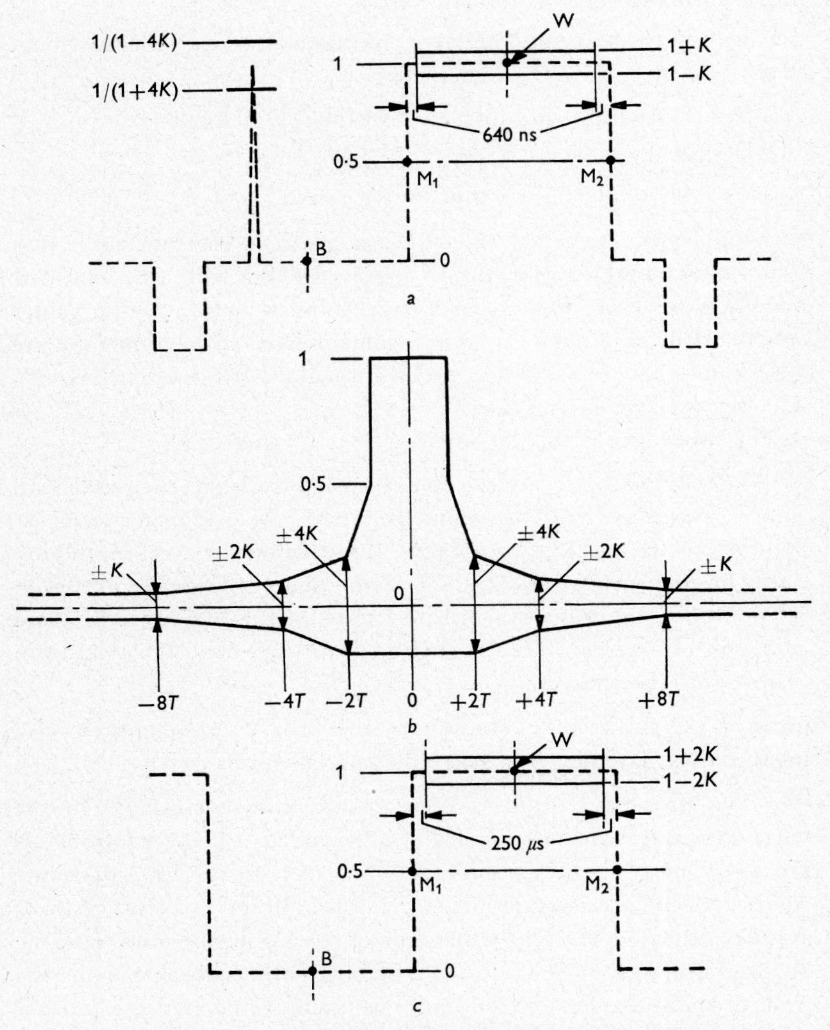

Fig. 6.6 Waveform response limits for sine-squared-pulse and bar signal

a K_{pb} and K_{bar}
b K_{2T}
c $K_{50\,\mathrm{Hz}}$

The K rating corresponding to the greatest deviation of the pulse from the undistorted condition is measured. Let this be K_{2T}.

In passing, it was at one time customary to attempt to measure the half-amplitude duration of the pulse, but this is no longer so. First, it is redundant, since the pulse/bar ratio expresses the same condition in a linear system, and secondly, the pulse duration is extremely susceptible to narrowing or widening as a result of even very small amounts of quadrature distortion arising from sideband asymmetry with amplitude-modulated signals, even when the other features of the waveform are not changed by any significant amount. On the other hand, by inserting an inverted sine-squared pulse into the top of the bar (Fig. 6.17) this property can be utilised as a test for quadrature distortion (Fig. 7.4).

(iv) The 50 Hz square wave is fitted to the points B, W, M1, M2 on the third graticule and the maximum deviation of the top of the bar (positive or negative) is measured as $K_{50\,\mathrm{Hz}}$. The first and last 250 μs of the top of the bar are excluded from this measurement so that only very low-frequency distortions are included.

These four rating factors K_{bar}, K_{pb}, K_{2T} and $K_{50\,\mathrm{Hz}}$ are noted, since they form an invaluable record of the precise form of the distortion eixsting at the time of measurement; and, if variations occur, it is possible to see which factor or factors have changed. The final K rating is obtained by taking the largest of the first three as the value. $K_{50\,\mathrm{Hz}}$ is normally omitted, since this type of distortion is removed, or at least greatly reduced, by the black-level clamp at the transmitter.

As an example, assume that in the course of a given routine test the following results are obtained: $K_{bar} = 1\%$; $K_{pb} = 1\frac{1}{2}\%$; $K_{2T} = 2\frac{1}{2}\%$; and $K_{50\,\mathrm{Hz}} = 3\%$. Although $K_{50\,\mathrm{Hz}}$ is the worst of the four readings, it is ignored for this purpose, and the K rating is said to be $2\frac{1}{2}\%$, the value of K_{2T}.

As has already been explained, the routine-test method has one weakness, namely that the shape of the spectrum of the waveform makes it impossible to examine at all stringently the behaviour of a system over the upper half of the videoband. If the television standard is one which is also employed for colour, 625 lines and

525 lines, this difficulty may be overcome very simply by replacing the simple monochrome sine-squared-pulse and bar signal by one of the composite luminance plus chrominance sine-squared-pulse and bar signals (Section 6.3.3). The chrominance component of the test waveform, with appropriately relaxed limits, then provides an adequate test for the upper portion of the video spectrum. In the author's opinion, this modification of the standard procedure would be desirable even where it is not intended to introduce a colour service in the immediate future. A further discussion of this is given later.

Where no colour-test waveforms are available, a means is suggested by the classical sine-squared-pulse and bar routine-test method for controlling to some extent the behaviour of the apparatus under test at the top of the band. This consists in replacing the $2T$ pulse-and-bar signal by a T signal. In general, the pulse will be distorted by the upper cutoff characteristic of the video channel. The following characteristics of the distorted T pulse are then measured:

(*a*) half-amplitude duration
(*b*) pseudofrequency of the 'ring'
(*c*) first (negative-going) lobe, leading or trailing
(*d*) second (positive-going) lobe, leading or trailing.

The figures are not used to obtain a K rating, but are noted, and successive readings are compared so that any significant shift of the upper video region becomes evident. A very similar procedure is adopted in the USA and Canada, but at the moment no other K rating is undertaken.

It may be desirable in addition to make a single-frequency measurement at or near the top of the videoband and impose the condition that this shall not differ by more than ± 1 dB with respect to the low video frequencies. Where equalisation is adopted, based solely on an examination of the $2T$ pulse shape, it is possible to end up unaware of a rise at the top of the band which, if repeated successively, could cause overloading. However, when the sine-squared-pulse and bar signal forms part of a combined luminance and chrominance test signal (Section 6.3), this precaution can be omitted since the chrominance test signal will furnish the requisite information.

6.2.6 Graticules

So far, for reasons of clarity, the description has been given of the routine-test method with respect to the three graticules of Fig. 6.7. However, it is inconvenient under operational conditions to have to change graticules during a measurement, not only because of the time involved but also on account of the possibility of damage to the graticules and the ease with which loose graticules can be mislaid.

One very obvious simplification is to omit the graticule for the 50 Hz square wave and use the limit lines for the top of the bar, doubling the value of the *K* rating as read. Even better is the use of a

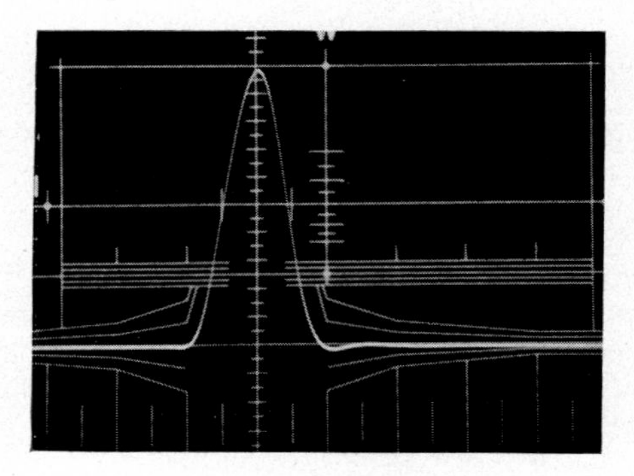

Fig. 6.7 Composite *K*-rating graticule

compound graticule such as is shown in Fig. 6.7. Although skeletonised to avoid confusion from too great a number of lines, it nevertheless has been found easy to interpret and use.

Another device which has been used for certain operational purposes by the BBC is a graticule engraved with a different pattern on each side. The edge-lighting lamps on the waveform-monitor bezel are divided into two groups and masked so that each set illuminates one of the patterns only, contrasting colours being used if desired. It is then very simple to switch from one graticule to the other.

All graticules should be accurately engraved, and used in such a way that parallax is minimised. They should preferably be bought from a reputable supplier; but if they have to be engraved, the first at least of the graticules from the template should be checked against a transparency made from the original scaled-up drawing by a specialist firm equipped for this kind of photography (see also Chapter 11.3.2).

6.2.7 Addition of *K* ratings

It is frequently required to know what the resultant *K* rating will be of a number of items of equipment or distribution links when put in tandem, or conversely, given an overall limit for the tandem connection, what limit should be placed on each individual item or link. No hard and fast rule can be given, since such a problem must always be statistical in nature and the numbers involved are usually too small for statistical methods to give an accurate forecast.

If the separate items are likely to have sine-squared-pulse and bar responses which are closely alike, e.g. video amplifiers made to close tolerances, the *K* ratings will add linearly. If the *K* ratings are random in nature, they would be expected to add on a root-mean-square basis. For distribution circuits of dissimilar types, it has been suggested (CCIR, 1966) that the law of addition is

$$K_{overall} = (K_1^{\frac{3}{2}} + K_2^{\frac{3}{2}} + \ldots + K_n^{\frac{3}{2}})^{\frac{2}{3}}$$

This is usually known as the 'three-halves power law'. On the other hand, extensive series of measurements carried out by Paddock (1970) suggest that the practical relationship may be closer to the square root of the sum of the squares of the individual values.

6.2.8 Small *K* ratings

The measurement of small *K* factors is one of the special problems of the method as a whole. When a lengthy and complex chain of equipment carrying video signals is given a reasonable overall limit, say 2 or 3%, and this overall limit is shared out among the component items of equipment, it is found that the smaller items must logically be allocated *K* ratings which are only a small fraction of 1%. Such a rating may correspond to a spacing of the limit lines on the graticule which is equal to, or even less than, the width of the trace of the waveform monitor.

If the item of equipment is such that it can be made available in relatively large numbers, all specimens of which should resemble one another quite closely, e.g. video distribution amplifiers, the most accurate method is undoubtedly to put, say, ten samples in series and measure the total overall distortion. Care must be taken, of course, to use the shortest possible connecting leads and to insert distortionless attenuators if and when required. The resulting *K* rating is then divided by the number of amplifiers in series. Since it is possible with a good waveform monitor using a 5 in cathode-ray tube or the equivalent rectangular tube to estimate, rather than measure, *K* ratings down to 0·25%, it follows that individual *K* ratings can be estimated by series measurement to about 0·03%.

Many waveform monitors are capable of considerable amounts of windowing with very little distortion; a selected portion of the waveform can be amplified without overloading from the rest of the waveform. This technique can be applied to the measurement of K_{pb}, K_{bar}, and K_{2T} when the errors are small. The measurement is made in the standard way except that the determination of the limits for the distortion is made with the gain increased by a suitable factor, say, three or ten times, and greater accuracy is obtained in the measurement of K_{pb} if the waveform monitor can be double-triggered so that the centre of the bar appears vertically above the peak of the pulse.

At first sight, this would seem to offer a means of increasing the accuracy by a considerable factor, but it imposes a severe requirement on the accuracy of the undistorted pulse-and-bar signal, apart from the difficulty of dealing with the inherent slight residual distortion of the generated pulse due to the Thomson network. In all but a very few instances, both the undistorted pulse and the distorted pulse are available simultaneously when extremely small *K* ratings are to be measured; so, this difficulty can be surmounted by examining both the input and output pulses under identical conditions and taking the differences. One well known technique is to take photographs from which the differences can be found by measurement, thus eliminating parallax errors; the various features of the waveforms can then be measured with a travelling microscope. The waveform-monitor gains must be standardised by using the centre of the top of the bar. Small distortions due to the waveform monitor are also eliminated provided that the two photographs are taken under identical conditions.

A further possibility is to subtract the input from the output waveform electrically in a high-grade difference amplifier. Unfortunately, this may come to grief on the difficulty of providing a signal delay with sufficiently low distortion. To measure the differences between the waveforms, the input waveform must be delayed by an amount precisely equal to the delay in the apparatus under measurement, which means that a virtually distortionless variable delay network must be available as well as a distortionless means of precisely equalising the gains in the two paths.

If the delay of the apparatus is very small, as is very often the case when the distortion is low, the method does become practicable, since a variable delay of some tens of nanoseconds with very low distortion can be constructed from a chain of prototype lowpass-filter sections whose shunt elements are supplied by a ganged capacitor (Fig. 6.8). The terminating resistors are preferably composed of ganged variable resistive networks or voltage-controlled resistors, so that the terminating impedances remain substantially equal to the impedance of the filter network as the capacitances are varied. Even for maximum delay, the cutoff frequency will be several times the highest video frequency; so the distortion should be negligible.

The basic circuit is shown in Fig. 6.8. The gain of the system must be adjusted to give zero resultant at a point near the centre of the pulse after the delay has been adjusted so that the peak of each pulse viewed separately is centred on the same graticule line. Assuming that the difference-mode gain of the amplifiers is known, the value of K_{2T} can be read from the graticule. The difference signal being so small compared with the test signal for the low-distortion conditions assumed, considerable amplification may be employed without fear of difficulty with the vertical amplifier of the waveform monitor. For example, with a gain of $10X$, a K_{2T} of 0·2% can be read against the

2% lines of the standard graticule. K_{pb} and K_{bar} may be measured similarly.

A very convenient variant of this technique for measuring K_{2T} is available if one has available a sine-squared pulse accurate and stable enough to serve as a master for measurement purposes. The output sine-squared pulse is then subtracted from this instead of from the input pulse, and the difference considered as the distortion component. Two advantages accrue from this: first, the ability to make

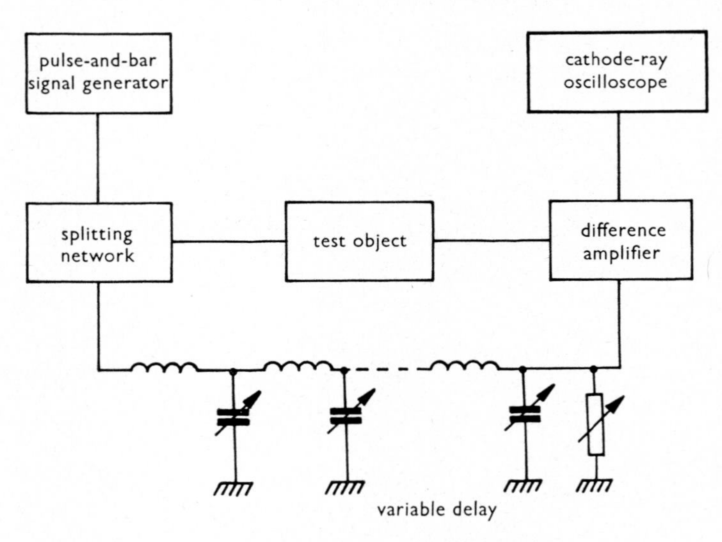

Fig. 6.8 Measurement of small K ratings

the measurement even when the input terminals of the apparatus under test are not accessible, and secondly, the facility of assuring the time coincidence of the two pulses by means of a trigger pulse with variable delay controlling the start of the master pulse. This can be timed from the leading edge of the received line synchronising pulse. The master pulse then remains undistorted, whatever the delay.

The measurement process is not materially different from a measurement against a graticule, except that one is operating with the difference or error waveform instead of the distorted waveform, and, of course, there is no possibility of parallax errors. Indeed, it is preferable to the conventional technique, in that one is no longer in any way concerned with the presence of the 'rings' following the pulse which normally cause a slight unbalance in the K ratings on the two sides of the pulse. Nevertheless, a more symmetrical sine-squared pulse than that furnished by the Thomson network would be advantageous in general for the measurement of small K ratings.

6.2.9 Routine test: bar only

A form of the routine test preferred by a number of administrations, particularly in Europe, makes use of the standard pulse-and-bar

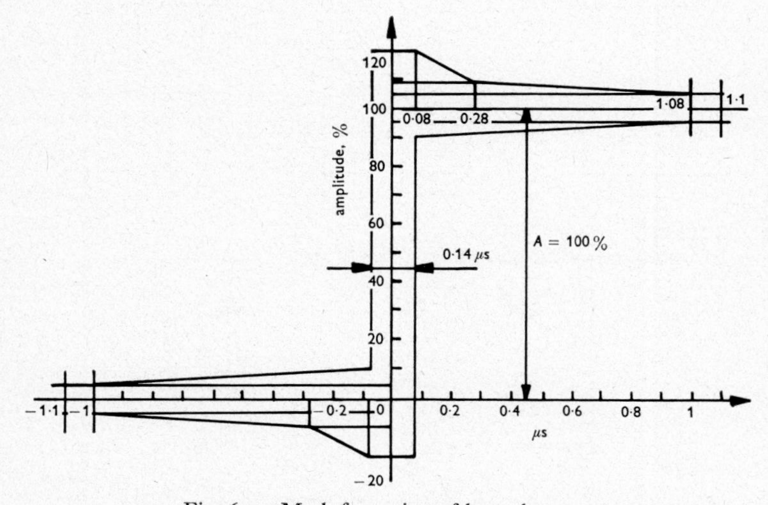

Fig. 6.9 Mask for rating of bar edge

signal with the pulse omitted, the transition being 1 T. A graticule of the type shown in Fig. 6.9 is used, and the test is carried out by observing whether the distorted signal fits within the graticule. In effect, this is the equivalent of the T pulse test, in which any transient resulting from the upper frequency band limitation of the system under test are rendered less objectionable by the use of a bar instead of a narrow pulse.

The advantage of this procedure is presumably the ease with

which it can be carried out and the much lower reliance placed on the technical skill of the operator. The disadvantages, on the other hand, seem to be considerable, the principal one being the abandonment of a well conceived and proven means for allocating to a distorted signal a unique rating factor which is a measure of the viewer's reaction to the corresponding picture.

As far as the test-signal generator is concerned, the saving when the pulse is omitted is relatively small, since the omission is only of a narrow-pulse generator and its accompanying trigger circuits. The rest of the generator, including the relatively expensive Thomson network, remains virtually unchanged.

While the use of the bar only is still to be preferred to steady-state measurements, this modification of the method is, in the author's opinion, markedly inferior to the standard sine-squared-pulse and bar system, and will no doubt become less frequently used as time passes.

6.2.10 Acceptance test: practical considerations

The acceptance-test method differs from the routine-test method essentially in that a T pulse is used instead of a $2T$ pulse, and a mathematical analysis is subsequently undertaken to obtain the rating value. Additional parameters are introduced to test the whole of the videoband more stringently while removing irrelevant information due to effects beyond the nominal upper limit of the videoband. The analysis cannot be described in detail in this limited space, but a lucid explanation of the mathematical procedures is given by Lewis (1954). Instead, the basic principles of the method, which appear to be little understood in general, are outlined briefly.

The first step requires accurate plots of the input pulse and the distorted output pulse from the system under test, using the same measuring equipment. Lewis recommends the analysis of the shape of photographs with a travelling microscope. Very much to be preferred, if it is available, is a waveform sampler whose sweep and output voltage are connected to the x and y scans, respectively, of an x–y plotter. The author has used a commercial sampling adaptor for cathode-ray oscilloscopes. The duration of the sampling pulse should not be greater than about 20 ns. The resolution easily obtainable with such a device is shown in Fig. 6.10, in which each division corresponds to 25 ns. Once the apparatus has been set up, such a curve can be traced in one minute or less.

From these two sets of information, i.e. the ordinates of the waveforms as a function of time, is derived the output waveform which would have been received at the output terminals if the input test waveform had been a $(\sin x)/x$ pulse corresponding to an ideal band

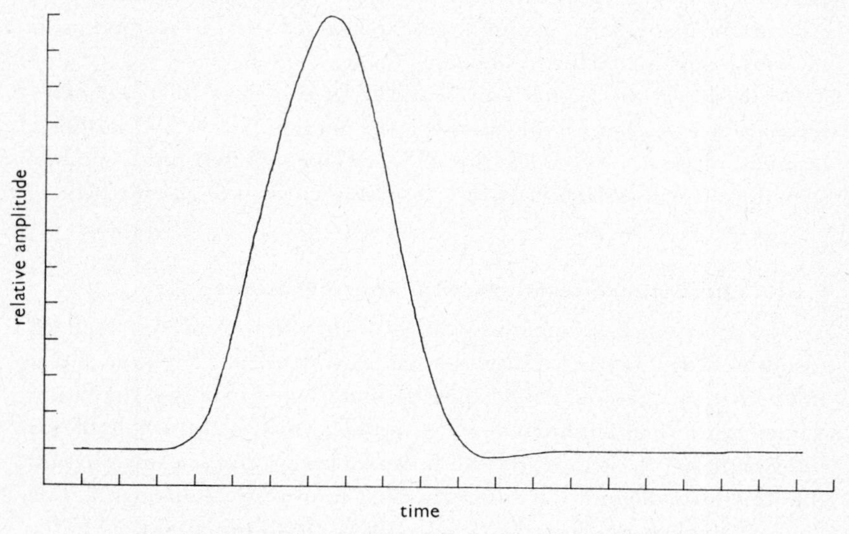

Fig. 6.10 T sine-squared pulse plotted by waveform sampler

limitation at a frequency of $1/(2T)$; this is the waveform which would have been obtained if an infinitely narrow pulse had been applied to the input terminals of the system under test via an ideal lowpass filter having its cutoff at that frequency. Such a waveform has a spectrum which is uniform between zero frequency and $1/(2T)$, and is zero beyond. This means that all frequencies within the nominal videoband are given precisely equal weighting, and no test of any kind is made at frequencies higher than the nominal band limit. In addition, the individual characteristics of the pulse-and-bar generator and the test apparatus at the output have both been allowed for.

The distorted hypothetical $(\sin x)/x$ output pulse obtained by computation from the output T pulse is finally used to derive by computation the output waveform of the system under test for an input waveform consisting of an ideal $2T$ pulse. The $(\sin x)/x$ and $2T$ waveforms then furnish four sets of conditions for which the K rating is finally derived.

The mathematical work would be very much more difficult if advantage could not be taken of the fact that all the waveforms concerned have spectra restricted to a finite frequency range. Shannon (1948) showed that a waveform which has no spectral components above some frequency f_c can be unambiguously described by a series of ordinates spaced at equal intervals $1/(2f_c)$. Hence the waveform can be written down and used for computation as a series of numerical terms known as a 'time series' (Lewis, 1952); e.g. the time series for the ideal $2T$ pulse can be written $\frac{1}{2}$, 1, $\frac{1}{2}$, in which the ordinates are spaced at intervals of T and all other terms are zero.

The evaluation of a convolution integral then reduces to a simple algorithm (serial multiplication); and the inverse process (serial division), although somewhat less straightforward, is not unduly difficult. These are the operations carried out on the two original waveforms, which for this purpose are required in the form of a time series with an interval T.

Let the time series for the derived $(\sin x)/x$ pulse be

$$B(rT) = B_{-r}, \ldots, B_{-1}, \quad 1, \quad B_{+1}, \ldots, B_{+r}$$

where the central term B_0 has been normalised to unity. By serial multiplication of $B(rT)$ with $\frac{1}{2}$, 1, $\frac{1}{2}$, we obtain the derived $2T$ time series

$$C(rT) = C_{-r}, \ldots, C_{-1}, C_0, C_{+1}, \ldots, C_{+r}$$

Then the conditions for a rating factor of K are

(a) $$\frac{1}{8}\left|\frac{C_r}{C_0} - \frac{1}{2}\right| \leqslant K \qquad r = \pm 1$$

$$\frac{1}{8}\left| r\frac{Cr}{C_0}\right| \leqslant K \qquad \begin{cases} -8 \leqslant r \leqslant -2 \\ +2 \leqslant r \leqslant +8 \end{cases}$$

$$\left|\frac{Cr}{C_0}\right| \leqslant K \qquad \begin{cases} r < -8 \\ r > +8 \end{cases}$$

(b) $$\frac{1}{4}\left|\frac{1}{C_0}\sum_{-8}^{+8} B_r - 1\right| \leqslant K$$

(c) $$\frac{1}{6}\left|\sum_{-8}^{+8} B_r - 1\right| \leqslant K$$

(d) $$\frac{1}{20}\left[\sum_{-8}^{+8} |B_r| - 1\right] \leqslant K$$

Conditions (*a*) and (*b*) are approximately equivalent to the limits laid down in the routine test, since $C(rT)$ represents the response of the system to an ideal $2T$ sine-squared pulse. The summation of the terms of $B(rT)$, that is $\sum_{-8}^{+8} B_r$, is equivalent to an integration of the pulse response and therefore gives a value proportional to the height of the bar. Condition (*b*) is accordingly equivalent to the derivation of K_{pb}. Condition (*c*) is likewise a restriction on the pulse/bar ratio of the $(\sin x)/x$ pulse response.

Condition (*d*) is difficult to interpret until one recalls that the $(\sin x)/x$ pulse corresponding to an ideal frequency limitation at $f_u = 1/(2T)$ should ideally have the time series...0, 0, 1, 0, 0,..., since the undistorted waveform passes through zero at intervals of T spaced symmetrically about the central term. Hence the time series represents the distortion terms introduced by the system under test, and condition (*d*) represents a restriction of the arithmetical sum of all these distortion terms.

6.3 Colour pulse-and-bar waveforms

6.3.1 General

Although there are now three systems of colour television in use, NTSC, Pal and Secam, they very fortunately have sufficient in common for it to be possible to use identical waveforms for testing the linear waveform distortion of all three, even though the allowable distortion may differ from one system to the other.

These common features are that the luminance information is transmitted as a video signal virtually identical with the video signal which would be delivered by monochrome cameras viewing the same scene, and that the chrominance information is transmitted as a narrowband signal modulated on a high-frequency carrier situated near the upper end of the videoband. The two signals are added linearly.

The luminance signal may therefore be tested with sine-squared-pulse and bar waveforms exactly as described above for monochrome purposes. The chrominance region must then be tested with a modulated signal so chosen that its sidebands occupy approximately the same frequency range as the coded chrominance signal; so the obvious choice for Pal and NTSC is a narrowband sine-squared-pulse and bar signal which amplitude-modulates to a depth of 100% the standard colour subcarrier. Although logically the Secam chrominance test signal should be frequency-modulated, an amplitude-modulated signal is found adequate, and is easier to use.

However, the testing of the chrominance channel cannot proceed completely analogously to the testing of the luminance (monochrome) channel, since, apart from distortion of the modulation envelope of the chrominance test signal, the mutual relationship between the chrominance and luminance channels is also of importance. The linear errors which can arise are luminance–chrominance gain inequalities and luminance–chrominance delay inequalities, the former affecting the saturation of the reproduced colours with Pal and NTSC and the latter affecting the registration of the colour picture. Experience has shown that these inequalities occur much more frequently than distortion of the chrominance pulse itself, and indeed the general distortion usually has to be fairly large before distortion of the chrominance pulse becomes at all evident. It is consequently necessary to build into the test signal a luminance amplitude reference signal, which is most conveniently the narrowband sine-squared-pulse and bar signal used in the generation of the modulated chrominance pulse-and-bar signal.

The accepted convention is that gain inequalities are expressed as a percentage (CCIR, 1966), and are defined as positive when the chrominance amplitude exceeds the luminance. Delay inequalities are expressed in nanoseconds, and are positive when the chrominance component lags the luminance.

6.3.2 Chrominance pulse-and-bar signal

The original signal for the measurement of chrominance-channel distortion was described by Macdiarmid and Phillips (1958). It consists (Fig. 6.11 *a*) of two distinct signals which are carried on adjacent lines. The first, the luminance reference, is a sine-squared-pulse and bar signal of one-half the normal amplitude whose half-amplitude duration depends on the standards of the system in use. For system *I*, a new parameter T_c (Weaver, 1965) has been defined corresponding to T for the luminance (monochrome) case, in which $T_c = 5T =$ 500 ns. The normal routine-test signal has accordingly a half-amplitude duration of $2T_c = 1\ \mu$s, and a video bandwidth of 1 MHz. It is sometimes also termed a '$10T$' signal. The T_c and $2T_c$ signals are generated entirely similarly to the T and $2T$ signals, the networks being identical except for the appropriate frequency scaling.

The second signal is generated from the luminance reference by using it to modulate chrominance subcarrier to a depth of 100%. For normal measurement purposes, it is preferable not to use a correctly locked subcarrier but to generate it from a crystal oscillator in the test equipment. The $2T_c$ pulse contains only eight or so cycles of subcarrier in all, and the T_c pulse only four; this results in the modulation envelope with locked subcarrier having a distinctly 'jagged' appearance, which interferes with the accuracy of measurement. When a reasonably stable but unlocked oscillator is employed, the inevitable slight changes of frequency give rise to a superposition of sinusoids with random phases, which effectively smooth the modulation envelope, as can be seen from the illustrations, without detriment to the accuracy.

Both the signals are mounted on a pedestal of amplitude equal to 50% of white level, which ensures that the lower half of the modulation envelope does not encroach too much, even under distortion conditions, on the synchronising pulse region.

Since both signals are generated from the same basic waveforms, the timings of the various portions of the modulation envelope correspond exactly to the timings of the same portions of the luminance reference. It is therefore possible to overlap them completely

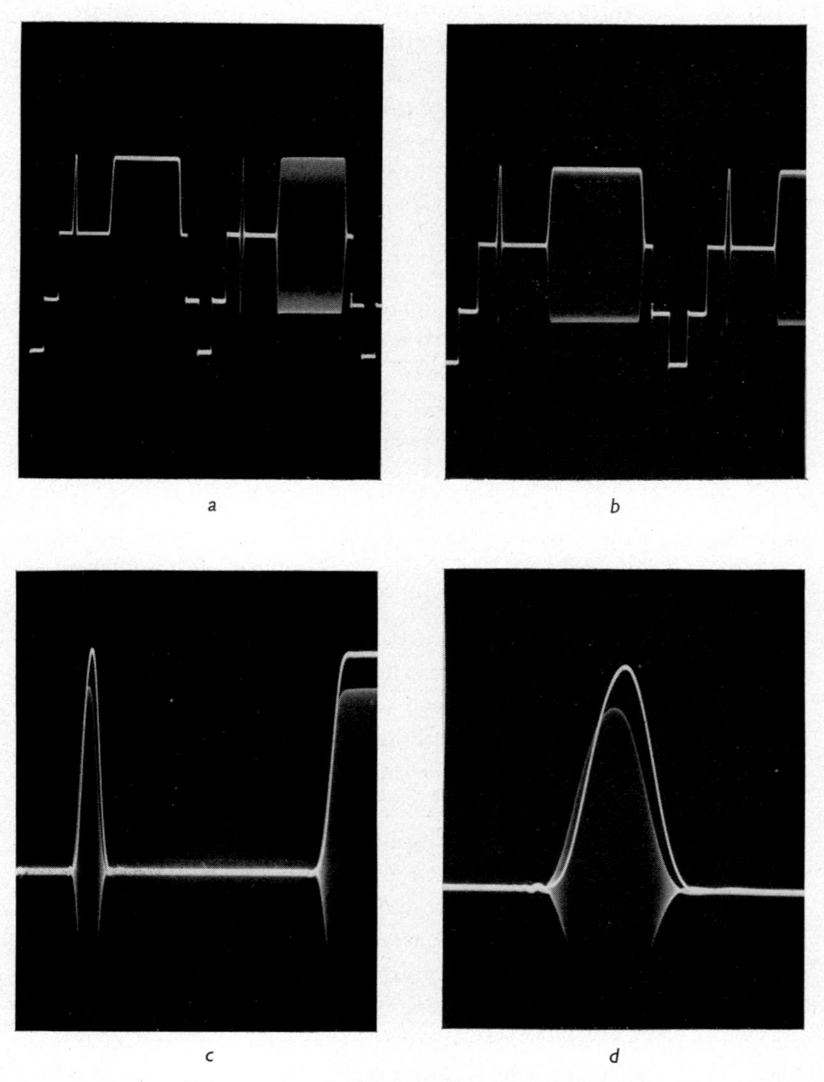

Fig. 6.11 Chrominance pulse-and-bar waveforms

a 2-line signal
b Adjacent lines superimposed
c Detail of relative gain and delay errors
d As (*c*), but in greater detail

when the waveform monitor is adjusted so as to trigger from each successive synchronising pulse. In fact, the only real indication (Fig. 6.11*b*) of the existence of two signals when no distortion is present comes from the brightening of the upper portion of the modulation envelope and black level.

It follows that any gain inequality or delay inequality between the low luminance frequencies and the subcarrier region will cause a characteristic loss of coincidence as in Fig. 6.11*c*, in which both the overlaid pulses and overlaid bar edges are shown for simultaneous gain and delay distortion. The delay error is most clearly indicated by the pulses and the amplitude error by the bars. Fig. 6.11*d* gives an enlargement of the superimposed pulses. The distortion of the modulation envelope is negligible.

It is possible to use the oscilloscope to measure the amplitudes of the luminance reference and chrominance signals from which the luminance–chrominance gain inequality may be calculated, and likewise the time intervals between the peaks of the luminance reference and the chrominance pulse furnish the luminance–chrominance delay inequality.

Apart from its simplicity, this procedure for measuring luminance–chrominance inequalities has little to recommend it. The accuracy is low and each reading is fussy and time-consuming. A considerably more convenient and accurate method, now standard in the BBC, is the use of a nulling device known as the 'colour gain and delay tester' (Fig. 6.12).

The output signal from the apparatus or circuit under test is applied to the input of this equipment, at which point it traverses two channels. The upper channel contains a phase-equalised bandpass filter which is distortionless over a range of ±2 MHz with respect to the subcarrier frequency. The luminance component of the test signal together with the synchronising pulses is thereby eliminated, and the chrominance component is recovered. A calibrated gain control allows the separated chrominance signal to be varied by ±30% with respect to its correct amplitude.

The lower channel contains a bandstop filter to eliminate the chrominance region without distortion and a variable delay, so that the signal at the output of the filter is the delayed luminance reference signal. The relative transmission times of the chrominance and luminance channels are adjusted to be equal when precisely one-half of the variable delay in the luminance path is inserted, so that the chrominance can be made either to lag or to lead the luminance. The function switch connects the output amplifier either to the chrominance signal only, the recombined chrominance and luminance signals, or the luminance reference only.

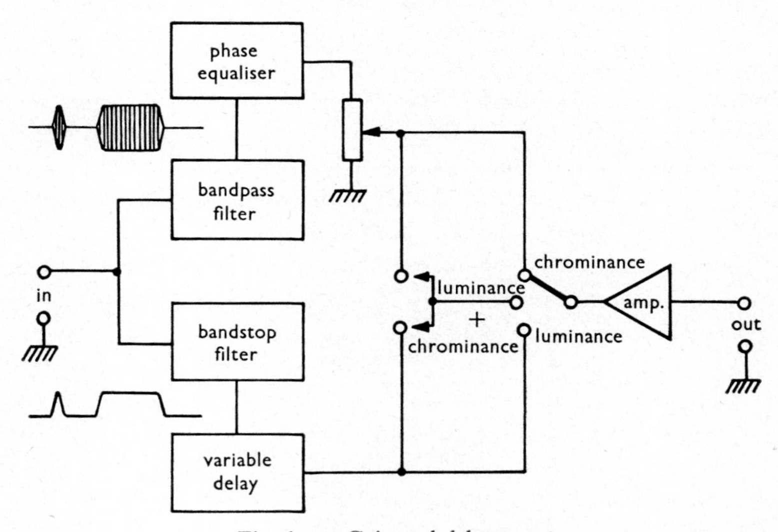

Fig. 6.12 Gain and delay tester

The method of use is very simple. The distorted test signal to be measured, preferably a $2T_c$ signal, is applied to the input, and the recombined signal is examined by connecting a suitable waveform monitor to the output of the test set. The gain control is varied until the relative amplitudes are correct, best carried out by adjusting the flat tops of the bars to coincidence, when the luminance–chrominance gain inequality can be read to the nearest 1% from the calibrated dial.

Adjustment of the delay then makes it possible to bring the two pulses to coincidence. The major section of the delay is adjustable in steps of 10 ns on a push-button switch up to a total of ± 100 ns with a continuously variable section providing an additional ± 10 ns. When the two component waveforms are undistorted by their passage through the apparatus under test, the coincidence can be effected with a setting accuracy of 2–3 ns; the magnitude and sign of the delay error are read directly from the switch positions. When the waveforms are distorted, the setting accuracy is reduced; but, on the other hand, the size of the error is normally comparatively large, so that the percentage error is not widely different. Furthermore, as the distortion increases the difference between the shapes of the two waveforms, the notion of transmission delay loses its precision, since one can no longer measure between completely corresponding points.

A rather more elaborate version of this instrument has since been designed by Coles (1970). Its operation depends on the cancellation of the amplitude and delay errors of the signal by the addition of pairs of equal-amplitude echoes of the input signal, of like polarity for the cancellation of luminance–chrominance gain inequalities, and of opposite polarities for the cancellation of luminance–chrominance delay inequalities. This will be recognised as an application of the theory of paired echoes (Wheeler, 1939). Such pairs give rise to cosine-shaped modifications of both the gain and the group delay characteristics, the magnitude and sign of which are determined by the magnitudes and signs of the echoes. Thus, with suitably calibrated controls, the instrument may be utilised in precisely the same way as the 'gain and delay tester' described above.

The instrument suffers in principle from the inherent weakness that, although pairs of like echoes affect the gain without modifying the group delay, pairs of unlike echoes affect both simultaneously. Fortunately, for delay adjustments up to 100 ns, the interaction between the delay and gain characteristics is small enough to be neglected for practical purposes. On the other hand, unlike the previously described instrument, this version is passive; and moreover, there are no discontinuities in its response, so it may be left permanently in circuit. It can accordingly serve both as a measuring instrument and as an equaliser.

The ease, accuracy and dependability with which readings can be taken with an instrument of this type compared with the use of a graticule make it very worthwhile to standardise such measurements for operational and other purposes. It represents an extra expense, of course, compared with a graticule; but in fact, the cost is quite moderate, since the performance demanded from its amplifiers and network is less stringent than might be thought.

The great weakness of this original test signal, in common with all signals having components on more than one line, is its vulnerability to errors from the low-frequency distortions of the video signal. These produce a difference in the 'sit' of each of the signals, and low-frequency interference causes a difference in shape between the two adjacent lines, both of which interfere with their correct superimposition. In some instances, this takes the form of an apparent delay error as a result of a difference in oscilloscope triggering between the successive lines. Their vulnerability to positional errors also makes such signals quite unsuitable for use with videotape machines. This subject is discussed at greater length elsewhere (Weaver, 1965). Consequently, although the chrominance pulse-and-bar signal is one of the test signals for linear waveform distortion which are specified by the CCIR for system *I*, its use is now restricted to short-time and line-time distortions occurring within the chrominance channel itself.

6.3.3 Composite pulse-and-bar signal

The most important objections to the '2-line' signal described immediately above can be overcome with additional advantages, by combining the two component waveforms on the same line (Weaver, 1965). The two components, the luminance reference and the chrominance signal, are generated precisely as described for the chrominance pulse-and-bar signal; but instead of being added to successive lines, they are added in time coincidence with the amplitude of the luminance reference precisely equal to the amplitude of either half of the modulation envelope.

The result is as shown in Fig. 6.13: the lower half of the modulation envelope appears to be cancelled by the addition of the luminance reference, and a sine-squared-pulse and bar signal results with twice the amplitude of the luminance reference but filled with subcarrier and having a flat base.

This is a unique condition depending on the time coincidence of the two component waveforms and their precisely correct relative amplitudes. It follows therefore that any change in the relative gains or the relative delays will give rise to some characteristic deformation of the

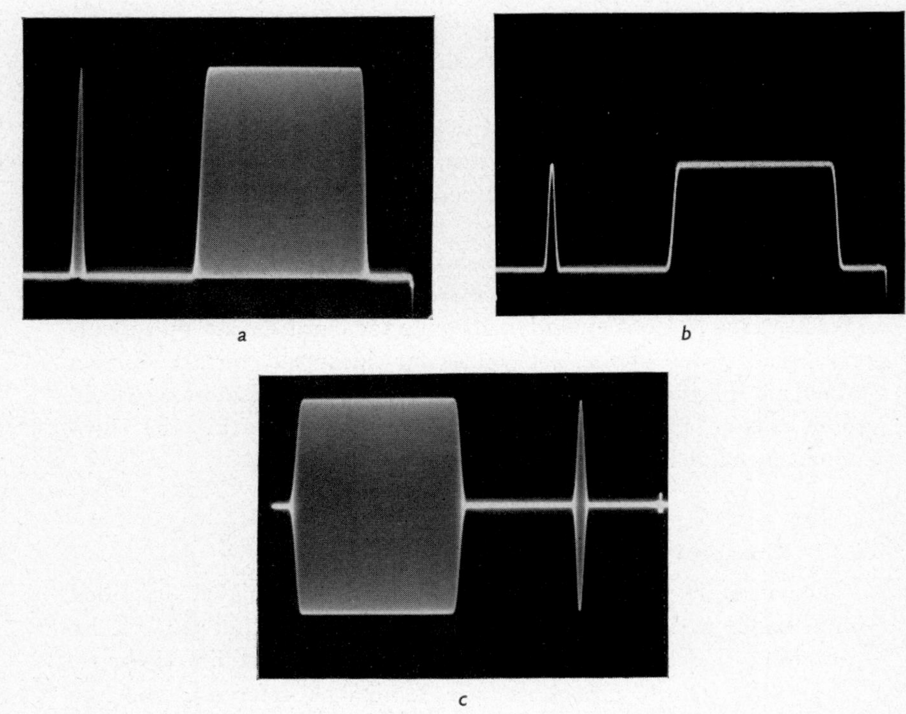

Fig. 6.13 Composite chrominance pulse-and-bar signals

a Composite waveform
b Luminance component of waveform
c Modulated chrominance component of waveform

composite waveform. When it is recalled that the sine-squared pulse is a single cycle of a sinusoid, and that the transitions of the 'sine-squared' bar have the shape of halfcycles of a sinusoid, a qualitative idea of the results of delay and gain errors is easily achieved.

Gain errors are very simple. The base of the composite signal is formed by the difference of two sine-squared pulses, whereas the signal itself is formed by the sum of the same two waveforms. Any change in the relative gain must consequently produce a deformation of the base having exactly the same shape as the waveform concerned,

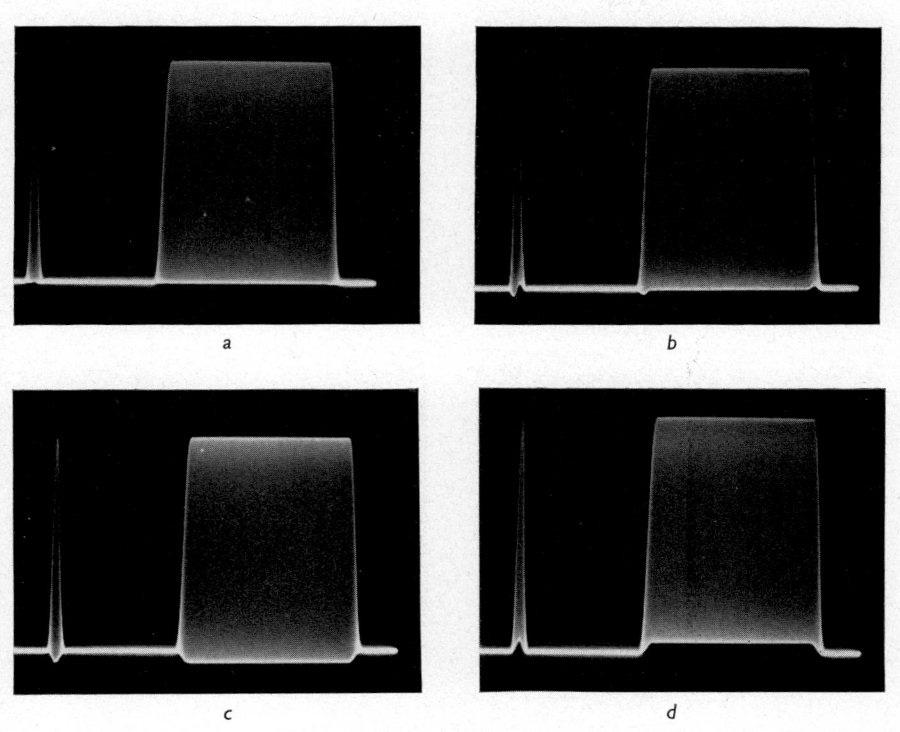

Fig. 6.14 Distortions of composite chrominance pulse-and-bar signal

a Undistorted signal
b Negative delay inequality only
c Positive chrominance–luminance gain inequality
d Negative chrominance–luminance gain inequality

positive for a decrease in the chrominance gain and negative for an increase (Figs. 6.14*c* and *d*). The amplitude of the waveform, whether the pulse or the bar, increases or decreases, respectively, by the same amount.

Delay errors present no difficulty if it is recalled that taking the difference between two identical waveforms with a small time displacement is equivalent to an approximate differentiation with respect to time. The base of the pulse consequently takes the form of

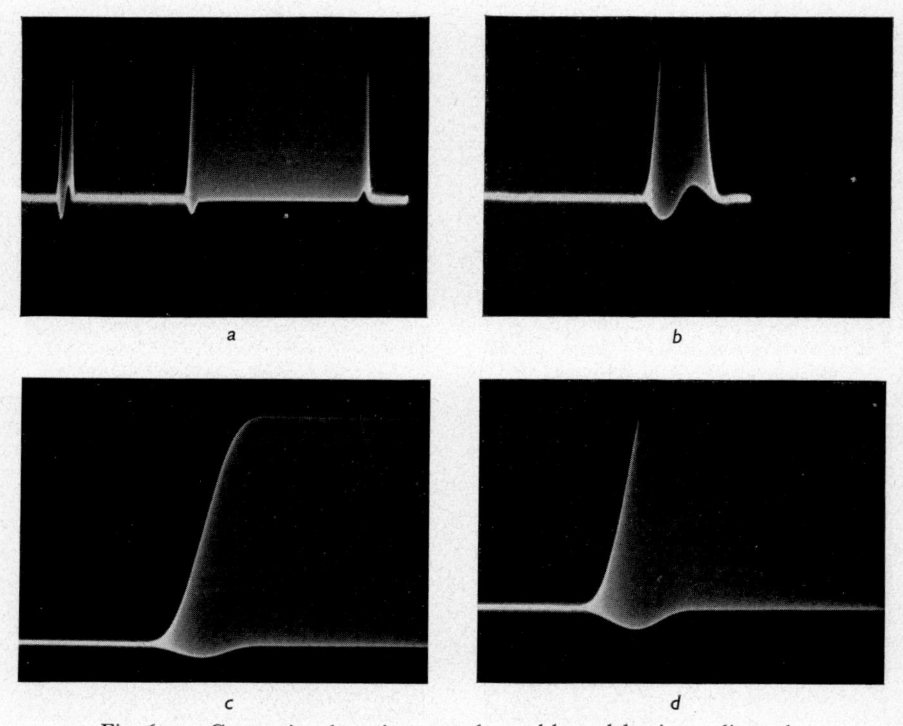

Fig. 6.15 Composite chrominance pulse and bar: delay inequality only

a Distorted pulse and bar
b Detail of base of pulse
c Distortion of bar edge
d As (*c*), but in greater detail

a sinusoid displaced by 90° with respect to the sinusoid forming the pulse, the polarity depending on the sense of the delay error. Likewise, the bar yields a differentiated bar, i.e. a sine-squared pulse, at each transition. These distortions are shown in Fig. 5.14*b* and, in greater detail, in Fig. 6.15.

The effect on the waveform itself, as distinct from the baseline, is far less evident. It consists of a slight widening of the waveform which is only really visible, even on the pulse, when the delay error is very great.

The exact expressions for the composite pulse are easily derived. If the half-amplitude duration of the pulse is T_0, the envelope of the modulated pulse for unit amplitude of the composite pulse will be

$$e_1(t) = \pm \tfrac{1}{2} \sin^2 \frac{\pi}{2} \frac{t}{T_0}$$

and the luminance reference will be

$$e_2(t) = \tfrac{1}{2} \sin^2 \frac{\pi}{2} \frac{t}{T_0}$$

The composite pulse is the sum of these two voltages, and evidently, in the undistorted case, the pulse amplitude is $e_u(t) = \sin^2(\pi/2)t/T_0$ and the baseline $e_L(t) = 0$.

Assuming that the ratio of the chrominance to luminance amplitudes is A and that the chrominance is delayed with respect to the luminance by a time τ,

$$2e_u(t) = A \sin^2 \frac{\pi}{2} \frac{t-\tau}{T_0} + \sin^2 \frac{\pi}{2} \frac{t}{T_0}$$

$$2e_L(t) = -A \sin^2 \frac{\pi}{2} \frac{t-\tau}{T_0} + \sin^2 \frac{\pi}{2} \frac{t}{T_0}$$

It is more convenient to replace A by $1 + \delta A$, so that the expressions are in terms of the errors δA and τ. The baseline then becomes

$$2e_L(t) = \cos^2 \frac{\pi}{2} \frac{t}{T_0} - (1 + \delta A) \cos^2 \frac{\pi}{2} \frac{t-\tau}{T_0}$$

and, by a simple transformation,

$$2e_L(t) = \sin \frac{\pi}{2} \frac{2t-\tau}{T_0} \sin \frac{\pi}{2} \frac{\tau}{T_0} - \delta A \sin^2 \frac{\pi}{2} \frac{t-\tau}{T_0}$$

The peak-to-peak amplitude of the ripple in the baseline caused by the delay error is only evidently $\sin(\pi/2)\tau/T_0$ for unity undistorted pulse amplitude, and the amplitude of the 'bulge' in the baseline for a gain error only is $-\tfrac{1}{2}\delta A$. For example, a delay error of 10 ns with a

$2T_c$ pulse will give a peak-to-peak baseline ripple of

$$\sin(90^\circ \times \tfrac{10}{100}) = 0{\cdot}016,$$

i.e. 1·6% of the undistorted pulse amplitude, which is quite measurable. Similarly a 1 dB chrominance–luminance gain error alone would result in a deformation of the baseline of 6% of the undistorted pulse amplitude.

The maximum and minimum of the baseline when both gain and delay errors are present have been found by Rosman (1967*a*). From these, the peak-to-peak deformation of the baseline is found to be, for unit amplitude of the undistorted pulse,

$$e_{pp} = \frac{1}{2}\left(1 + A^2 - 2A\cos\frac{\pi\tau}{T_0}\right)^{\frac{1}{2}}$$

which, for small distortions, can be approximated to

$$e_{pp} = \frac{1}{2}\left\{\sqrt{\left(\delta A^2 + \frac{\pi^2\tau^2}{T_0^2}\right)} + \delta A\right\}$$

Rosman (1967*b*) has attempted to make use of the above expressions to design a graticule for the direct estimation of the magnitude of the combined distortion from the distorted composite chrominance pulse, to avoid the use of measuring apparatus such as the colour gain and delay tester. Although this seems to be practicable for small distortions it has not yet been elaborated into an operational procedure. This work has been further extended by Siocos (1968).

Another approach has been made by Wolf (1967), who used an ingenious method for obtaining an approximate K rating from the peak deviation of the baseline of the $20T$ pulse and attempted to relate this to the luminance K rating, although at the moment this should be treated with some reserve, since the precise establishment of the relationship between Wolf's chrominance-pulse K rating and the routine-test K rating requires further experimental data. It also has the disadvantage of being applicable only to the $4T_c$ ($20T$) pulse, whose bandwidth is insufficient for the UK and some other standards.

Both the methods just described run into difficulty when chrominance–luminance crosstalk is present, since the baselines of the pulse and the bar are shifted vertically from their correct positions. An attempt must accordingly be made to allow for this shift.

6.3.4 Measurement with chrominance bar

So far, the measurement of luminance–chrominance gain and delay inequalities has been described in terms of the deformation of the base of the pulse, but the bar is also distorted and, in principle, can be used in a similar way to the pulse.

The measurement of the gain inequality is certainly even easier to carry out when the bar is used, whether the measurement is made against a graticule or whether a nulling device such as the 'colour gain and delay tester' is employed, and is less likely to be in error due to deformation of the envelope of the chrominance signal from linear waveform distortion in the chrominance channel. The routine to be preferred is the adjustment of the centre of the lower half of the bar envelope of the black level of the signal by a nulling device.

The measurement of the delay error with the bar offers a lower sensitivity, since it is shown (Weaver, 1965) that, for the same delay error, the distortion of the baseline is $1/\pi$ of the distortion of the baseline of the pulse. On this basis, the use of the pulse is to be preferred.

Nevertheless, Hill (1966) suggested a useful criterion for the measurement of the delay inequality when the chrominance envelope is distorted. In this proposal, the nulling device is used to adjust the gain error to zero at the centre of the bar, after which the relative delay is varied until the negative half of the bar envelope passes through zero at a point vertically below the half-amplitude point of the leading transition of the bar. A corresponding reading is obtained by using the trailing transition of the bar, and the average of the two readings is taken as the luminance–chrominance delay inequality. Although this criterion may afford a clearer method of specifying the measurement procedure when the envelope is distorted, in the author's experience an estimate of the setting for the optimum flatness of the base of the pulse is quicker and as accurate. Moreover, at least one instance has been encountered where the distortion of the envelope was such that no crossover point was available under the leading edge of the bar. It must also be borne in mind that, where there is more than a very small disparity between the waveforms of the luminance-reference signal and the corresponding modulation envelope, it ceases to be possible to define clearly what is meant by delay inequality in the context of picture registration.

6.3.5 Augmented pulse-and-bar signal

It is known that the standard sine-squared-pulse and bar signal is very suitable for the measurement of the linear transmission properties of the luminance channel, whereas the composite pulse-and-bar signal is well adapted to the measurement of luminance–chrominance gain and delay inequalities. The original modulated chrominance pulse-and-bar signal is evidently capable of being used for the measurement of the remaining linear chrominance parameters, and also for the measurement of chrominance–luminance crosstalk (Section 6.3.6).

This trio of signals has accordingly been recommended by the CCIR (1966*a*) for system *I* linear waveform measurements. It is sometimes found surprising, therefore, that various broadcasting authorities have for so long investigated alternative signals, and even in some instances adopted signals differing from those recommended. This overlooks the fact that the CCIR recommendations apply only to the transmission of television signals over long distances, and not to the large number of other purposes for which such test signals are required, e.g. the testing of studio circuits, studio equipment, transmitters and signal-origination equipment. For these applications, test signals in which the luminance and chrominance components are combined in a single line offer considerable advantages, such as speed and convenience of operation. In particular, the simultaneous presentation of the luminance and chrominance errors makes it possible to establish a reasonable compromise between them during the alignment of equipment and circuits, and to make measurements in the presence of positional instabilities such as occur with videotape recorders.

The first proposal for a signal of this type was made by Wolf (1965), who took advantage of the fact that, provided no nonlinearity distortion is present, the luminance bar can serve equally well as a white-level reference for both the luminance and chrominance test waveforms. The chrominance pulse, as has already been shown, can serve for the measurement of both gain and delay relative errors.

The way in which the two test signals are disposed in Wolf's proposal is shown in Fig. 6.16. The luminance bar is inserted on every other line in approximately the same position as in the luminance pulse-and-bar signal. On the intervening lines are inserted a $2T$ luminance pulse and a $20T$ composite chrominance pulse, so that, when the waveform monitor viewing the signal is triggered on every

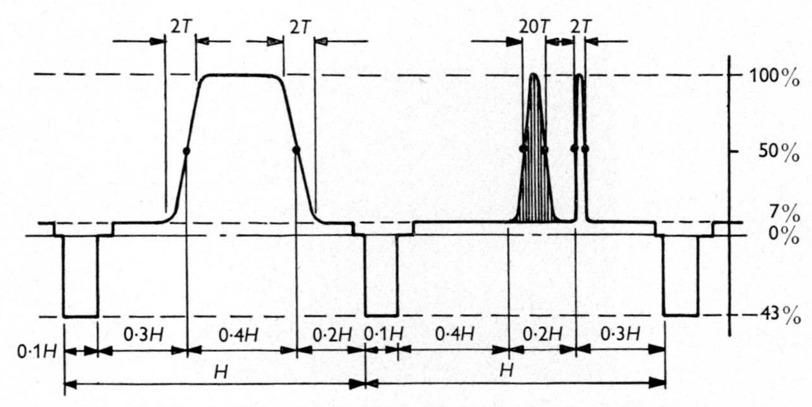

Fig. 6.16 Institut für Rundfunktechnik 2-line signal

line, the two pulses appear to be superimposed on the bar and symmetrically disposed about its centre point. The normal measurements on the chrominance pulse and the luminance pulse-and-bar signal may now be carried out.

Although this signal is very easy to use, it is inevitably liable to errors occurring as a result of the difference between the signals contained on successive lines. In particular, the considerable disparity between the energy in the signals on alternate lines makes it especially susceptible to waveform distortion attributable to amplitude and phase errors in the neighbourhood of halfline frequency, which can give rise to errors in measurement.

However, other combinations of the luminance- and chrominance-component test signals are readily available. One of these (Fig. 6.17) was proposed by the author (Weaver, 1965) under the name of the 'augmented pulse-and-bar' signal, where the reference is to the increase in the field of application of the original signal. It has been used extensively in the BBC, and has been found highly satisfactory, particularly for those in technical areas associated with studios.

6.3.6 Chrominance–luminance pulse-and-bar signal

The very considerable use to which the augmented pulse-and-bar signal has been put for the testing of circuits and equipment has demonstrated its very great utility; but, at the same time, it has

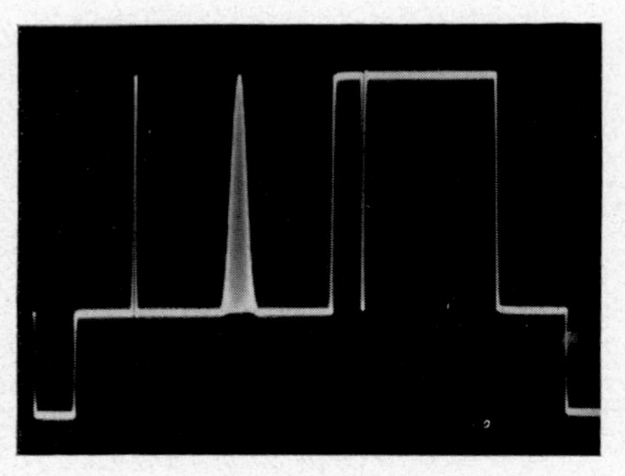

Fig. 6.17 Augmented pulse-and-bar signal

revealed difficulties in making an accurate measurement in cases where the circuits under test involve amplitude or frequency modulation of the video signal.

The reason for this is a distortion known as chrominance–luminance crosstalk (or intermodulation), which manifests itself as a vertical shift of the axis of a modulated chrominance signal as a result of the appearance of a spurious component having the approximate shape of the modulation envelope. For this reason, it is known in the USA as axis shift. It may have either polarity. The similarity of the generation of this component to the demodulation of an amplitude-modulated signal by an envelope detector has given rise to the alternative name of 'subcarrier rectification', although this is to be deprecated since it appears to imply that a considerable degree of amplitude nonlinearity is present. In fact, this may be quite untrue.

Chrominance–luminance crosstalk has two principal causes. The first is distortion in frequency-modulated systems, such as radio links, due to inadequate group-delay correction of the intermediate-frequency stages aggravated by the use of a quite high degree of pre-emphasis, and may also be associated with discriminator nonlinearity at high deviations. The second is quadrature distortion in amplitude-modulated systems with sideband asymmetry, such as radio re-broadcast links.

The position of the chrominance region in the video signal ensures that it is always transmitted as a single-sideband signal. It can be shown (Cherry, 1949) that sideband asymmetry has the effect of introducing a spurious quadrature component into the modulation envelope, and at the same time, of shifting the axis of the modulated signal by an amount proportional to the instantaneous depth of modulation (Weaver, 1965), which is the chrominance–luminance crosstalk.

The effect on a picture of chrominance–luminance crosstalk is a change in the saturation of colours due to a change in the amplitude of the accompanying luminance signal as a function of the sub-carrier amplitude. In a sense, it may be said to be the converse of differential-gain distortion which modifies the subcarrier amplitude as a function of the luminance amplitude. It is fortunate that the eye does not appear to be greatly sensitive to this distortion, since a certain amount always occurs in the television receiver as a result of envelope detection of the sideband asymmetry of the radiated signal. Nevertheless, measurements need to be made so that the total distortion can be kept within acceptable limits.

Another consequence of this distortion is its effect on the composite pulse-and-bar signal, which is so modified that misleading measurements are possible unless care is taken.

The special test signal which is particularly effective for the measurement of chrominance–luminance crosstalk is shown at the left-hand side of Fig. 6.18*a*. It is a chrominance bar, similar to that which forms a component part of the composite pulse-and-bar signal, but somewhat shorter. It is mounted on a 50% pedestal which is some microseconds longer than the chrominance bar, the front edges of the bar and the pedestal being coincident, so that a reference is available for the original position of the axis of the chrominance bar.

Fig. 6.18*b* and *c* show what happens when negative chrominance–luminance crosstalk is present. The negative-going luminance cross-

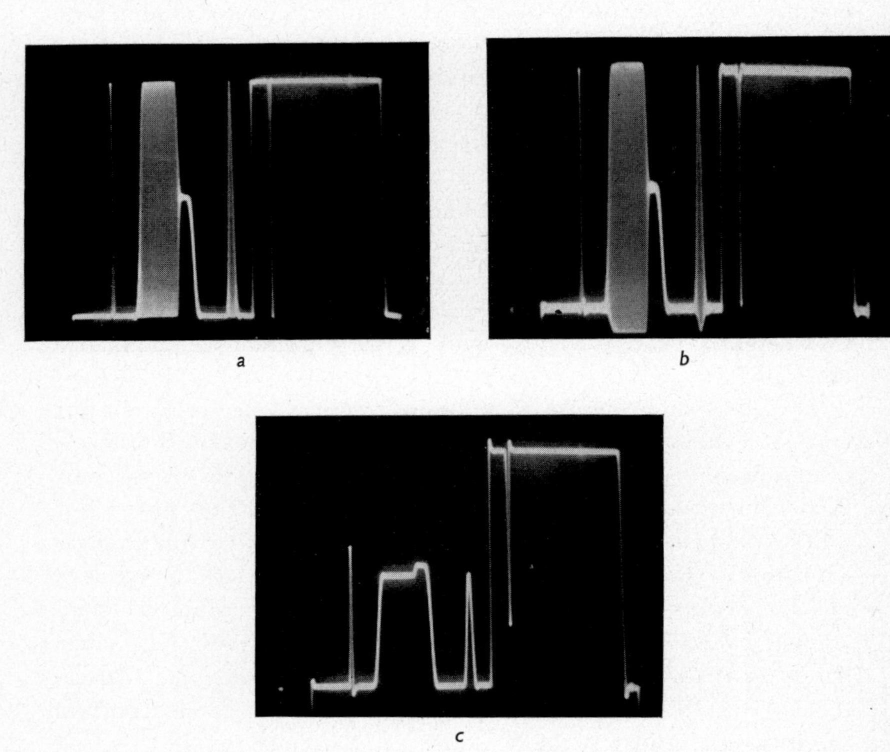

Fig. 6.18 Chrominance–luminance pulse-and-bar signal

a Undistorted signal
b Negative chrominance–luminance crosstalk: positive inequality
c Signal of (*b*) after filtration

talk component, which is generated as a result of the distortion, creates an effective downward shift of the axis of the chrominance signal, so that, when the subcarrier is filtered off, a step remains in the pedestal whose amplitude is equal to that of the crosstalk component.

The peak-to-peak amplitude of the chrominance bar in the undistorted signal is equal to the amplitude of the luminance bar, and therefore equal to the black-level/white-level signal amplitude. Let this be a. Then, if the amplitude of the step is d, the chrominance–luminance crosstalk is defined as $d/a \times 100\%$, which in the present example should have a negative sign since the axial shift of the chrominance signal is downwards.

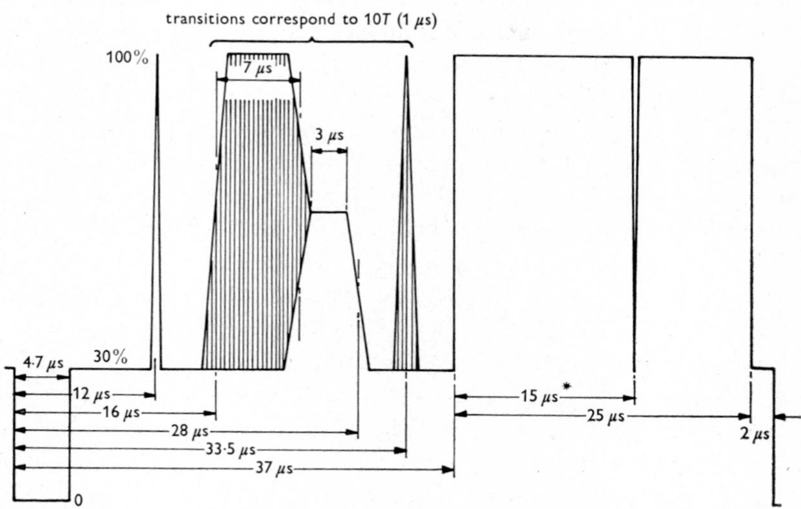

Fig. 6.19 Chrominance–luminance pulse-and-bar signal
Luminance transitions correspond to 1 T or 2 T (100 ns or 200 ns)
* The inverted pulse is optional

As will be seen from the photographs, it has been found possible to add this chrominance-bar component to the augmented pulse-and-bar signal, with a certain rearrangement of timings, to form an extremely versatile and useful signal for the general measurement of linear waveform distortion, particularly on long links intended for colour. It had been called the 'chrominance–luminance pulse-and-bar signal' (Fig. 6.19), where the name refers not only to its function for measuring both chrominance and luminance regions of the video spectrum but also to its ability to measure chrominance–luminance crosstalk.

The distorted signal of Fig. 6.18*b* was deliberately adjusted so as to exaggerate the apparent discrepancy between the indications of luminance–chrominance gain inequality as given by the top and the bottom of the chrominance waveforms. It makes the point that, where chrominance–luminance crosstalk is present, it is not sufficient just to consider either of these singly, but the peak-to-peak chrominance amplitude must be measured and compared with either the luminance bar or the luminance reference amplitude. Evidently, in Fig. 6.18*c*, the luminance–chrominance gain inequality is positive.

When the augmented pulse-and-bar signal is used, it is possible to detect from this kind of anomalous behaviour that crosstalk is present. If, however, this indication is ignored and a measurement is made using the base of the pulse or bar only, an error will result. The presence of the chrominance bar makes it possible to measure both the crosstalk and the gain inequality simply and without ambiguity.

The chrominance-pulse envelope is *K* rated in the normal way, except that care has to be taken in interpreting any overshoots, because the 100% modulation depth used overmodulates any part of the waveform which goes below the baseline, although no error is likely in the *K* rating since the limit lines are symmetrical about the lower reference line. Moreover, the 100% modulation depth means that no simple demodulation can be carried out to simplify the display, since an unacceptable amount of distortion would be introduced thereby.

The question of the acceptable value of the chrominance-pulse *K* rating for given circumstances has not yet been answered, and further work needs to be carried out on the subject. The present suggestion is to use the same *K*-rating limit as for the luminance pulse, but this must be regarded as quite tentative.

Let us sum up the properties of the chrominance–luminance pulse-and-bar signal and compare it critically with the augmented pulse-and-bar signal.

The chrominance–luminance pulse-and-bar signal contains not only the luminance but also chrominance pulse-and-bar signals, so that it can be used for the standard measurements on the chrominance channel as well as on the luminance channel. The presence of the bar, although not essential where chrominance–luminance crosstalk is absent, is nevertheless useful for the measurement of gain inequalities. Although its duration is rather less than one-third of the normal bar duration, enough of the horizontal portion remains, even when the waveform distortion of the chrominance channel is severe, to make the measurement quite simple. Furthermore, when the gain inequality has been removed with a nulling device, the adjustment of the relative delays on the pulse to the optimum position in the presence of waveform distortion is appreciably simplified, since one of the two possible variables has already been eliminated.

On the other hand, the chrominance–luminance waveform is rather crowded, since so much is included in a single line; and, whereas the spacing is just about tolerable for $2T_c$ ($10T$) transitions, it is definitely only marginally safe for $4T_c$ transitions, at least for the timings shown in Fig. 6.19. However, the durations of the chrominance bar and the unmodulated part of the pedestal cannot be reduced appreciably without losing the advantage of being able to make measurements under poor signal/noise conditions. Figs. 6.21 *a* and *c* demonstrate the reduction of the 'step' to a vestige when the cut-off frequency of the filter is made as low as practicable for noise-reduction purposes. If any modification is to be made in the timings to provide more room for the $20T$ chrominance pulse and bar, it is probably better to shorten the luminance bar, even though this makes the waveform non-standard. This expedient is adopted in the case of insertion test signals (Chapter 7).

The augmented pulse-and-bar signal is excellent for all situations in which no chrominance–luminance crosstalk is to be expected, and preferably where the chrominance envelope distortion is not too severe; in other words, for almost every use which does not involve the testing of links, transmitters and radio rebroadcast systems. Where a $4T_c$ ($20T$) pulse is required, this can be accommodated without difficulty.

The lack of the chrominance bar does not really make itself felt for the kind of use just mentioned, since the distortion of the chrominance is likely to be small and there is then no great difficulty in finding the correct adjustment of the nulling device for the measurement of gain and delay inequalities.

6.3.7 Other combined waveforms

For the sake of completeness, it would be useful to describe very briefly two experimental waveforms which have been tried out for the measurement of luminance and chrominance parameters and also for the measurement of chrominance–luminance crosstalk.

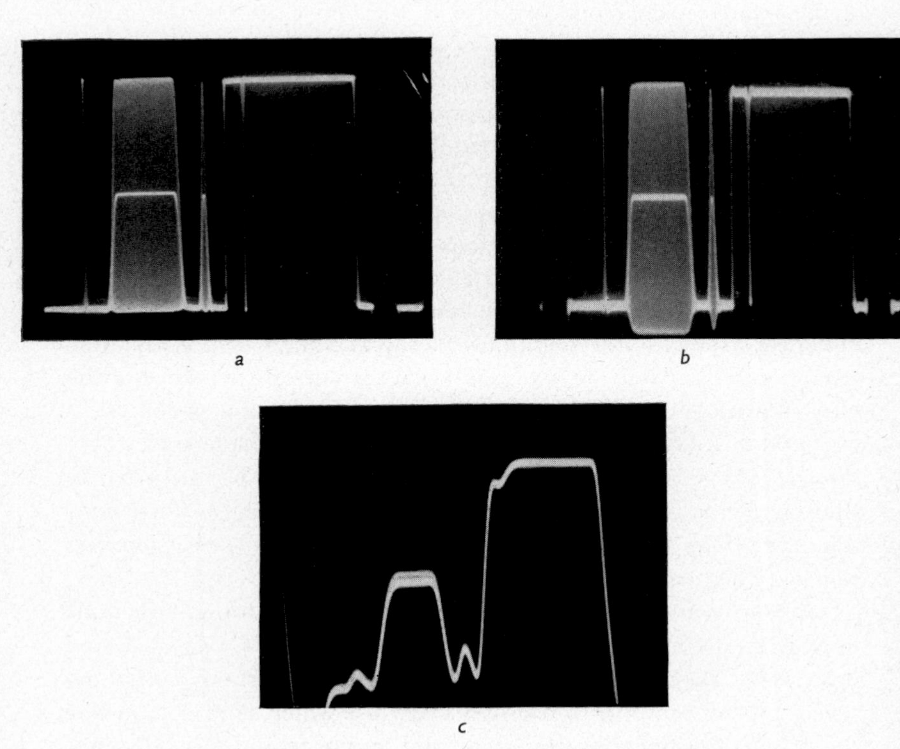

Fig. 6.20 'Keyed chrominance' pulse-and-bar signal

a Undistorted signal
b Negative chrominance–luminance crosstalk: positive gain inequality
c The signal of (*b*) after filtration.

The first, the 'keyed chrominance' signal, has the form of the chrominance–luminance signal with the 'minibar' replaced by a shortened composite chrominance bar. The chrominance component of the bar and the pulse is suppressed every fourth or fifth line so that, when the waveform monitor is triggered normally for a sequential display, the waveform appears as in Fig. 6.20*a*, in which the luminance reference bar, obtained whenever the subcarrier is suppressed, is seen superimposed on the composite bar. The unequal mark/space ratio introduces a deliberate difference between the brightnesses of the two superimposed displays, and also makes the waveform less susceptible to distortions occurring in the vicinity of halfline frequency.

When chrominance–luminance crosstalk is present, the axis of the chrominance bar is shifted, downwards in Fig. 6.20*b*. Removal of the subcarrier with a lowpass filter then gives the superimposed display of Fig. 6.20*c*, in which the upper of the two traces is the zero-distortion reference. Visually, it is definitely less bright than the other, although this is less obvious in the photograph than in actuality.

The great advantage of this waveform lies in the extent to which the cutoff frequency of the lowpass filter can be lowered compared with the chrominance–luminance waveform, to minimise the masking effect of random noise. This is clearly indicated in Fig. 6.21, in which the keyed subcarrier waveform and the chrominance–luminance waveform were given precisely identical treatment. In the left-hand pair of waveforms corresponding to the chrominance–luminance signal,the step in the pedestal has become a just recognisable vestige, whereas, in the right-hand photographs, it is clear that the available duration of the flat portions of the bar is so much greater that the cutoff frequency of the lowpass filter used could have been reduced at least to one-third of the value employed, with a corresponding reduction in the visual amplitude of the random noise.

The principal disadvantage of this waveform was discovered, after practical tests, to arise as a consequence of the superimposed hum voltages. It was found indispensable either to clamp the signal or to introduce the hum filter.

The second experimental signal was known colloquially as the 'butterfly' pulse-and-bar signal, for reasons evident from Fig. 6.22. It is really constructed on the same principle as the chrominance–luminance signal, with the exception that the unmodulated portion of the pedestal is placed in the middle of the chrominance bar instead of at one end, and it is used in precisely the same way.

It really has an advantage over the chrominance–luminance signal only in those instances where the edges of the bar are used for

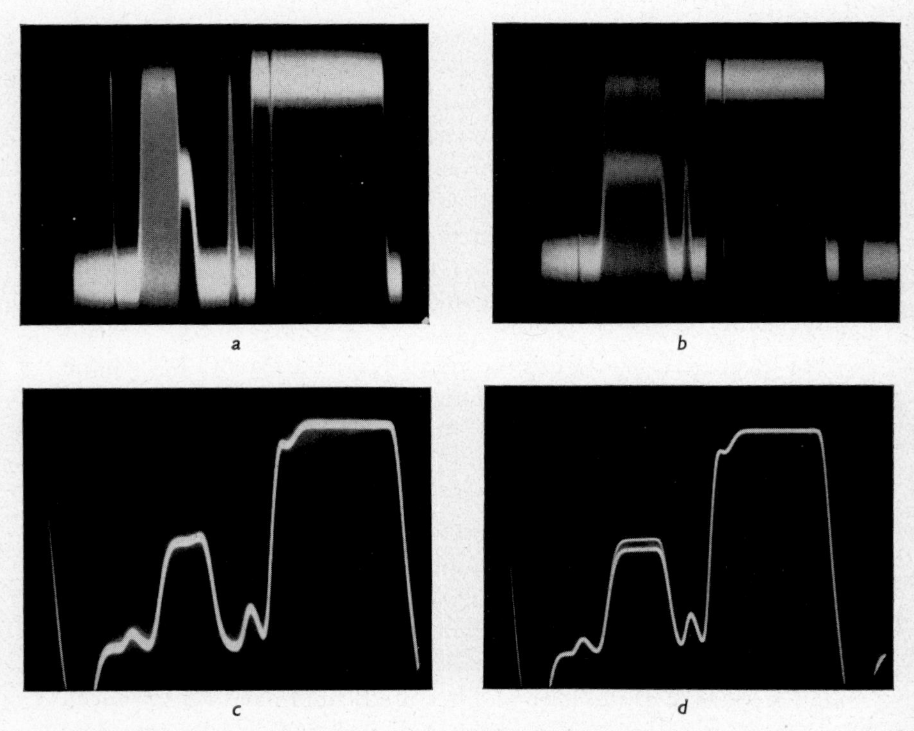

Fig. 6.21 Comparison of chrominance–luminance and 'keyed chrominance' waveforms

a Chrominance-luminance waveform
b 'Keyed chrominance' waveform
c Signal of (*a*) after filtration
d Signal of (*b*) after filtration

measurement, as in Hill's (1966) proposal for the measurement of delay inequalities. On the other hand, the circuitry required is slightly more complicated.

Another combined chrominance and luminance test waveform of a rather different kind has been proposed by the Federal German Post Office (1966). The object here is to produce a test signal which is the simplest possible kind and at the same time consistent with the practice of that administration.

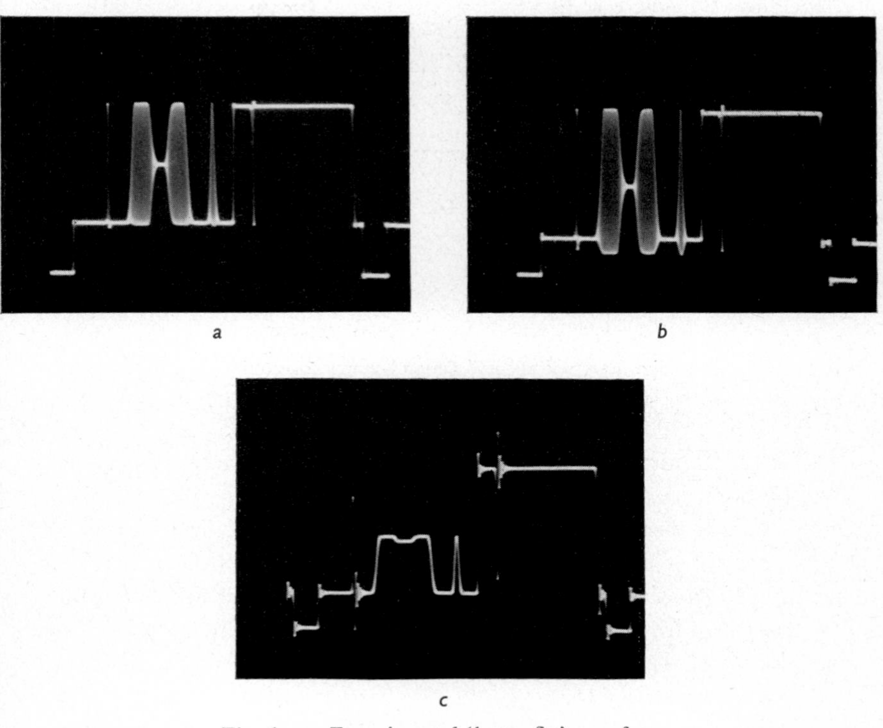

Fig. 6.22 Experimental 'butterfly' waveform

a Undistorted signal
b With chrominance–luminance crosstalk
c Signal of (*b*) after filtration

The proposed test signal is shown in Fig. 6.23. It consists of a 50% pedestal bearing a full amplitude of subcarrier, followed by a square wave where transitions are shaped by a $2T$ Thomson network.

The subcarrier on a pedestal can very obviously be used as the minibar described above for the measurement of luminance–chrominance gain and delay inequalities and chrominance–luminance crosstalk. The $2T$ bar fulfils the remaining requirements for the luminance region. The day-to-day checking of links is carried out mainly in terms of the 'ringing' pattern following the bar transitions.

This signal is evidently tailored to the requirements of a particular administration and to fit in with existing practice; furthermore, it is specifically intended for use on links rather than in studio areas, for example, where the errors should be much smaller. For that reason, it is not to be recommended for general use.

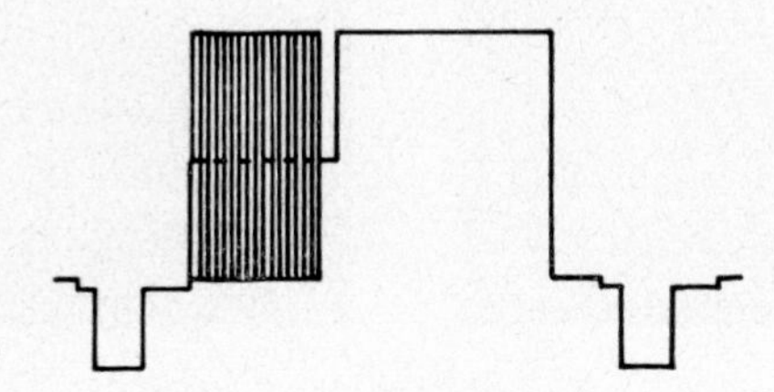

Fig. 6.23 Simplified general-purpose test signal

6.3.8 Fullfield signals

For many purposes, it is sufficient to have the above signals generated on line rate only; and normally, line synchronising pulses will be provided. However, for many television purposes, it is desirable to have test signals available which contain both line and field synchronising information. The reasons are various. The equipment under test may require field information for correct operation, the waveform monitor may not trigger normally with line-synchronising pulses only, or it may be required to record the results of tests on videotape. Where picture monitors are provided in operational areas, these may 'flash' in an annoying manner if no field-synchronising pulses are present.

Signals which contain full synchronising information are known as 'fullfield' test signals in north America, and this seems a convenient term to use. Fullfield signals can be generated in two ways: either full synchronising information must be provided within the generator, or arrangements must be made to supply it from an external synchronising-pulse generator.

The first solution is normally far too expensive to be entertained other than in exceptional circumstances. The second is generally the better solution. In administrations responsible for programme generation, a supply of standard pulses is available in many locations; two signals, 'mixed synchs.' and 'mixed blanking', usually suffice. Otherwise, a separate synchronising-pulse generator must be provided which can supply a number of test-signal generators. In instances where the availability of a standard composite video signal can be relied on during testing periods, it may prove cheaper to use a simple regenerator which reconstructs the required pulses from the synchronising information contained in the video signal.

6.3.9 Extension of use of colour test signals

It seems to have been first pointed out by Wolf (1965) that the addition of the chrominance test waveforms to the sine-squared-pulse and bar signal transforms it into a much more effective test of the videoband than it was originally.

The reason can be understood from Fig. 6.24, which gives the spectra of the $2T$ pulse and the modulated chrominance pulse on the same linear scale. The amplitudes of the spectral components tail away quite rapidly after about 3 MHz, with the result that the pulse is far from sensitive to distortions occurring within the upper third of the videoband.

The answer to this so far has been the use of the T instead of the $2T$ pulse, taking advantage of its much more uniform spectrum as far as the videoband is concerned (Fig. 6.25). However, a considerable proportion of the energy of the T pulse lies outside the nominal limit of the videoband; and, if the response of this region is not ideal, as is almost always the case, distortion of the type shown in Fig. 6.26*b* may result and obscure any irregularities resulting from inband distortions. As we know, the information may still be derived from the output waveform by the type of mathematical processing used in the acceptance test, but this is not at all convenient for routine testing.

Reference again to Fig. 6.24 reveals that the combined luminance and chrominance pulse spectra very largely complement one another, in that, when the spectrum of the $2T$ pulse is approaching zero, the

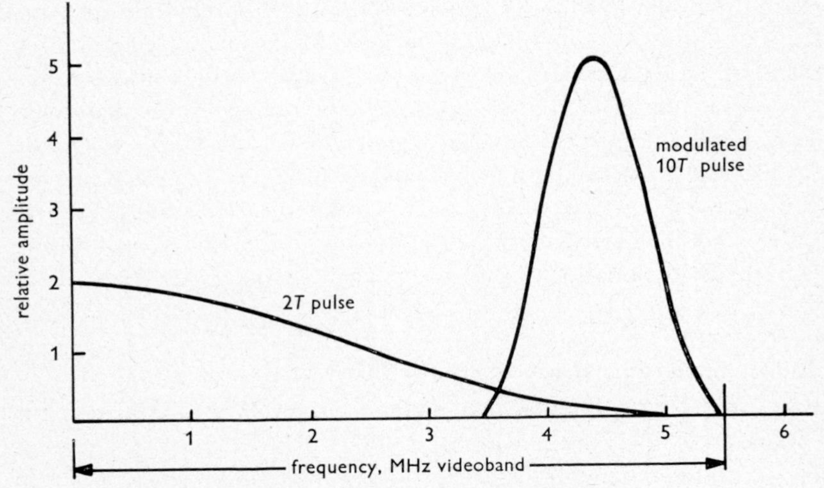

Fig. 6.24 Spectra of $2T$ pulse and $10T$ pulse

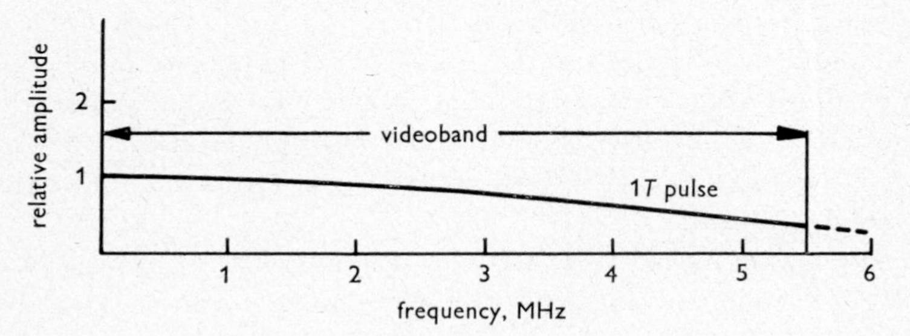

Fig. 6.25 Spectrum of T pulse

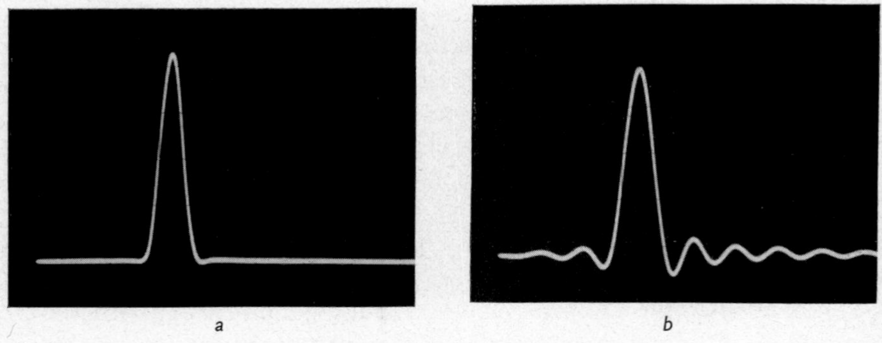

Fig. 6.26 Effect of band limitation on T pulse

a Undistorted T pulse
b Band-limited T pulse

$10T$ pulse adds large-amplitude components from the colour sub-carrier and its sidebands, thus providing a sensitive indication of amplitude/frequency distortion in that region.

This is demonstrated very well by Fig. 6.27*a*, illustrating what happens when a chrominance–luminance pulse-and-bar signal is passed through a delay-corrected lowpass filter with a loss of 6 dB at 3·6 MHz. It is very clear that the insensitivity of the $2T$ pulse to frequencies above the filter cutoff has resulted in a drop in the

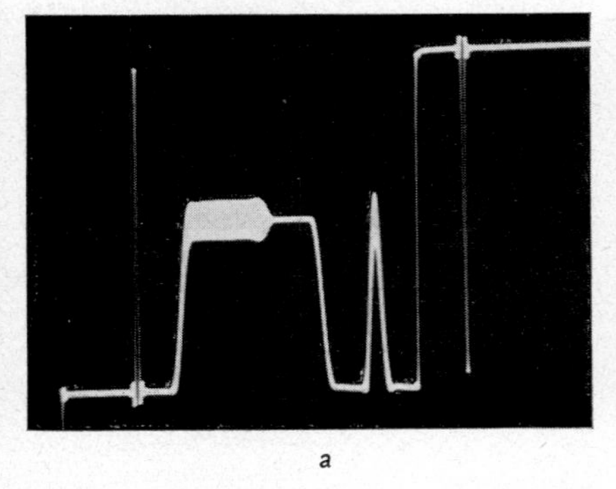

a

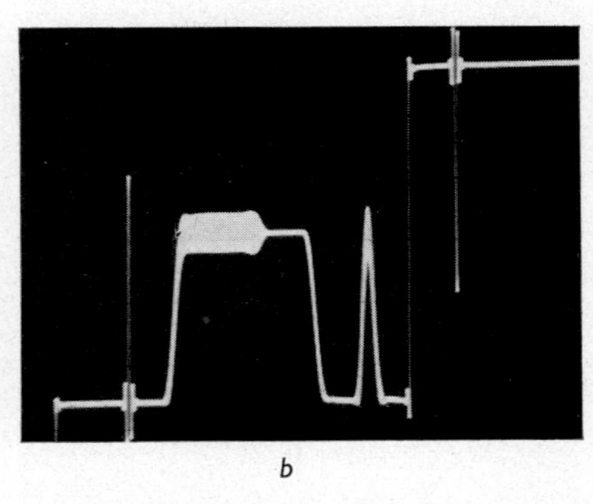

b

Fig. 6.27 Effect of band limitation on chrominance–luminance pulse-and-bar waveform

a $2T$ pulse
b T pulse

pulse/bar ratio of only 4%, whereas the effect on the chrominance component has been drastic, in fact a reduction of 90%. To make the demonstration more convincing, the experiment has been repeated with the $2T$ pulse replaced by a T pulse; the result is shown in Fig. 6.27*b*. As would be expected, the T pulse is much more markedly reduced in amplitude than the $2T$ pulse, while large 'rings' of the kind shown in Fig. 6.26*b* are quite evident at the base of the pulse.

A further remarkably useful property of the combined waveform is the fact that the complete spectrum is contained within the limits of the videoband, which implies that the signal could be transmitted through an ideal lowpass filter with a cutoff at 5·5 MHz without distortion. Ideal filters are not available, but high-grade video filters with excellent group-delay correction can be found. Figs. 6.28*a* and *b* show the result of passing the same chrominance–luminance signal through a filter of this type, in fact an example which was especially designed for incorporation in picture-origination equipment (Weaver,

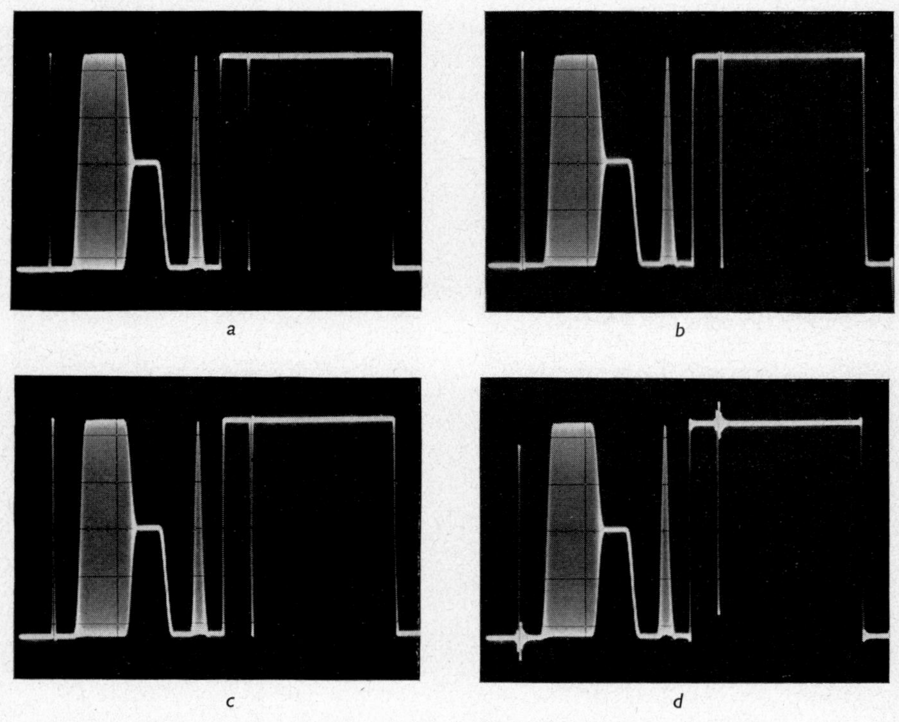

Fig. 6.28 Chrominance–luminance pulse-and-bar signal limited to nominal video bandwidth

a 2*T* pulse before filtration
b 2*T* pulse after filtration
c *T* pulse before filtration
d *T* pulse after filtration

1958). Fig. 6.28*a* gives the input signal, and Fig. 6.28*b* the signal at the output of the filter. A comparison of the two reveals that the distortion is so slight that it can only just be discerned at the bases of the pulses. On the other hand, in Figs. 6.28*c* and *d*, in which the 2*T* pulse has been replaced as before by a T pulse, the distortion is very clear.

A potential means for making use of these two useful properties is suggested by the acceptance-test method (Section 6.2.10), in which, for a given standard of performance, four restrictions are placed on information derived from the distorted output pulse by mathematical processing. The first two are equivalent to tests carried out with the $2T$ pulse, and can be considered as potentially known since the $2T$ test waveforms are available in any of the combined luminance and chrominance signals. The fourth restriction is a function of the arithmetical sum of the principal terms in the time series representing the distortion component of the waveform corresponding to the output distorted T pulse. It can only be measured directly if one has at one's disposal a lowpass filter with a sharp cutoff, accurately corrected for group delay with an equaliser whose basic delay is at least 1 μs. This is not impossible, and indeed the author has designed such a filter; but it cannot be considered as an operational tool, and the measurement is, in any case, not easy to carry out. However, if one is principally concerned with studio measurements, this factor is likely to be quite small.

The situation is quite different with respect to the third restriction, which is a measure of the pulse/bar ratio of the pulse corresponding to the distorted T pulse, and is thus an indication of the effective bandwidth of the circuit or equipment under test. This again cannot be measured directly; but, if we assume that the nonunity pulse/bar ratio is due to amplitude–frequency distortion having an approximately Gaussian form, it can be derived indirectly, since it is related to the T-pulse/bar ratio and a connection can be found between the latter and the reduction in chrominance subcarrier amplitude in the test signal. An experimentally derived curve showing the relationship between these two quantities is given in Fig. 6.29, which enables an immediate conversion to be made where, as is very commonly the case, a T pulse is used for studio testing instead of the $2T$ pulse.

It is not suggested that the procedure just outlined should replace the acceptance test, but it could form the basis of a measurement technique for use in areas where the inband distortion is relatively low, and where the standard *K* rating method does not provide a sufficiently searching test. Further work is still proceeding on this subject. It should be added that more information than has already been mentioned is derivable from the chrominance–luminance waveform, e.g. the group delay difference between low video frequencies and the chrominance region, and this might also be utilised.

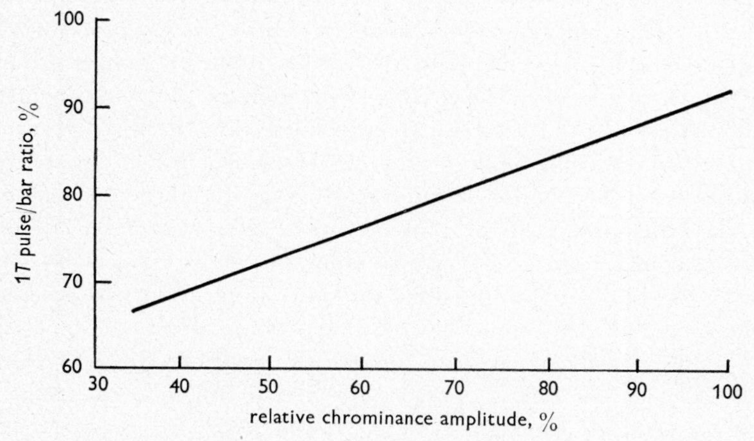

Fig. 6.29 Relationship between *T* pulse/bar ratio and the relative gain at subcarrier frequency

6.3.10 Compact test signals

The continual increase in the number of hours of television broadcasting in most countries has inevitably brought with it a decrease in the time available for the testing of circuits and equipment. A very effective answer to this is the use of insertion test signals (Chapter 7), but these have certain limitations and, where the highest possible accuracy is required, there is no doubt that one would always prefer to employ line-repetitive or fullfield test signals.

A special problem arises where circuits or equipment can be made available for testing with such signals but the time available for the tests is very strictly limited, since the tests must be planned so that the maximum of information can be derived in the time allowed. A very convenient solution is the use of insertion test signals generated repetitively, since they have been designed with great care to ensure that they provide the greatest number of useful measurements in the most compact form possible; and where they are normally in use, equipment must be available for the evaluation of the distortion. The provision of the generating apparatus for these signals need not raise any particular problems if the possibility of their employment is taken into account when the insertion-test-signal generator is designed, because it is by no means difficult to include the facility for originating the signals either in 'single shot' or in any desired sequential form.

The suggested procedure has an incidental advantage, in that it makes possible a unified system of testing embracing both insertion and repetitive test signals, which eliminates certain inconsistencies between the two types of test, e.g. the problem of dealing with the measurement of long-term distortions as revealed by the top of the bar when the length of the bar top differs between the insertion and repetitive waveforms.

The evaluation of the distorted output waveforms can be varied according to circumstances and the equipment available. Fairly standard measuring apparatus can be employed with suitable switching, and a separate feed of the signal is provided to each of the individual devices; or, if time is at a premium, sequence switching to a waveform monitor makes it possible to record the output waveforms photographically for subsequent evaluation. In more sophisticated methods, a mixture of sampling and digital techniques are used to obtain output signals capable of being recorded or of being transmitted to a remote point.

There is no doubt of the efficacy of this technique for the rapid assessment of the quality of circuits and equipment, and it will certainly increase very considerably the opportunities for running fast tests on these either during out-of-service periods or immediately prior to service. For example, it has been found possible to carry out a check of the quality of a videotape recording and the quality of the reproducer combined from as little as two minutes of recorded compact signals at the end of a programme tape. This application is discussed in Section 13.3.4.

6.4 Steady-state methods

Although steady-state testing methods are rapidly giving place to waveform-testing methods over the greater part of the television field, there are nevertheless certain considerations which make their use preferable. In particular instances, conditions have to be satisfied which must necessarily be stated in terms of frequency or phase, e.g. in transmitters where the shape of the channel is so specified. In investigational work also, it is often very useful to transfer from the time domain to the frequency domain to obtain additional information. Another special case is the measurement of networks, e.g. filters and equalisers.

6.4.1 Amplitude/frequency characteristics

In general, the measurement should be made in terms of insertion loss or gain (Chapter 2) unless some very special circumstances prevail, such as measurements at test points on apparatus specified to be treated thus.

Since some of the measurements are likely to be on circuits the two ends of which are not simultaneously available, it is advisable to make use of a sine-wave generator especially designed for this type of measurement. A good commercially available example furnishes 1 V pk–pk into a 75 Ω termination over the frequency range 10 KHz–10 MHz. An internal variable 75 Ω attenuator is provided.

A valuable feature of this instrument is the constancy of the output level over the range of output frequencies, but even more important is the fact that the level is held constant at a point in the output circuit where the impedance is nominally 75 Ω. Any small variations of this impedance from the correct value are minimised by a 10 dB attenuator pad in series with the output, which improves the output impedance by a factor of 20 dB. This arrangement ensures that the output circuit of the oscillator is closely equivalent to a generator with a fixed output level which feeds the apparatus under test via a series 75 Ω resistance, which is one of the correct methods of measuring insertion loss.

The output from the apparatus under test must be measured by a voltmeter which has an exact 75 Ω input impedance, and preferably also a good approximation to a square-law characteristic for both positive and negative peaks of the sinewave, so that no significant error is introduced by a reasonable harmonic content in the output waveform. Since a calibrated reading is only required for ± 1 dB around the reading corresponding to 1 V pk–pk, it is possible to construct a very simple meter utilising fullwave semiconductor diodes at a low level where their voltage/current curve is approximately parabolic. Larger output levels can be measured by inserting an attenuator, and lower levels by means of a calibrated amplifier with good 75 Ω input and output impedances. A typical circuit is described in Section 2.5.

A sweep generator will provide a panoramic display of amplitude characteristics, but the accuracy is all too often rather poor. A great improvement can be made by using the sweep generator to drive an *x–y* plotter. After the desired curve has been traced, the network is replaced by a series of known losses from a variable attenuator, thus calibrating the vertical scale. Frequency markers can be superimposed on one of these, from which vertical lines may be drawn to form a frequency scale, as in Fig. 30. Apart from the advantage of a permanent record, quite accurate readings are possible.

A quantised form of sweep generator which is often employed for video testing is the multiburst generator, which furnishes a signal consisting of a series of rectangular bursts of sinusoidal oscillations, each burst having a different frequency (Fig. 6.31). These bursts are all of equal amplitude at the output of the generator, and are mounted on a pedestal of 50% signal amplitude on a standard television-line waveform. Field synchronisation can be provided as required. Such a multiburst signal may be utilised in the same way as a normal video test signal, with the additional advantage that it can also be displayed on a picture monitor or receiver and used as a test pattern for checking overall resolution.

The multiburst signal is usually measured with the implicit assumption that each burst is a continuous sinusoidal signal. This, of course, is far from the truth. Each burst can be expressed mathe-

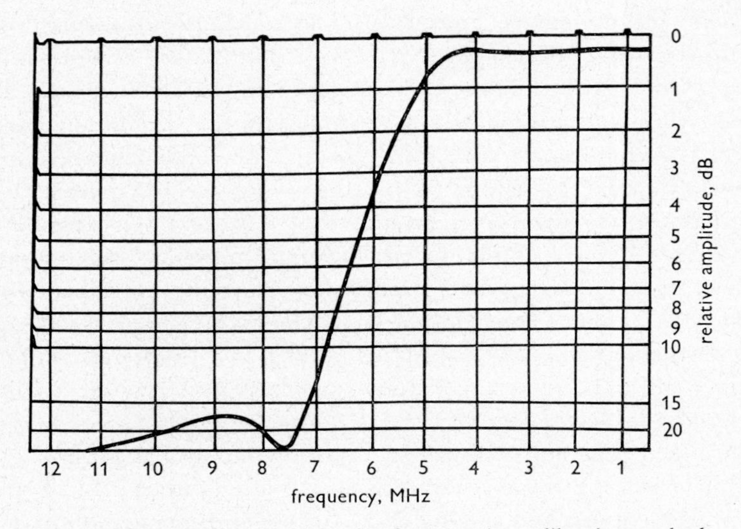

Fig. 6.30 Automatic plot of insertion loss showing calibration method

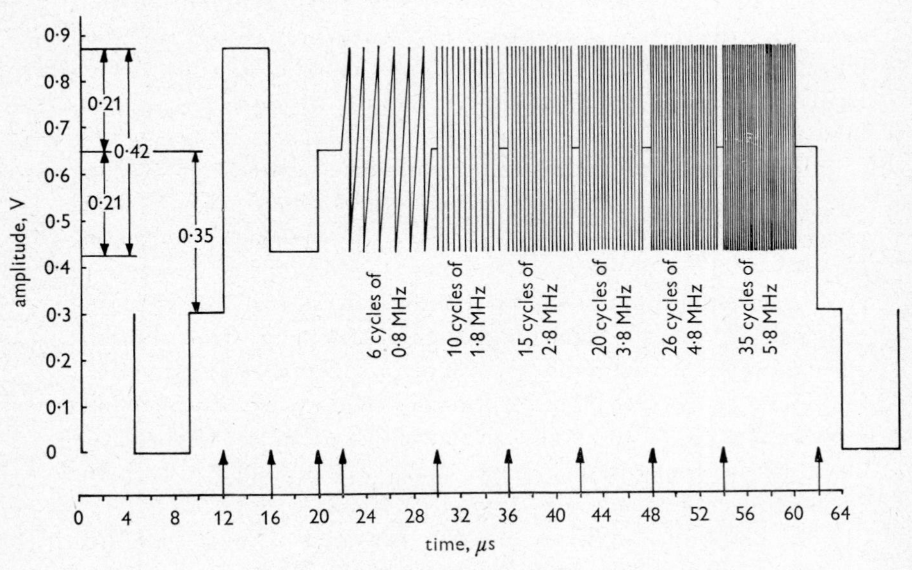

Fig. 6.31 Typical multiburst signal
Vertical arrows denote waveform-transition points

matically as a sinusoid multiplied by a rectangular repetitive pulse, and hence as a Fourier series of symmetrical sidebands of the sinusoidal frequency. Furthermore, to obtain adequate coverage of the whole frequency band, as many bursts as conveniently possible are usually included, with the result that the mark/space ratio of the shaping pulse is probably something like 1:10, and its risetime is kept short, so that the amplitudes of the sidebands do not decrease very rapidly with increasing order. This is very clearly illustrated in Fig. 6.32, in which the spread of the sidebands should be noted. It

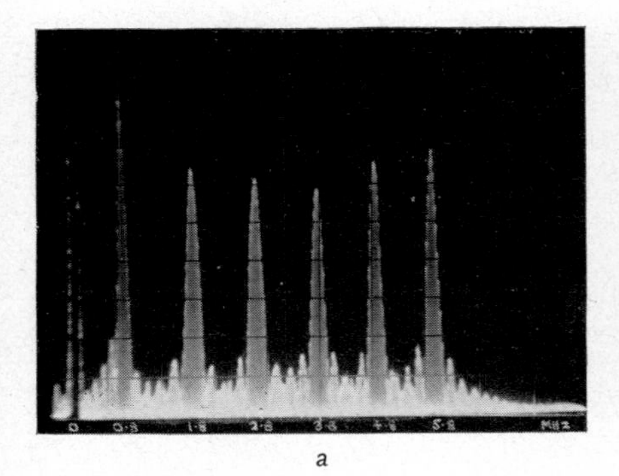

a

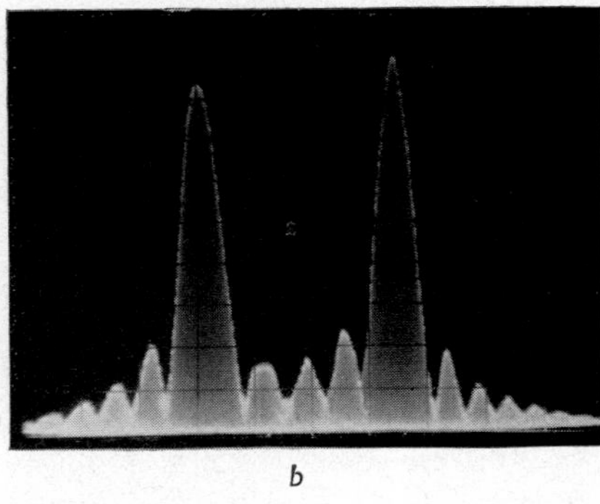

b

Fig. 6.32 Spectrum of multiburst signal of Fig. 6.24

a Overall view
b Detail of two highest-frequency bursts

follows that the higher-frequency bursts are liable to envelope distortion owing to the attenuation of elimination of the upper sidebands, which in a practical case may well be combined with quadrature distortion from vestigial sideband amplitude modulation or the characteristic distortions from the frequency-modulation systems of radio links and videotape machines. Where phase distortion also exists near the upper end of the videoband, it is possible to obtain the higher-frequency bursts with highly curious shapes. In all cases, there is interaction between the bursts as a result of the overlapping of the sidebands. Some examples are shown in Fig. 6.33. It is worth noting that the subcarrier notch filter used for Fig. 6.33*d* gave no perceptible distortion on pictures.

Certain other possible errors can arise from typical shortcomings in the multiburst generator, which should be designed with great care if its indications are not to be misleading. Obvious faults are asymmetry of the bursts with respect to the pedestal and distortion of the burst

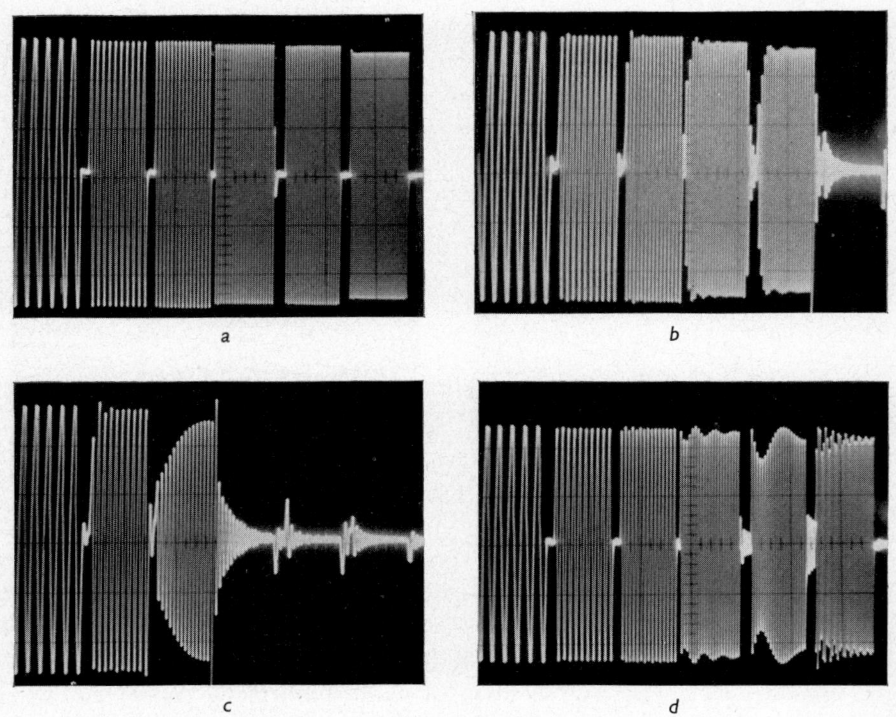

Fig. 6.33 Distorted multiburst signal of Fig. 6.24

a Amplitude response drooping towards high frequencies
b Through delay-corrected 5 MHz lowpass filter
c Through delay-corrected 3 MHz lowpass filter
d Through subcarrier notch filter

envelope. Less obvious sources of error are frequency modulation of the sinusoids, a lack of flatness of the burst top, and harmonic distortion of the component sinusoids. In this last respect, it is advisable to select a set of frequencies which are not harmonically related. Such distortion can cause a change in burst shape as soon as any band limitation is introduced, as well as an apparent distortion of the lower-frequency bursts as a result of severe upper-frequency gain or phase distortions (Fig. 6.33*c*). It is also evidently necessary for the burst amplitudes to remain very stable with time, to avoid the need for frequent readjustment.

Some of these errors can be lessened by reducing the spread of the sidebands associated with the bursts, either by suitably shaping their envelopes or by increasing, as far as possible, the duration of each burst. It is also advisable to measure a burst amplitude at the centre of the envelope, which is less likely to be affected by transient distortion, and to employ a line selector to minimise the influence of field-frequency distortions and hum.

A burst frequency can be most quickly and conveniently adjusted against a notch filter centred on the nominal burst frequency. The notch filter is connected in series with the generator and the frequency adjusted for minimum output on the burst. In maintenance areas, a set of switchable notch filters can be contrived very simply, with which the successive burst frequencies furnished by the generator can be checked extremely quickly.

Clearly, no great accuracy should be looked for with multiburst measurements even where the generator is well designed. The best general advice is to be very wary of using the outline of the group of bursts as a quantised amplitude/frequency curve, particularly since some of the bursts may be shifted vertically, and, wherever possible, to measure each sinusoid in terms of the peak-to-peak amplitude at the centre of a burst.

6.4.2 Group delay

The measurement of group delay is, for all practical purposes, never required for the day-to-day checking of apparatus and links, since, when steady-state methods are used in preference to waveform testing methods, the amplitude/frequency curve is sufficient to define the performance. The simplification introduced by this is perhaps one reason why steady-state methods were able to resist the introduction of waveform testing for longer than might have been expected. The reason is quite simply that a very large proportion of the component elements of the signal chain have minimum-phase or essentially minimum-phase properties, so that the probability is very great that any change in the transmission characteristics takes place in such a manner that the resulting variation in group delay is dependent on the variation in gain. Consequently, the network can be maintained within limits by controlling one of these quantities only, and the amplitude/frequency characteristic is by far the easier to measure.

A measurement of group delay is, however, required for the alignment and checking of equipment involving frequency modulation, for transmitters and for experimental work and certain types of fault finding. On the other hand, the alignment of precision networks is often best carried out by sine-squared-pulse methods (Weaver 1958), since this automatically ensures the best waveform response.

Group-delay measuring sets are available for both point-by-point and sweep measurements. For uses where adjustments have to be made to optimum flatness or to a predetermined curve, the sweep presentation is to be preferred; even though the accuracy is reduced, it is still possible with a good instrument to obtain an accuracy within a few nanoseconds.

Let us look very briefly at the theory of the most popular method of measuring group delay, the Nyquist method (Nyquist and Brand, 1930; Sandeman and Turnbull, 1939), to see how errors can arise from limitations in the measuring equipment.

The search frequency p at which the group delay is to be measured is modulated by an incremental frequency q. For a modulation depth m, the resultant voltage is given by the well known expression

$$e_0 = \cos pt(1+m\cos qt) = \cos pt + \frac{m}{2}\{\cos(p+q)t + \cos(p-q)t\}$$

Now, assume that the phase shift at p is ϕ, and at $p+q$ and $p-q$ it is $\phi \pm \Delta\phi$, respectively. It is assumed further that the gain is constant over the region between $p+q$. The expression for e_0 then becomes

$$e_0' = \cos(pt+\phi) + \frac{m}{2}[\cos\{(p+q)t+\phi+\Delta\phi\} + \cos\{(p-q)t+\phi-\Delta\phi\}]$$

which can be put back into the form

$$e_0' = \cos(pt+\phi)[1+m\cos(qt+\Delta\phi)]$$

Consequently, if e_0' is demodulated, a comparison of the incremental frequency thus recovered with a reference feed of the incremental frequency in a phase comparator gives the angle $\Delta\phi$; and, since this is

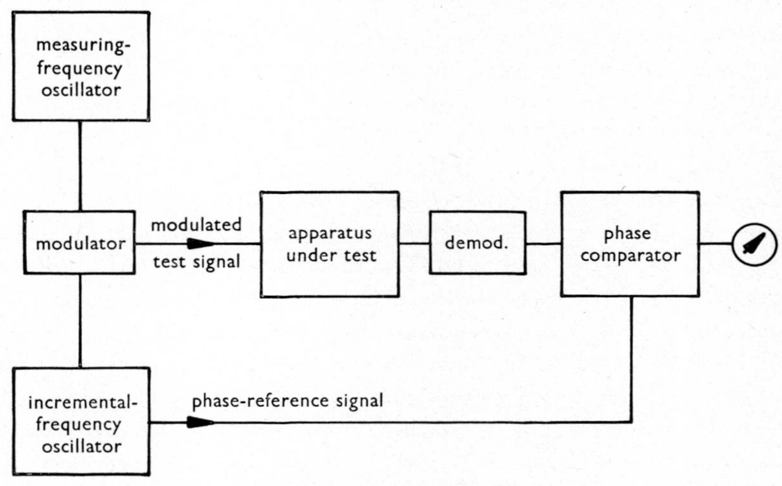

Fig. 6.34 Nyquist method of group-delay measurement

the increment of phase for an increment in angular frequency q, the group delay is approximately given by

$$t_g = \frac{\Delta\phi}{q}$$

It is approximate, since the incremental frequency q is finite whereas it should be vanishingly small. A block diagram of the method is given in Fig. 6.34.

In practice, of course, q will depend on the desired sensitivity of measurement. If, for example, it is required to measure down to 2 ns, and the smallest phase angle which can be measured by the phase comparator is 0·1°, we have

$$2\times 10^{-9} = \frac{0{\cdot}1}{360Q}\times 10^{-6}$$

where Q is the incremental frequency in megahertz, so that

$$Q = \tfrac{5}{36}\,\text{MHz} \simeq 140\,\text{kHz},$$

which means that the measurement is actually performed over a range of some 280 kHz, since the sidebands of the modulated search frequency are spaced at $2Q$. This imposes a lower limit of about 200 kHz on the measurement, apart from the other disadvantages.

The measurement of group delay by the Nyquist method involves a number of assumptions. The first is that the slope of the chord on the phase/frequency curve connecting the points corresponding to $p-q$ and $p+q$ is the same as the slope of the tangent to the curve at the frequency p, which is the true group delay. Furthermore, it is assumed that the increment of phase angle between $p-q$ and p is the same as the increment between p and $p+q$, and finally that the gain is the same at $p-q$ and $p+q$. Failures to meet both these conditions will result in errors in the measurement, although, for small degrees of asymmetry, the errors are not serious.

The smaller the incremental frequency, the smaller these errors are likely to be, but, on the other hand, the smaller is the phase angle which has to be measured; so, the temptation is always great for the manufacturer to choose a high incremental frequency to increase the sensitivity, particularly with group-delay sweeps where less refined methods can be put into service than with point-by-point measurements. It is therefore well to assure oneself that the incremental frequency is small enough to give the desired accuracy of measurement, taking into consideration the greatest likely slopes of the group-delay/frequency and amplitude/frequency curves.

This point should be borne very much in mind when one is considering using for video measurements a group-delay measuring set which has been designed for very different conditions, e.g. for the measurement of very wideband systems. It is then possible for the search frequency to be quite high, e.g. 1 MHz or so.

The proportion of the frequency spectrum adversely affected by a high incremental frequency is much greater than might be supposed. Both sidebands of the modulated measuring frequency need to be accommodated, but a source of error arises also from the need to include a trap filter in the output of the modulator to ensure that the amount of the incremental frequency at this point is extremely small under all conditions. The delay characteristic of this filter then appears as part of the measured delay characteristic. A bandstop type of network is usually chosen, since the delay errors are then kept as far as possible to the region immediately surrounding the suppression frequency, but frequencies well away from the incremental frequency itself are affected to an extent depending on the tolerable error. This limits the lowest frequency of measurement to rather more than twice the search frequency.

The immediately obvious solution of delay-equalising the filter entails an equalisation to an accuracy better than the required measurement accuracy over the whole range of frequencies. An alternative, which avoids the difficulty altogether, generates the modulated measuring waveform by a double-modulation process (Gareis, 1963); e.g. measuring frequencies up to 5 MHz might be generated first by modulating a 20 MHz carrier wave with the incremental frequency and then further modulating this with frequencies of 20–25 MHz. The distortionless selection of the lower sideband of this second modulation presents no great difficulty; neither does the separation of the 20 MHz carrier from the incremental frequency. So far, it has been assumed that a local measurement is being made in which both ends of the equipment under test are available. The Nyquist method can also cope where only one end of the equipment is available, e.g. in the measurement of a distribution link. The preferred method is to make use of another circuit to send the reference frequency for phase-comparison purposes from the transmitting to the distant end. This may be a protection circuit or, if the incremental frequency is sufficiently low, a speaker or music circuit, but whatever is used must have an adequate phase stability.

An attractive alternative is to mix the reference frequency linearly with the modulated search frequency and to transmit them together as a single system; but, for accurate measurements, high amplitude linearity is demanded from the equipment under test. Amplitude nonlinearity, even in very small amounts, causes intermodulation between the reference frequency and the search frequency which modifies the phase angle between the reference and the modulation envelope. On the other hand, the circuit used to transmit the reference frequency must have high phase stability, which may not be easy on a long circuit.

The group-delay characteristic may depend to some extent on the level of signal at which measurements are made; and, if this is suspected, spot checks should be made at selected frequencies, particular attention being paid to the higher video frequencies. Augustin (1962) discusses a number of errors which can arise with the Nyquist method.

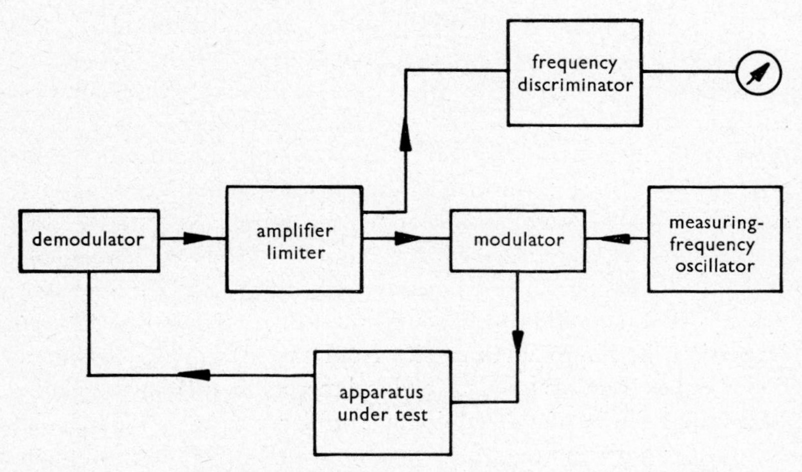

Fig. 6.35 de Boer and van Weel method of group-delay measurement

An instrument operating on a quite different and very ingenious system (Fig. 6.35) has been described by de Boer and van Weel (1954). The apparatus under test is connected in a regenerative loop with a modulator, a detector and an amplifier containing a phase-shift network so that oscillation is maintained at about 25 kHz when the modulator is fed with a carrier frequency which is the desired frequency of measurement.

The frequency of oscillation is determined by the total phase shift around the loop; so that, if the carrier frequency is altered to a point at which the group delay is different, the frequency of oscillation must change by an incremental amount sufficient to restore equilibrium. de Boer and van Weel show the group delay to be the effective parameter in determining the magnitude of the frequency shift, which can therefore be measured to give a value from which the group delay can be derived.

This method has the virtue that $d\beta/d\omega$ is measured, and not $\Delta\beta/\Delta\omega$ as with the Nyquist method; but there are certain practical difficulties, in particular the provision of an amplitude limiter which introduces no phase error; furthermore, only loop operation is possible.

6.5 References

AUGUSTIN, E. (1962): 'Probleme der Gruppenlaufzeitmessung nach der Spaltfrequenzmethode', *Tech. Mitt. RFZ*, **6**, (4), pp. 166–176

CCIR (1966*a*): Documents of the XIth Plenary Assembly, Oslo, Recommendation 421–1

CCIR (1966*b*): *ibid.*, Recommendation 451

CHERRY, C. (1949): 'Pulses and transients in communication circuits' (Chapman & Hall)

COLES, D. A. (1970): 'A set for annulling chrominance/luminance gain and delay inequalities', *IERE Conf. Proc.* 18, pp. 209–221

DE BOER, H. J., and VAN WEEL, A. (1954): 'An instrument for measuring group delay', *Philips Tech. Rev.*, **15**, (11), pp. 307–316

FEDERAL GERMAN POST OFFICE (1966): Technical Report 5311, Oct.

GAREIS, W. (1963) 'Messung der Gruppenlaufzeit im Videobereich', *Radio Mentor*, **29**, (5), pp. 418–419

HILL, M. W. (1966): 'Measurement of gain and delay inequalities between the luminance and chrominance channels in colour television', *Electron. Lett.*, **2**, pp. 57–58

LEWIS, N. W. (1952): 'Waveform computations by the time-series method', *Proc. IEE*, **99**, Pt III, pp. 294–306

LEWIS, N. W. (1954): 'Waveform responses of television links', *ibid.*, **101**, Pt III, pp. 258–270

MACDIARMID, I. F. (1952): 'A testing pulse for television links', *ibid.*, **99**, Pt IIIA, pp. 436–444

MACDIARMID, I. F. (1959): 'Waveform distortion in television links', *Post Off. Elect. Engrs. J.*, **53**, Pts 2 and 3, July and Oct.

MACDIARMID, I. F., and PHILLIPS, B. (1958): 'A pulse-and-bar waveform generator for testing television links', *Proc. IEE*, **105 B**, pp. 440–448

NYQUIST, H., and BRAND, S. (1930): 'Measurement of phase distortion', *Bell Syst. Tech. J.*, **9**, pp. 522–549

PADDOCK, F. J. (1970): 'The relationship between individual link and chain distortions in a television network', *IERE Conf. Proc.* 18, pp. 127–141

ROSMAN, G. (1967*a*): 'Interpretation of the waveform of luminance–chrominance pulse signals', *Electron. Lett.*, **3**, p. 128

ROSMAN, G. (1967*b*): 'Use of simple graticules with luminance–chrominance pulse signals', *ibid.*, **3**, pp. 302–303

SANDEMAN, E. K., and TURNBULL, I. L. (1929): 'The Nyquist method of measuring time delay', *Elect. Commun.*, **7**, (4), pp. 327–330

SHANNON, C. E. (1948): 'A mathematical theory of communication', *Bell Syst. Tech. J.*, **27**, pp. 379–623

SIOCOS, C. A. (1968): 'Chrominance/luminance ratio and timing measurements in color television', *IEEE Trans.*, **BC-14**, (1), pp. 1–4

SPRINGER, H. (1963): 'Die Vorteile des Impuls- und Sprung-Signales bei der Messung linearer Übertragungsverzerrungen', *Rundfunktech. Mitt.*, **7**, pp. 25–41

THOMSON, W. E. (1952): 'The synthesis of a network to have a sine-squared impulse response', *Proc. IEE*, **99**, Pt III, pp. 373–376

VILLE, J. (1951): 'Décomposition d'un signal en composantes en cosinusoide surélevée', *Câbles & Transm.*, **5**, p. 126

WEAVER, L. E. (1958): 'The design of a linear phase-shift low-pass filter', BBC Engineering Monograph 17, April

WEAVER, L. E. (1965): 'Sine-squared pulse and bar testing in colour television', BBC Engineering Monograph 58, Aug.

WHEELER, H. A. (1939): 'Interpretation of amplitude and phase distortion in terms of paired echoes', *Proc. Inst. Radio Engrs.*, **27**, pp. 359–385

WOLF, P. (1965): 'Eine zweckmässige Erweiterung des Impuls- und Sprungsignals', *Rundfunktech. Mitt.*, **9**, pp. 19–25

WOLF, P. (1967): 'Bewertung von Fehlern im Amplituden- und Laufzeitgang bei der Übertragung von Farbfernsehsignalen', *ibid.*, **11**, pp. 176–179

7 INSERTION SIGNALS

7.1 General

In all television standards, the field synchronising pulse in the video waveform is followed by a number of empty lines to allow sufficient time for the field scan of a receiver or monitor to return to the top of the screen. It seems first to have occurred to Fröling (1955) that these lines might be utilised for signals which serve some purpose other than picture information, e.g. for the measurement and monitoring of transmission quality during programme hours, for the automatic measurement and control of distribution-network parameters, for identification of signal sources, for control purposes such as network switching and so on.

These signals have been variously known in the past as 'test-line signals' and 'vertical-interval test signals (v.i.t.s.)', but it is strongly recommended that preference should be given to a proposal by Working Party M of the European Broadcasting Union that all such signals should be generically called 'insertion signals,' and that, if it is required to indicate the function, a further word can be added, e.g. 'insertion test signals (i.t.s.)' for measurement and control purposes, 'insertion communication signals (i.c.s.)' where operational instructions or data are to be transmitted, and 'insertion switching signals (i.s.s.)' where executive action is to be taken.

The name 'v.i.t.s.' seems to be strongly entrenched in the USA, and is not infrequent elsewhere, but no attempt seems so far to have been made to develop in connection with it a system for indicating the signal function. In Europe, in any case, 'vertical interval' is non-standard; so, in that area it is to be deprecated.

Since insertion signals are likely to be radiated from the transmitter, precautions should be taken when devising them to minimise the annoyance to the viewer. Regions of high luminance should be of minimum duration and confined as far as possible to the start and finish of a line. The signals in the two fields should be as similar in content as possible, to avoid flicker at picture rate, which is particularly annoying. Countries having a 60 Hz field rate are particularly fortunate, since the magnitude of the flicker can be several times that with a 50 Hz field rate for the same picture impairment. Any chrominance subcarrier component of the signal must be locked to the colour burst, to stop annoying 'rainbow' effects at the top of the screen, and the phase angle between the inserted subcarrier and the burst should be chosen to give a low-luminance colour such as blue.

To the purist, it is intolerable that the picture should be marred by signals added purely for the convenience of engineers, and this is a view with which the author has a great deal of sympathy. Nevertheless, insertion signals undeniably have a very important part to play in the maintenance of signal quality and are doubtless destined to become indispensable before long; so, it is essential to adopt all possible means to reduce picture impairment to the absolute minimum.

With receivers having mean-level a.g.c. systems, the chief impairment occurs when the insertion signals are clearly within the displayed area, since the change of picture drive with picture content alters the contrast and draws the viewer's attention to the pattern at the extreme top of the picture. On the other hand, when a monochrome receiver with black-level control is overscanned, a glow extending downwards from the top of the tube becomes evident whenever large areas of black occur in this region; white captions on a black ground often show the effect very clearly. With mean-level a.g.c., such areas are never black, but always a midgrey which effectively masks the glow.

Experiments by the author have shown that the impairment caused by insertion signals in pictures displayed on monochrome receivers having d.c. restoration depends only to a minor extent on the lines used for the signals; but, on the other hand, it does depend greatly on the vertical-scan amplitude. As the amount of overscan increases, the brightness of the glow at the top of the tube increases; and with further overscan the glow becomes distributed over the whole of the picture area, but with reduced brightness.

It seems probable that the phenomenon is due to multiple reflection in the glass of the tube along the region of high curvature, aggravated by the presence of the phosphor material around this region. This seems to be confirmed by the considerably smaller degree of impairment obtained with shadow-mask tubes, in which the phosphor is masked so that it does not extend beyond the normally scanned area; but electron scatter may also play a part.

From this, the most effective method of reducing or eliminating the impairment due to insertion test signals seems to be some action in the receiver, e.g. a more effective blackout of the cathode-ray tube during the field-blanking interval, with particular attention to the region immediately preceding the start of the picture.

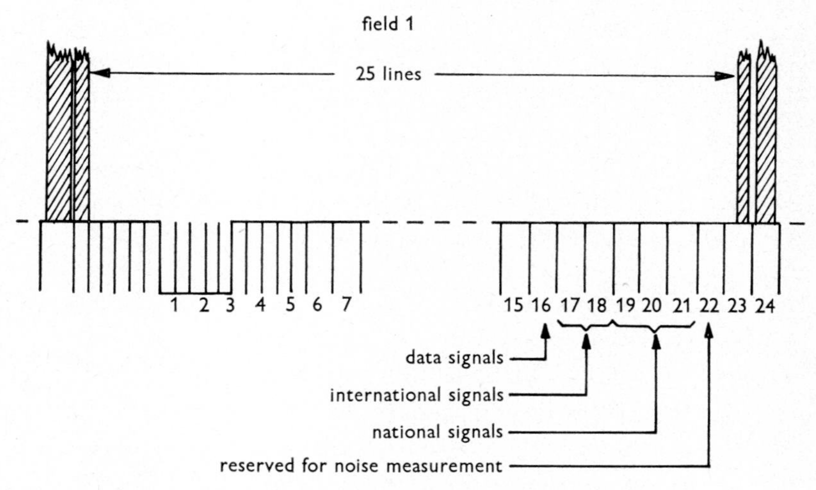

Fig. 7.1 Recommended locations for insertion test signals
Analogue in field 2 of line n is line $n+313$

7.2 Location of insertion signals

The CCIR has done much to establish standards for the location of insertion test signals in the field-blanking interval, as an obvious prerequisite for an international system of test and measurement practice. The proposals (CCIR, 1969) for 625-line systems (Fig. 7.1) go hand in hand with a further recommendation that the duration of the field-blanking interval should correspond to 25 lines, to allow room for the insertion signals at present envisaged and the special synchronising signals required by the Secam system. An ambiguity has also been removed, in the definition of the line in field 2 corresponding to a line n used for insertion in field 1, since evidently there are two possibilities when the same signal is to be inserted in both fields. The line in field 2 corresponding to line n in field 1 is now standardised as line $n+313$, so that it is no longer necessary to specify both lines unless different signals are inserted in the two fields.

Line 16 is devoted to insertion communication and data signals, which will normally only concern the country of origin, although not always. Lines 17 and 18 are reserved for international insertion test signals, such a signal being inserted by the country originating the program. It is mandatory on all countries to preserve these lines intact in the signal handed to the next country downstream. Lines 19, 20 and 21 are allocated to national insertion test signals, which, being a purely internal affair of the user country, may be inserted and erased as desired.

The remaining recommendation is the reservation of line 22 for the measurement of random noise, which is the necessary complement to the other measurements made (CCIR, 1969*d*). This line must be maintained intact as received in the signal handed on during international exchanges, and a similar arrangement will be made by any country using this line for noise measurements internally. This is variously termed the 'quiet line' or 'noise line', the nomenclature not having been fixed. The methods of measurement are discussed in Section 3.13.

7.3 Insertion test signals: monochrome

The signal originally recommended by the CCIR for monochrome testing (CCIR, 1963) consisted simply of a 10 μs bar with well shaped transitions corresponding to T = 100 ns, located at the extreme left-hand end of the test line. It provided a reference for white level, and also permitted the measurement of linear waveform distortion from the bar edge by countries using this criterion (Chapter 6).

A later recommendation (CCIR, 1965) amplified this signal by the addition of a $2T$ sine-squared pulse and a 5-step staircase waveform to give the signal shown in Fig. 7.2. It was based on a signal which had been used extensively by the BBC for both 405 and 625 lines and had amply proved its worth. It is inserted on one line only in each field. A photograph of the 405-line signal is given in Fig. 7.3; the

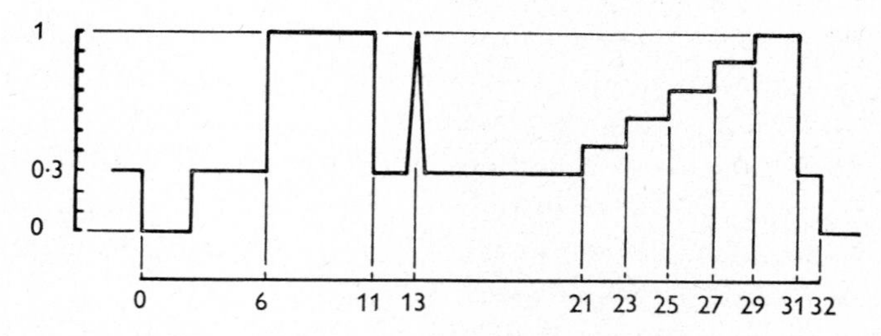

Fig. 7.2 Monochrome insertion test signal

Horizontal-scale unit = $H/32$
White-bar rise and fall times = 100 ns, or derived from sine-squared shaping network
$2T$ sine-squared pulse half-amplitude duration = 180 ± 20 ns
Amplitude of staircase steps = 140 mV

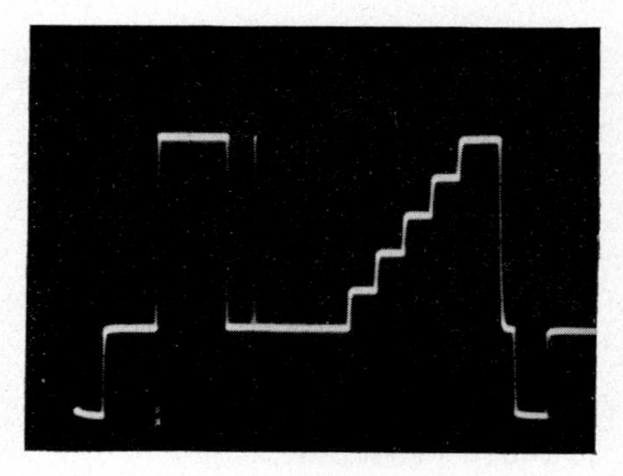

Fig. 7.3 Photograph of monochrome insertion test signal

625-line version has since been replaced by a further modification containing chrominance information described below, which is preferable even for monochrome purposes.

In the later recommendation (CCIR, 1969), the timings of the component waveforms are based on a 2 μs module, i.e. 1/32 of the

64 μs fixed line period. This has the advantage that the waveform timings are defined with greater precision and less jitter, since the modular points can be derived by 'clock pulse' techniques; i.e. they are defined by the periods of an oscillator whose frequency may be made as accurate and stable as required. For this purpose, the reference timing points of the test waveform itself are defined as the peak of the pulse component and the half-amplitude point of any bar component, including the steps of the staircase waveform. However, the precise timings have to be achieved with the aid of a number of auxiliary delays, on account of the inevitable small delays in the triggering circuitry and, most important of all, of the delays through the waveform-shaping networks. For instance, the peak of the $2T$ pulse is delayed by 200 ns with respect to the initiating pulse at the input of the sine-squared network. As a result, it becomes necessary to time a waveform component from the modular timing point immediately preceding the one with which its reference timing point should coincide, and to make up the difference by the introduction of additional delay. Since these auxiliary delays will be very small, obviously much less than 2 μs, the additional jitter introduced should be negligible.

The facilities provided by the monochrome i.t.s. of Fig. 7.2 can be classified as follows:

(*a*) white-level reference for level measurement and control purposes
(*b*) sine-squared-pulse and bar measurements (K rating)
(*c*) line-time nonlinearity.

Facility (*b*) needs some qualification with respect to the measurement of K_{bar}, since, in this instance, the length of the bar is only 10 μs against the specified length of 25 μs in the standard sine-squared-pulse and bar signal, so that any long-term distortion will appear to be reduced to less than one-half of the expected value, whereas some shorter-term distortions will be substantially unaltered.

It is impossible to circumvent this difficulty completely. The most acceptable expedient seems to be that standardised by the BBC and the UK Post Office, where the 10 μs bar is treated as though it were one-half of the full 25 μs bar, the 20% difference in length being ignored to avoid unduly complicating the graticule employed for the assessment. The standard graticule is provided with an additional reference mark M3 to the right of the vertical centre line and symmetrical with the left-hand mark M1 (Section 6.2.4.) The time-base speed is adjusted so that the half-amplitude points of the bar flanks coincide with the marks M1 and M3, and one end of the bar top is brought into coincidence with the zero line of the K_{bar} scale. The value of K_{bar} is then found from the largest deviation of the bar top from the zero line, the first and last 640 ns being ignored, as usual. This is evidently analogous to the normal procedure employing the full width of the bar in which the centre of the bar top coincides with the zero line and the larger of the two deviations of the bar top is taken as the value of K_{bar}, but the assumption is made that the distortion can be obtained from rather less than half the bar. This device seems reasonably satisfactory, but the readings it provides cannot always be in agreement with those furnished by the line-repetitive waveform.

The problem of reconciling the two methods of measurement is a universal one, since so much information is required from modern insertion test signals that no room is available for a bar with a duration of half the active line period, particularly when colour waveforms are included, as will be seen below. In view of the ever increasing importance of insertion test signals, the author suggests that some of the anomaly could be removed by modifying the duration of the luminance bar in the 625-line sine-squared-pulse and bar signals to 20 μs. This would make the duration of the luminance bar in the insertion test signal precisely half that of the line-repetitive bar, instead of 20% less as it is at the moment.

It is not thought that this step would appreciably prejudice line-repetitive measurements, for which a considerable duration of bar would still be available. In one respect, there would even be an advantage. It is believed that the original reason for the choice of the bar duration was the wish to make the average picture level close to 50%, so that measurements would be made on a.c.-coupled equipment at the centre of the transfer characteristic where the linearity is likely to be optimum. The average picture level is increased by the luminance components of the colour test waveforms in the latest sine-squared-pulse and bar signals, but the reduction in the duration of the luminance bar offsets this; in the case of the luminance–chrominance pulse-and-bar signal, the compensation is almost perfect.

The determination of the line-time nonlinearity presents no particular problem when it is carried out with one of the shaping networks described in Section 5.2.2, provided of course that the waveform monitor can be suitably triggered; other methods are likely to prove less satisfactory, particularly in the presence of appreciable random noise or interference. A typical result of passing an insertion test signal containing a staircase through such a shaping network is shown in Fig. 7.7*d*. The noise on the original signal has been smoothed somewhat by the integrating effect of the relatively long photographic exposure, so that the differences between the successive steps are somewhat clearer than in the original display; nevertheless, this particular measurement is not often found difficult to carry out.

One difference between the practical waveform and the standard version is the presence of an inverted $2T$ pulse in the centre of the bar. This was introduced by the BBC some years ago as a supplementary check on quadrature distortion, which may be introduced in transmitters, translators and radio rebroadcast systems, in fact anywhere where amplitude modulation may be followed by envelope demodulation. The presence of the quadrature component has the effect of either broadening or narrowing the pulse according to the polarity of modulation (Cherry, 1949). In addition, phase distortion around the vestigial-sideband region may give rise to asymmetry between the two pulses of a different kind. An indication of the severity of these distortions may be obtained by triggering the waveform so that the two pulses are displayed overlaid or side by side (Figs. 7.4 and 7.7*c*). The presence of this inverted pulse can occasionally cause some difficulty in the measurement of K_{bar}, and it is recommended that it should be omitted unless specially required.

7.4 UK colour insertion test signals

The colour insertion test signal must, of course, contain luminance as well as chrominance test waveforms, since neither channel can be considered in isolation, and the relationship between the two channels is quite as important as the individual behaviour of each.

The signal described below is that agreed between the broadcasting authorities in the UK. The signal on which it was based had been in use for a number of years; and, after some modifications dictated by experience, it has now been standardised as shown in

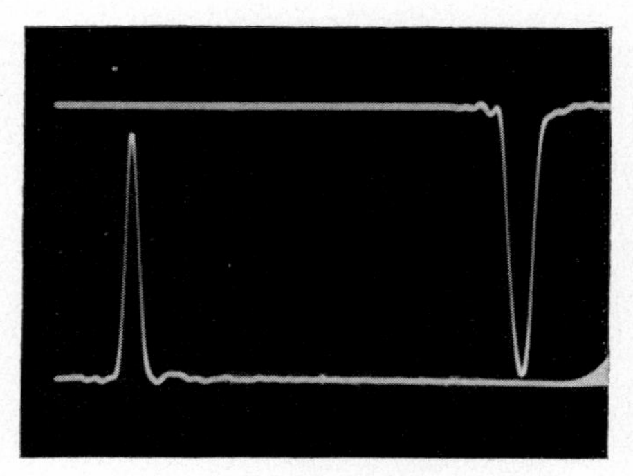

Fig. 7.4 Use of inverted pulse

Fig. 7.5. It consists of two full lines of test information, i.t.s. 1 and i.t.s. 2, respectively, inserted on lines 19 and 20 in field 1, and their analogue lines 332 and 333 in field 2. The effective repetition rate is consequently 50 Hz if one takes into account the possibility of removing a halfline difference in timing between the two fields during the display or analysis or display of the i.t.s.

7.4.1 Insertion test signal 1

The original version of this signal was a CCIR-recommended single-line signal derived from the one described in Section 7.4 by the addition of a $20T$ chrominance pulse, at first as an option and later made mandatory (CCIR, 1966*b* and 1967). This addition

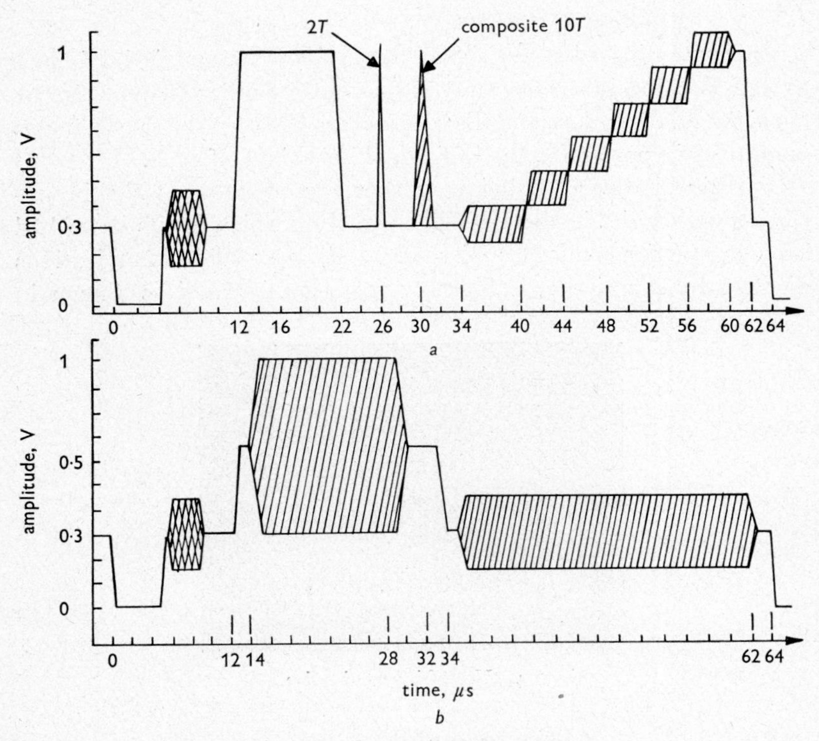

Fig. 7.5 UK national insertion test signal

Timings are within ±1% of time from leading edge of line-synchronising pulse

Burst is only present on colour transmissions

Added subcarrier is locked along axis which is nominally 60° to $B-Y$ axis when burst is present

a Insertion test signal 1

Width of last staircase step may vary slightly from nominal value

Signal is inserted on lines 19 and 332

Timings are nominally with respect to half-amplitude positions of bar and staircase components and to peak amplitude of pulse components

b Insertion test signal 2

Transitions correspond to $10T = 1\ \mu s$

This signal is inserted on lines 20 and 333

Timings are nominally with respect to half-amplitude positions of bar components increases the usefulness of the waveform by allowing measurements to be made of luminance–chrominance gain and delay inequalities and chrominance-envelope distortion; paradoxically, it also converts the signal into a considerably improved test for linear waveform distortion for monochrome purposes. The reason is discussed in Chapter 6; briefly, the improvement is a consequence of the extra information provided by the chrominance pulse in the region in which the sensitivity to distortion of the $2T$ pulse is considerably reduced by the low amplitudes of the spectral components. The chrominance pulse complements the $2T$ pulse, but in such a manner that their combined spectra do not extend beyond the limit of the videoband.

It was later realised that the superposition of a constant level of subcarrier on the staircase component would allow almost all the necessary measurements to be carried out on a single line of test signal. The final form of the waveform with a subcarrier amplitude of 140 mV was subsequently recommended by the CCIR for colour testing, and, with a slight modification, it has been adopted as the UK insertion test signal 1 (Figs. 7.5 and 7.6).

The facilities provided by this i.t.s. 1 can be summarised as follows:

(*a*) white-level reference
(*b*) sine-squared-pulse and bar measurements
(*c*) line-time nonlinearity measurement
(*d*) luminance–chrominance delay-inequality measurement
(*e*) luminance–chrominance gain-inequality measurement
(*f*) measurement of chrominance-pulse envelope distortion
(*g*) measurement of differential gain
(*h*) measurement of differential phase.

However, this list needs qualification. Strictly speaking, the measurement of line-time nonlinearity can only be carried out without error if there is no interaction between the chrominance and luminance regions, i.e. if chrominance–luminance crosstalk is absent (Chapter 6). Otherwise, the measured nonlinearity will contain a component derived from the crosstalk as the price of the economy effected by forcing the staircase component to perform a dual role. The error is not found to be serious, first, because of the close control maintained over chrominance–luminance crosstalk in the UK distribution network, and secondly, because of the low subcarrier amplitude. In addition, sine-squared-pulse and bar measurements are subject to the remarks made in Section 7.3, and differential-gain and -phase measurements to those made in Section 7.4.2.

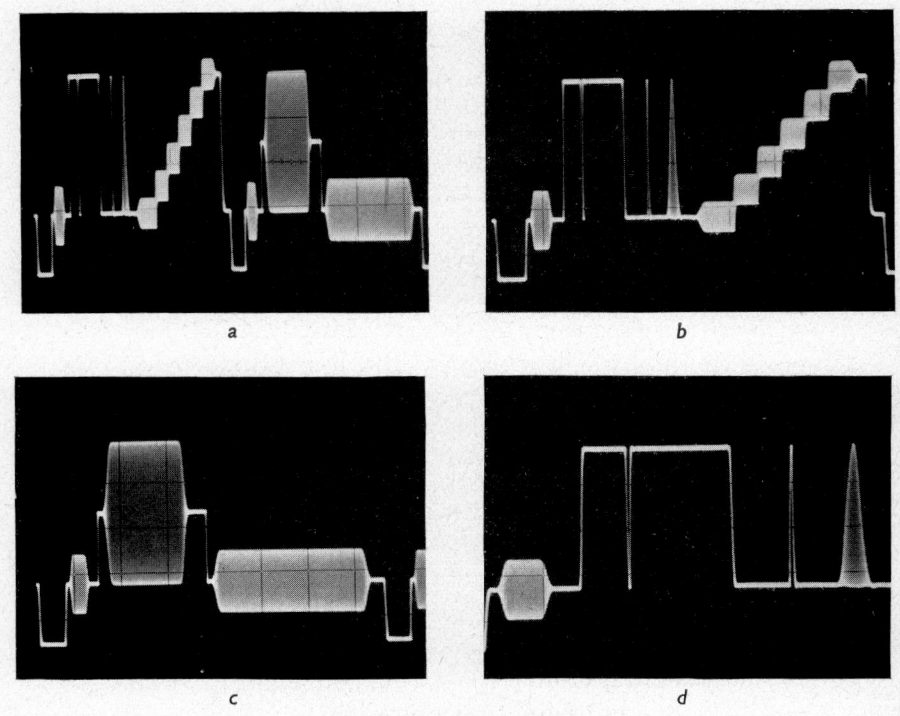

Fig. 7.6 UK national insertion test signal

a Complete waveform
b I.T.S. 1
c I.T.S. 2
d Detail of i.t.s. 1

One important difference between the CCIR-recommended waveform and the i.t.s. 1 is the replacement of the $20T$ chrominance pulse by a $10T$ pulse, since the wider chrominance bandwidth of the UK system-I standards would not be adequately tested by the spectrum of the $20T$ pulse, extending as it does over a total range of only ± 500 kHz. Moreover, the sensitivity of luminance–chrominance gain- and delay-inequality measurements from the baseline of the chrominance pulse is increased by a factor of two when the $20T$ pulse is replaced by a $10T$ pulse, a very worthwhile improvement.

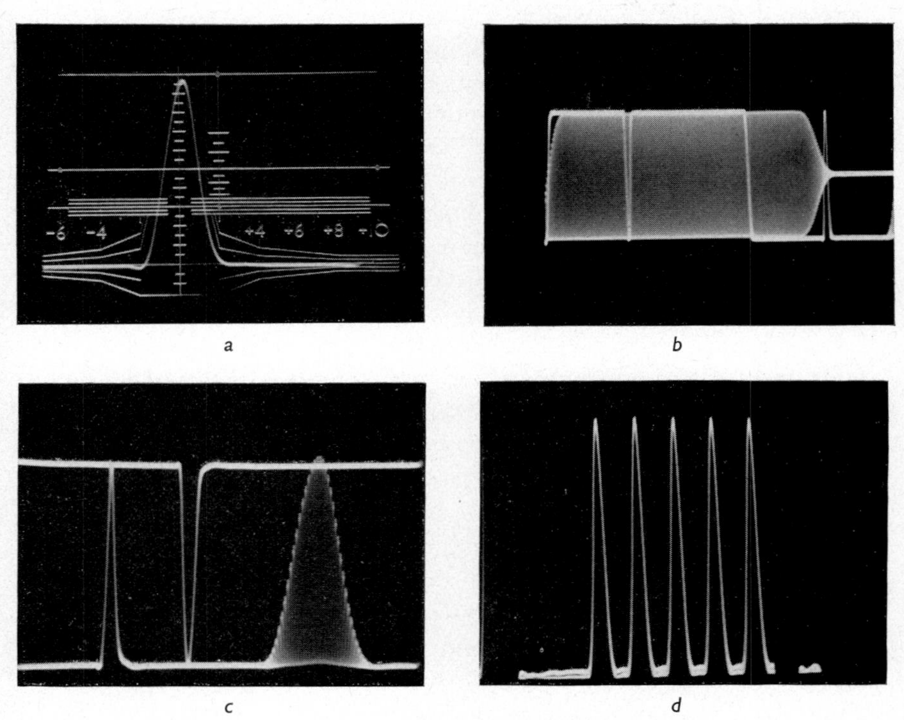

Fig. 7.7 Evaluation of UK national insertion test signal

a K rating of $2T$ pulse
b Measurement of chrominance-bar height
c Measurement of chrominance pulse and comparison of erect and inverted pulses
d Measurement of line-time nonlinearity

Typical waveforms during some of the measurements available with i.t.s. (Figs. 7.7*a*, *c* and *d*) were obtained at the end of a distribution network, as might be surmised from the visible distortion, with a waveform monitor especially designed for this purpose. The clarity of the displays is noteworthy, although, admittedly, the integrating effect of the photographic emulsion has been of assistance.

7.4.2 Insertion test signal 2

If no other source is available, the phase reference for differential phase measurements carried out with i.t.s. 1 must be derived by regenerating the subcarrier from the burst carried by the signal. This implies the use of a burst-locked oscillator, which with Pal must be of high quality if phase errors over the field-blanking region are not to be introduced by the burst blanking. Such a regenerator is relatively expensive, and a cheaper and more rugged method is desirable.

This can be achieved by following i.t.s. 1 with a further insertion test line containing a suitably wide and properly located uniform area of subcarrier (Figs. 7.5 and 7.6*a* and *c*). This is employed for the measurement of differential phase in a technique described in Section 5.4.2. Briefly, it consists in utilising a standard Pal delay line of the type incorporated in receivers to delay the subcarrier component of i.t.s. 1 by precisely 64 μs, thereby making it simultaneously available with the reference subcarrier signal of i.t.s. 2 so that a measurement becomes possible. The equipment required for this is simple, cheap and stable.

The length of the subcarrier reference signal need only be equal to the duration of the subcarrier component of the staircase, so that the use of a further line has been additionally justified by the incorporation of a signal for the measurement of chrominance–luminance crosstalk (Chapter 6). It takes the form of a block of subcarrier with $10T$ transitions mounted on a 50% pedestal, an important feature being the longer duration of the pedestal compared with the subcarrier block. The presence of chrominance–luminance crosstalk is indicated by the superposition of a positive or negative pedestal equal to the length of the subcarrier block on the 50% pedestal, and is very clearly indicated when the signal is passed through a lowpass filter which removes the subcarrier.

The generating equipment might have been somewhat simplified if the reference subcarrier signal had also been mounted on the same 50% pedestal as the chrominance bar, but this must be weighed against the greater visibility of i.t.s. 2 when this is done as a result of the higher luminance component. In the design of the signal, it was considered that the visibility of the signal should be given preference over the slight increase in the equipment cost.

Incidentally, if no chrominance–luminance crosstalk is present, the chrominance bar may be utilised to indicate the magnitude of any chrominance–luminance gain error, which is rather clearer than the baseline of the composite chrominance pulse when chrominance–luminance delay errors are simultaneously present. In use, the two waveform components may be overlaid (Fig. 7.7*b*).

Finally, the reference subcarrier signal in i.t.s. 2 becomes unnecessary if one is equipped to measure differential phase by the method of Voigt (Section 5.4.4), although the chrominance bar is still required for the measurement of chrominance–luminance crosstalk. Even this could be eliminated, which would mean the removal of i.t.s. 2, if one were prepared to agree to the use of luminance bar in, say, field 1, and a chrominance bar in field 2. Whether such considerations will eventually affect the form of the UK i.t.s., it is much too early to say.

7.4.3 International insertion test signal: colour

I.T.S. 1 and 2 form just one of a number of markedly different kinds of insertion test signal devised for purely national use by various countries, largely as a result of the widespread demand for insertion test signals for operational purposes ahead of any international agreement.

However, as long as utilisation of these signals is purely internal to a given organisation or country, no great harm is done; but, when international programme exchanges are concerned, it becomes much more important for a common signal to be employed with common methods for measuring and interpreting the distortion. Such international signals are by agreement inserted at the point of origin of the programme, and it is obligatory for all user countries to leave the original signal intact until the final country 'downstream' is reached.

Nevertheless, as a result of a great deal of dedicated work by international committees, an insertion test signal has now been recommended for international programme exchanges (CCIR, 1969*b*). Even so, it is indicative of the difficulty of reconciling so many differing points of view that the form of the signal is still not unique; as will be seen below, various options have been left open to accommodate national preferences.

The recommended signal (Figs. 7.8 and 7.9) consists of four different test signals inserted, respectively, on the four lines made available for this purpose, namely lines 17, 18, 330 and 331. The component test waveforms are as follows:

Line 17:

(*a*) luminance bar with an amplitude of 0·7 V pk–pk $\pm 1\%$; apart from its use in measuring medium–low frequency distortions, it also serves as the white-level reference of the signal

(*b*) $2T$ sine-squared pulse

(*c*) composite $4T_c$ ($20T$) pulse

(*d*) 5-step luminance staircase waveform

Line 18:

(*a*) multiburst reference bar; composed of two portions, the first with an amplitude $80 \pm 1\%$ of the reference white-bar amplitude, and the second either a bar of amplitude $20 \pm 1\%$ of the white bar, or one complete period of a 200 kHz sine wave whose peak-to-peak amplitude is $60 \pm 1\%$ of the white-bar amplitude

(*b*) multiburst signal with frequencies as indicated, each burst starting from a zero point in the sine wave.

All the above signals are mounted on a pedestal with an amplitude $50 \pm 1\%$ of the white bar.

Line 330:

identical with line 17, except that the composite chrominance pulse is removed and the staircase is overlaid with colour subcarrier, the room left by the removal of the composite pulse being utilised to lengthen the first chrominance step; the amplitude of the subcarrier is 280 mV $\pm 1\%$, and its transitions have a risetime of 1 μs.

Line 331:

(*a*) either a chrominance bar with the amplitude of the white bar

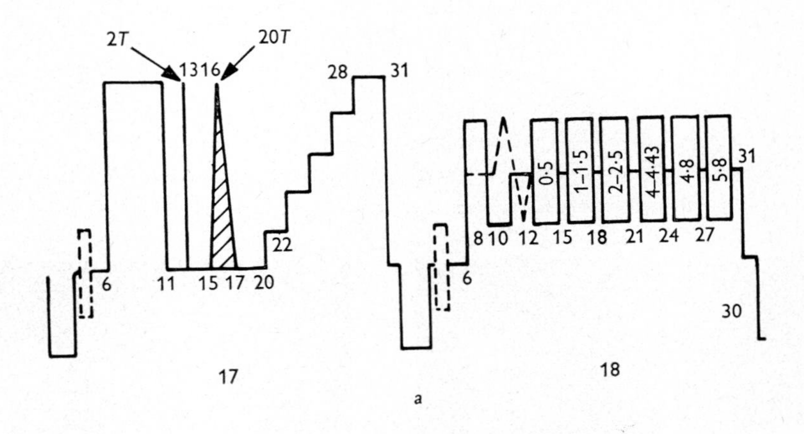

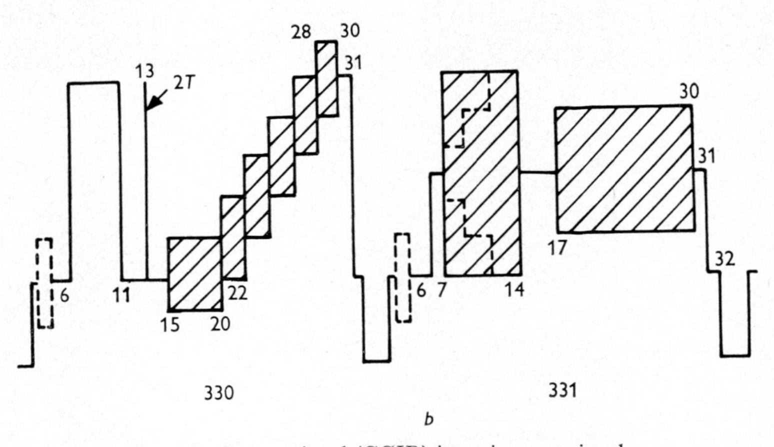

Fig. 7.8 International (CCIR) insertion test signal

a Field 1
b Field 2

Horizontal-scale unit $= H/32$; timings refer to leading edge of synchronising pulse

Waveform timings are taken from peaks of pulses and midpoints of transitions of bar waveforms

With Pal systems, subcarrier phase angle is 60° to $B-Y$ axis or a 3-level staircase with successive amplitudes $20 \pm 1\%$, $60 \pm 1\%$ and $100 \pm 1\%$ of the white-bar amplitude.

(*b*) a reference colour subcarrier signal with an amplitude $60 \pm 1\%$ of the white bar.

This rather complicated signal provides the same facilities as the UK signal, but with the advantage that quite separate measurements of luminance nonlinearity and differential phase and gain are possible. In addition, chrominance–luminance crosstalk may be measured, not

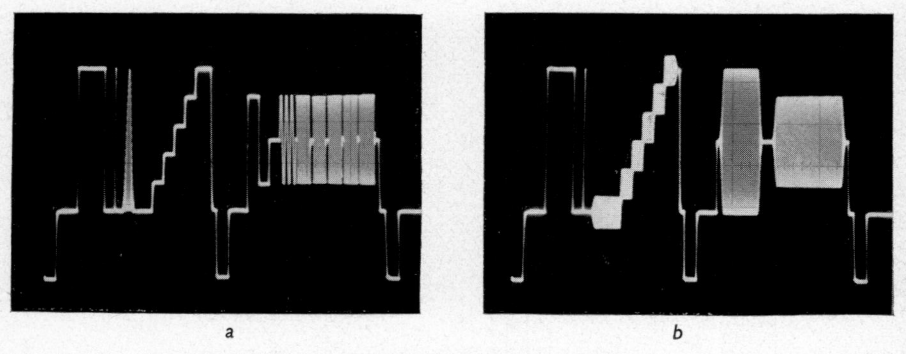

Fig. 7.9 International (CCIR) insertion test signal
a Field 1
b Field 2

solely at one level, but at three levels if desired. The multiburst is something of an anomaly here, since it is not otherwise a CCIR-recommended signal for circuit testing. It is not a reliable guide (Section 6.4.1) to the gain–frequency response of a circuit, although it can occasionally provide a rapid indication of certain types of fault.

The price paid for the additional information available with the international i.t.s. is the halving of the rate of presentation of information, since each individual signal occurs only at a 25 Hz rate. This means reduced brightness of the waveform monitor, with added flicker and increased difficulty with automatic measurement and control equipment. More elaborate triggering facilities are required with any apparatus used with this signal. Nevertheless, with the right equipment, and under the best conditions, measurements can be carried out with an accuracy not much lower than that possible with line-repetitive signals.

7.5 Erasure

It is necessary from time to time to erase an i.t.s. to insert a fresh signal, and by the same token it is prudent always to erase the line before the addition of the i.t.s., to ensure that nothing is left on the test line or lines which will later prejudice the use of the signal for measurement purposes. Another reason is to avoid double insertion, e.g on contribution signals which already contain an i.t.s.

Some general remarks on the design or erasure equipment may prove useful. First, all equipment concerned with the i.t.s., whether for erasure or for insertion purposes, is almost always installed in a main signal path. Consequently, not only must it be as reliable as possible but, in case of failure, it must cause the least possible damage to the signal.

Two main principles are employed for erasure, which may be termed the series and the shunt method, respectively. In the first, a line or group of lines is gated out or selectively amplified and clipped, after which the black level is replaced by a locally generated signal. This can give a high degree of suppression of existing signals; but, since one is acting, as it were, in series with the erased signal, any fault in the equipment can have serious consequences.

In the second, a shunt circuit extracts any existing signal from the test line or lines and adds it back to the signal in antiphase. The result of any malfunctioning is then less severe; and, in particular, the equipment can be switched off without interrupting the signal passing through it. However, it is not easy to ensure a high degree of suppression, so that, for colour working, the series method is likely to be preferred to ensure good suppression of chrominance components even though the risk from malfunctioning may be rather greater.

It is sometimes assumed that one may erase and reinsert almost without limit; but, in fact, there is always an element of risk from each operation, either from partial or complete failure of the eraser, or from a progressive degradation of its performance with time resulting from component aging. The latter chiefly affects the high-frequency end of the videoband of the erased signal to give incomplete suppression of the subcarrier frequency, usually accompanied by small 'spikes' derived from the waveform transitions. Another unwelcome effect is a variable pedestal on the erased line or lines as a consequence of drift in the reinserted black level. In bad cases, this has been known to cause false operation of succeeding clamps.

7.6 Utilisation

Experience with i.t.s. measurements over a period of some ten years has shown quite conclusively that insertion test signals can maintain a high-quality distribution network even without the aid of out-of-service testing. It is not easy to provide meaningful figures for measurement accuracy since it is so greatly affected by factors such as the signal/noise ratio at the point of measurement and the quality and suitability of the equipment used; but, in one series of comparative tests on the intercity network with a signal/noise ratio of about 50 dB, the median value of the difference between the results for the i.t.s. and a repetitive signal tended to be less than one-third of the tolerance allowed for the test in question, and more refined equipment seemed likely to provide an even closer agreement.

In making comparisons, a problem arises particularly with differential phase and gain; both are a function of the average picture level, which is constantly changing during measurements on the i.t.s. under inservice conditions, whereas the conventional specification is framed in terms of two fairly extreme values of average picture level (Section 5.2.4). However, experience shows that if the measurement is taken over a long enough time for a complete range of values of average picture level to have occurred, the highest reading is likely to agree fairly closely with that given by the conventional repetitive measurement.

It should be feasible to devise apparatus to take a reading during any sufficiently long duration of a given average picture level; but, in either case, one might have to wait a considerable time, and under operational conditions this is usually unacceptable. One will probably be compelled to admit that the insertion test signal furnishes nonlinearity-distortion figures, luminance and chrominance, which are invaluable for monitoring and maintenance purposes but which cannot necessarily be reconciled with the figures obtained with repetitive testing methods. As a corollary, performance limits need to be considered in the light of this fact.

The precise method of implementing a system of measurement and control of the quality of a distribution network in any given instance will depend on so many local factors that it is proposed to confine the discussion to some general comments and observations which a lengthy experience with insertion test signals suggests might be desirable.

As far as the insertion of the i.t.s. is concerned, the earlier the point in the network at which this can be done the better, preferably at the central mixer or whatever location corresponds to the completely assembled programme signal in the form in which it is to be distributed, allowance being made for locally originated signals which may be introduced at intermediate points along the network. Only thus can one ensure that any distortion observed on the i.t.s. is equated with the distortion undergone by the programme signal. This also implies that the original i.t.s. should preferably be retained as long as possible, and any erasure and reinsertion should only be carried out with good reason.

However, this policy raises a serious difficulty whenever it becomes necessary to check the contribution to the overall distortion of any individual link or section of the network, which can only be done by a comparison of the waveforms of the two distorted signals at the ends of the given section, making it necessary for the complete information on both these waveforms to be available simultaneously at a common point. While this is by no means impossible, it is nevertheless operationally inconvenient.

A modification of an original Australian proposal (CMTT, 1965*a*) gets round this problem. The waveforms inserted on the corresponding lines in two consecutive fields are identical, but are inserted at two different points in the network. The i.t.s. in field 1, say, is inserted at the normal point, i.e. at or very near the start of the network. The i.t.s. in field 2, however, may be inserted at any intermediate point at which it is required to carry out a local test. Hence any desired section can be measured with an input signal known to be immaculate, and the need for any comparison of the input and output signals is obviated.

The ingenuity of this proposal is that the i.t.s. is made to carry two separate pieces of information without any prejudice to the viewer. The price paid, of course, is the 25 Hz rate at which the information about either signal becomes available, and the slight complication introduced by the need to identify either field positively. This device is not applicable, of course, to the international i.t.s., which has different signals in the two fields.

Another way of achieving the same object is to make temporary use of some other line in the field-blanking interval for the insertion of the local i.t.s., making sure that it is removed or erased after use. Provided that another suitable line is available, which might not be the case under all circumstances, this variant has the advantage that the information is received from the i.t.s. at the normal rate.

So far, the measurement problem has been discussed without any reference to the means for acquiring the information from the i.t.s. If engineers are available the solution is straightforward, but tedious and time-consuming unless the work is suitably organised. This at least requires the equipment to be permanently located in the measurement area with provision of triggering signals and switching arranged so that the minimum number of operations is carried out during an evaluation. A more elaborate arrangement has a special test console to analyse the i.t.s. and display the resulting waveforms time-multiplexed in the space previously occupied by the i.t.s. The magnitudes of the various distortions may then be read off almost at a glance, or alternatively, the complete display can be photographed for record purposes or for later evaluation. This device has been successfully employed for the routine measurement of videotape machines when the time available for signal evaluation is extremely restricted (Section 13.3.4). See also Savage and Carter (1970).

An attractive alternative is an electronic means for signal evaluation. Work in this direction has been carried out by a number of organisations in the UK and abroad (CCIR, 1963*a*; CMTT, 1965*b*; and Fasshauer, 1967), but not all of it has so far been published. The universal basic principle is the conversion of the i.t.s. waveform, or a form of time-multiplexed analysed signal, into an analogue signal by waveform-sampling methods utilising a very narrow pulse. For colour i.t.s. waveforms, an already analysed signal must be used, since, although the sampling system can be made to reproduce the cycles of the subcarrier waveform, it is not really feasible to utilise the result.

The method is not without its difficulties, not so much in principle as in the design of the equipment to provide sufficient accuracy at reasonable cost and with adequate stability; the number of equipments required is likely to be relatively high, and they must operate correctly and reliably over long periods without frequent attention.

A common difficulty with sampling systems which must translate into analogue form the fine detail of the original i.t.s. waveform is the choice of the duration of the sampling pulse itself. If it is too wide, the resolution will be poor, and errors will result, particularly in the amplitude of the sine-squared pulse, which can be shown to be the most critical feature of the waveform. If the pulse is unnecessarily narrow, the charge extracted from the video signal during each sampling operation is so minute that the corresponding storage capacitor has to be extremely small and the total effective shunt resistance has to be extremely high if errors due to noise and other factors are not to be introduced.

An investigation by the author has shown the duration of sampling pulse needed for a given error in the measurement of the peak amplitude of the sine-squared pulse (Weaver, 1968*a*). It is to some extent a function of the way in which the gating is performed; but, for a $1T$ pulse, a typical expression for the error in amplitude dA would be $dA = r^2$, where $r = t/2T$, t in this instance being the half-amplitude duration of the sampling pulse. Thus, for a 625-line $1T$ pulse, allowing a maximum error of 1% in amplitude, the pulse duration required turns out to be 20 ns.

A fairly standard method of analysis of the i.t.s. is to move the sampling pulse along the signal at a very slow rate, which accordingly yields an analogue signal completely equivalent to the input signal but with a greatly diminished frequency band. The slower the rate, the more smoothing can be introduced to lessen the effect of the random noise component (Section 6.2.10). Finally, the analogue signal is converted into some form convenient for storage and transmission, most usually into an f.m. signal whose bandwidth may now be so low, say of the order of 1 kHz, that tape recording and transmission over a telephone pair are perfectly feasible; indeed, at least two such modulated signals, or a modulated signal and a single speech channel, can still be transmitted over a single circuit.

A variant of this technique recognises the fact that a great deal of information in the sampled i.t.s., and more especially in the signal analysed before sampling, can be considered as redundant and may be replaced by samples taken at strategic points along the waveform; thus, not only more sophisticated methods of data transmission, but also features such as out-of-limit alarms, become possible. Alternatively, the measurements required can be extracted from the waveform by refined circuit techniques and made available either in analogue or digital form, or data-coded for transmission to a central point. An instrument of this type is described in detail by Shelley and Williamson–Noble (1970).

Once the information has been converted into a form in which it can be transmitted where required, many possibilities for utilising it become available. It may, for example, be returned from all monitoring points to a central supervisory area where the control of the complete network is carried out. It then becomes possible to envisage the processed data being fed into a computer which analyses it on a statistical basis and makes use of the output to take charge of the quality control of the network, including executive action where required. In any given instance, plans for the automatic or semi-automatic control of a distribution network based on data extracted from insertion test signals suggest themselves readily, and there is no doubt that they can be made to work; whether the results would justify the enormous capital cost would have to be decided for each individual case.

Alternatively, the actual values of distortion may be regarded as less important operationally than the knowledge whether or not any given part of the circuit is approaching a value of distortion considered excessive under the circumstances. This allows the equipment to be simplified somewhat, since the conversion of the measured distortion to analogue form may then be carried out by whatever means happens to be the most suitable. Preferably, this would be evaluated in terms of an overall quality factor obtained by summing the individual impairment values for the various distortions (Weaver, 1968). Unfortunately, at the time of writing, not all distortions can be treated in this manner. It would also be necessary to lay down a maximum limit for any individual distortion, since this might well indicate an incipient fault condition.

In an equipment used experimentally by the BBC, the actual values of the distortions are furnished by the test set, which also provides two types of alarm at appropriate locations, a nonurgent alarm which announces that a predetermined distortion limit has been reached, and an urgent alarm which indicates a fault condition. In the latter case, it is proposed to transmit an inhibit signal to all automatic monitors downstream to prevent them from also raising an alarm; this has the additional advantage that the area of fault occurrence is identified.

Automatic monitors of the type described immediately above are evidently most appropriate where the distribution network is highly stable and reliable and moreover not unduly extensive. Where the network is less stable it might be more profitable to attempt to equalise the distortion automatically within reasonable limits by means of information derived from measurement of the i.t.s. This approach may also be worth while in certain other instances, e.g. when the composition of the network is frequently modified. In such a situation, it may not be practicable to maintain the component links within such tolerance limits that no combination formed will not show one or more types of distortion at an undesirably high level on certain occasions. This can be met by the use of a 'mop up' equaliser controlled by the i.t.s.

A very simple application of this technique is the control of the video-signal amplitude at the input to a transmitter by comparing the amplitude of the white bar in the i.t.s. with a local standard amplitude and utilising the error to control a motor-driven potentiometer or electronic attenuator to set the amplitude correctly. It was first described by Springer (1959), and the increasing use of unmanned transmitters has encouraged its widespread adoption, most usually nowadays in completely electronic form.

By far the most powerful and flexible technique for the correction of linear waveform distortions is the transversal equaliser of Kallmann (1940), in which both gain and delay characteristics are equalised simultaneously by direct operations on the waveform itself, and it happens to be one, moreover, to which electronic control is particularly applicable. Very briefly it depends on the fact that the distortion can be shown to be equivalent to the superposition on the undistorted waveform of a pattern of echoes of itself with delays and polarities which are a function of the nature of the distortion. If then the sine-squared pulse in the i.t.s. is sampled at the minimum Shannon intervals, i.e. 100 ns or less for a 5 MHz bandwidth, the echo pattern corresponding to the distortion present can be found, and accordingly the distortion can be equalised by feeding back into the video signal a pattern of its own echoes having the appropriate delays, amplitudes and the inverse polarity.

In a version of this studied by the author for the equalisation of colour video signals, it was proposed to supplement the data derived from the sine-squared pulse corresponding to the higher video frequencies with data derived from the composite chrominance pulse, in view of the reduced sensitivity of the $2T$ pulse in that region and the close tolerances required in the chrominance region. An alternative would have been the replacement of the $2T$ pulse by a $1T$ pulse, but it would not have been so satisfactory. It was concluded that an automatic equaliser of this type was perfectly feasible, but expensive and unnecessary with the relatively stable distribution circuits available in this case.

It became clear that a highly simplified version of such an automatic equaliser which only corrected for luminance–chrominance gain and delay errors would be much more appropriate, the correction being carried out as before with echo patterns but of much lesser complexity.

Equipment of this type has been designed and put into service experimentally. An automatic equaliser for the correction of differential-gain and -phase errors has also been designed and likewise put into service.

7.7 Local insertion test signals

Insertion signals have so far been discussed in terms of the measurement and control of distribution networks, but other measurement applications are possible. One in particular which has been found extremely useful in colour-signal generation takes the form of three colour-separation components which are linearly added to the appropriate colouring channels of a colour-signal source, e.g. a colour camera, in such a way that, after coding has taken place, they result in a form of colour-bar signal situated on one or more lines in the field-blanking interval. This then serves as a sensitive check of the accuracy of the coding process and as a means of detecting any distortion which may have been introduced in the studio circuits and equipment. No name appears to have been given so far to this signal; so it is suggested that 'studio insertion test signal (s.i.t.s.)' would be appropriate.

As an example, an experimental application of this technique by the BBC will be described. The test signal in this instance was the 100% (100–0–100–0) colour-bar signal (Section 10.8.3). It was chosen because it is the standard for the operational checking of the coded signal, the apparatus for generating it was readily available, and the staff were accustomed to analysing it and diagnosing faults from it. The signal was inserted on the two adjacent lines which are normally allocated to the national i.t.s.; two lines because of the properties of the Pal system, but with NTSC only one would be required as a minimum. This signal was erased after use, to allow the national i.t.s. to be inserted.

Fig. 7.10 shows the generation and insertion equipment. The basis is a standard colour-bar generator which fortunately had originally been designed on a 'one-shot' basis; i.e. a line of test signal is only generated on the receipt of a trigger pulse. Suitable triggers were derived from a unit initially intended for use with a waveform monitor.

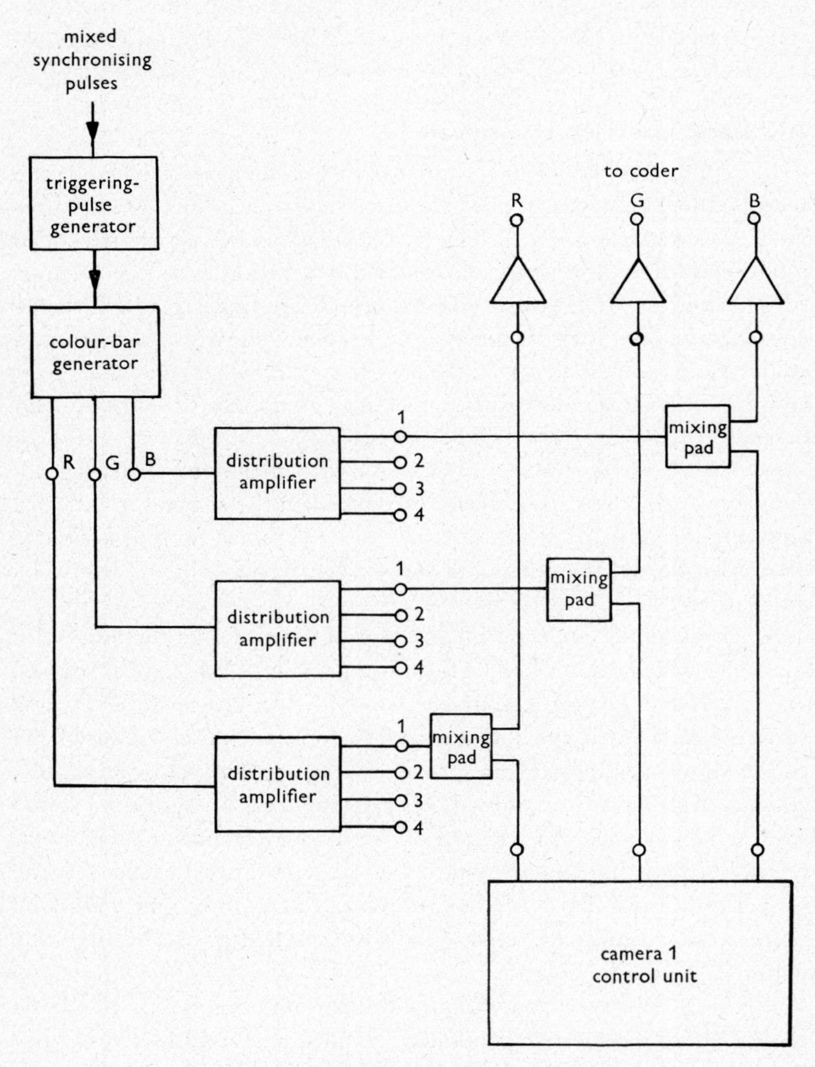

Fig. 7.10 Insertion test signal for studio use

The colour-bar generator then supplies, for the period of the two selected lines only, the three colour-separation components of the colour-bar signal (Figs. 7.11*a*, *b* and *c*; Fig. 7.11*d* shows the blue waveform in greater detail). Each of these is passed through a distribution

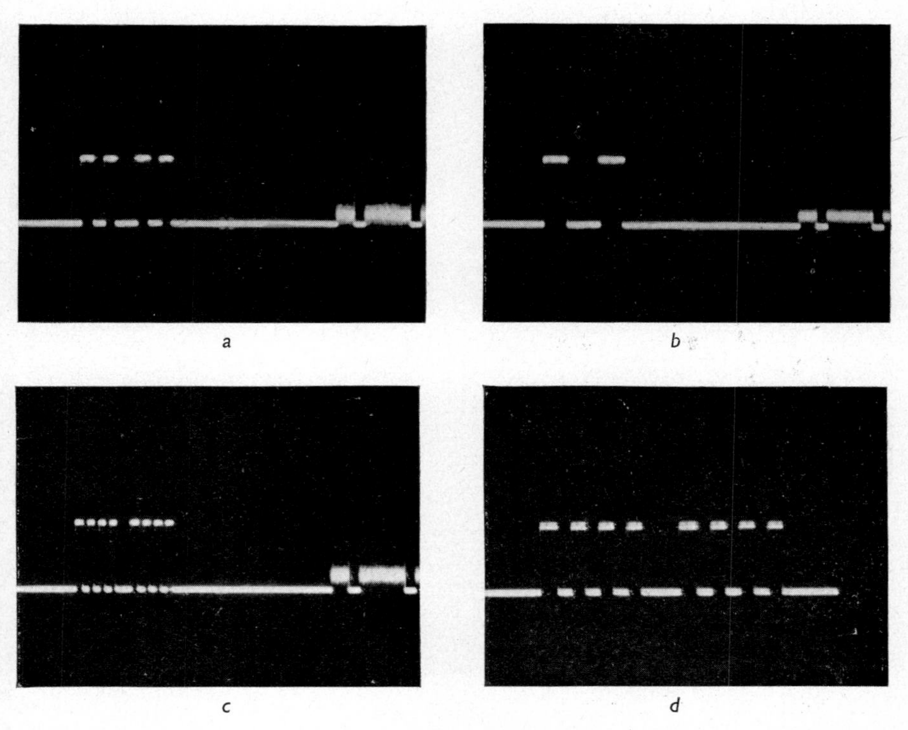

Fig. 7.11 Colour-separation components of studio insertion test signal
a Red channel
b Blue channel
c Green channel
d Detail of blue-channel waveform

amplifier, so that a number of signal sources may be fed from a single colour-bar generator if required, and is then linearly added to the appropriate channel of the signal source in a resistive pad. Each pad is preceded by an amplifier with a gain of 6 dB to restore the correct level.

The studio i.t.s. components now form part of the colouring-channel outputs of the signal source, so that, at the output of the coder, they should have the appearance of Figs. 7.12*a* and *b*, the former showing the location of the two lines of colour bars in the field-blanking interval, and the latter the detail of the bars. They may now be analysed, with a good vector monitor, and accordingly an assessment may be made of the quality of the coded signal. The studio i.t.s lines will eventually be erased to clear the two lines for the insertion of the studio i.t.s.; so the colour-signal quality may be monitored anywhere up to that point.

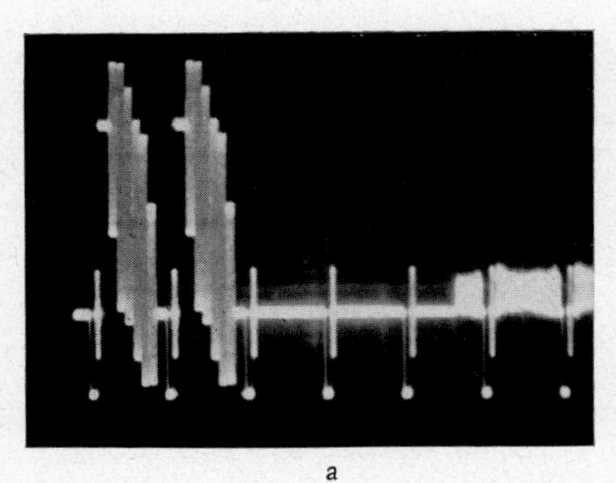

a

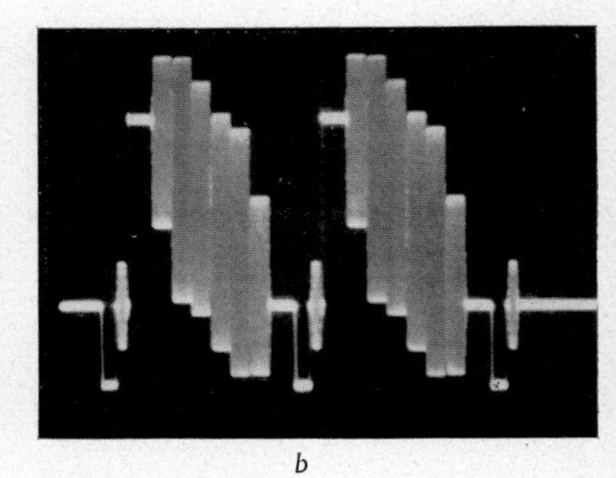

b

Fig. 7.12 Coded studio insertion test signal

a Location in field-blanking region
b Detail of coded waveform

A comment should be made on the method employed for mixing the studio i.t.s. components into the signal-source colouring channels. The method of Fig. 7.10 was employed, not merely because it is simple and convenient, but more fundamentally because it is essential for the success of the technique, in that it automatically ensures that the d.c. components of the individual signals are correct. Any shift of these d.c. components will modify the coded colour-bar signal and give the appearance of a nonexistent distortion. Shifts of any separately generated pedestals, which, although small, are nevertheless significant under the circumstances, are practically impossible to prevent.

A rather different version of the studio i.t.s has been proposed by Auld and Rao (1968) termed the 'simplified single-line colour-bar (s.l.c.b.)'. It is recommended for use with colour cameras operating on NTSC standards. Apart from the fact that it occupies only one line in the field-blanking interval, as might be anticipated in view of the restriction to NTSC practice, it differs from the signal described above in that it takes the form of the standard US colour-bar signal less the green and magenta bars. The practical advantages of this simplification are not entirely clear, since a special generator is necessitated and the resulting signal is nonstandard as far as operational staff are concerned. If a colour-bar signal familiar to the operational staff is utilised, there will be fewer difficulties in interpretation, and the cost of installation may be reduced by largely utilising equipment already standardised.

7.8 References

AULD, J. S., and RAO, G. V. (1968): 'A continuous simplified single-line color bar (SLCB) test signal facility in color cameras', *J. Soc. Motion Picture Televis. Engrs.*, **77**, (3), pp. 228–232

CCIR (1963*a*) Plenary Assembly, Geneva, Document 256 (OIRT)

CCIR (1963*b*): *ibid.*, Recommendation 420

CCIR (1966*a*): Document 85E, July

CCIR (1966*b*): XIth Plenary Assembly, Oslo, Recommendation 420–1

CCIR (1967): Draft Recommendation to 420–1, Oct.

CCIR (1969*a*): Document CMTT 1017E, Oct.

CCIR (1969*b*): Document CMTT 1025E, Oct.

CCIR (1969*c*): Document 1028E, Oct.

CCIR (1969*d*): *ibid.*, Section 208

CHERRY, C. (1949): 'Pulses and transients in communication circuits' (Chapman & Hall)

CMTT (1965*a*): Document 26 with Annex II/89, Geneva

CMTT (1965*b*): Document CMTT 11, Jan.

FASSHAUER, P. (1967): 'Auswertung von Prüfzeilensignalen nach dem Abtastverfahren', *Rundfunktech. Mitt.*, **11**, (5), pp. 242–250

FRÖLING, H. E. (1955): 'Das Prüfzeilenverfahren beim Fernsehen', *Tech. Hausmitt. NWDR*, **7**, pp. 129–138

KALLMANN, H. E. (1940): 'Transversal filters', *Proc. Inst. Radio Engrs.*, **28**, pp. 302–310

SAVAGE, D. C., and CARTER, D. A. (1970): 'Application of insertion test signals to television transmission chain operation', *IERE Conf. Proc.* 18, pp. 143–155

SHELLEY, I. J., and WILLIAMSON–NOBLE, G. E. (1970): 'Automatic measurement of insertion test signals', *ibid.*, pp. 159–170

SPRINGER, H. (1959): 'Andwendung und Weiterentwicklung der Prüfzeilentechnik', *Rundfunktech. Mitt.*, **3**, pp. 40–50

WEAVER, L. E. (1968*a*): 'The error due to sampling a sine-squared pulse', BBC Designs Department Technical Memorandum 9.83 (unpublished)

WEAVER, L. E. (1968*b*): 'The quality rating of colour television pictures', *J. Soc. Motion Picture Televis. Engrs.*, **77**, (6), pp. 610–612

8 RETURN LOSS

8.1 General

Television equipment, in common with most communication equipment, is normally designed to operate between definite impedances, usually 75 Ω. If these impedances, or the input and output impedances of the apparatus concerned, are not correct, one result will be the changes in gain considered in Chapter 2. Another may in certain instances be the generation of delayed echoes, which can form a serious signal impairment (Section 6.2.4).

Consider, for simplicity, a very long, uniform cable, cut at a given point and terminated in the characteristic impedance of the cable, i.e. with $Z_0 = V_i/I_i$ in terms of the Thévénin equivalent circuit of Fig. 8.1*a*. Since the terminating impedance is equal to the cable impedance, the cable is 'unaware' that it has been truncated, and all the transmitted energy is absorbed in the terminating impedance Z_0.

Now, if the terminating impedance is altered to $Z_0 + \Delta Z_0 = Z_T$, by the compensation theorem (Shea, 1929), a most useful device for dealing with impedance changes in networks, the effect is exactly equivalent to the insertion in series with Z_T of an ideal generator with an e.m.f. $I_i \Delta Z_0$. This gives rise to a reflected current I_r (Fig. 8.1*b*). The voltage developed by I_r across the impedance facing the termination Z_T, in this instance Z_0, is $(I_i \Delta Z_0 Z_0)/(Z_T + Z_0) = \{(Z_T - Z_0)/(Z_T + Z_0)\} V_i$, which is therefore the echo voltage generated by the mismatch. The ratio of this to the incident voltage V_i is consequently $(Z_T - Z_0/Z_T + Z_0) = r$, known as the reflection coefficient. In general, neither Z_0 nor Z_T is a pure resistance, so that r is complex, and the echo is a somewhat distorted version of the incident signal. Since the echo energy is derived from the input signal, the latter is reduced in amplitude accordingly and also suffers some distortion.

Although only the receiving end of the cable has been considered above, it should be evident that echo generation takes place at the

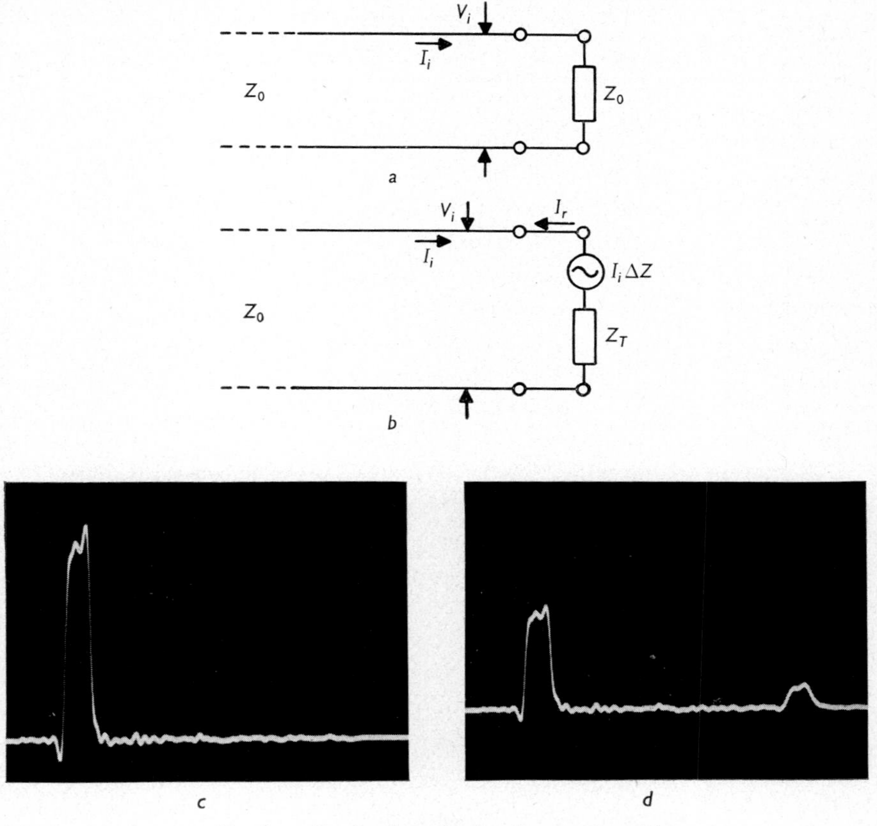

Fig. 8.1 Production of reflection at mismatch

a Correctly terminated cable
b Misterminated cable
c Test pulse at input of correctly terminated delay line
d Test pulse at delay-line input with open-circuited output

sending end in a completely analogous fashion, so that any primary echoes produced at the receiving end will give secondary echoes, which are now much more important, since they are travelling in the same direction as the original signal and will therefore add to it and form part of it. Furthermore, they are delayed, compared with the original signal, by twice the transmission time along the cable; and, as is known (Allnatt and Prosser, 1965) the impairment corresponding to an echo increases, up to a certain point, with increasing delay. All even-order echoes add to the signal as distortion; but, in practice, with most cables and other bilateral transducers, such as filters, the rapid increase in attenuation with echo order reduces considerably the importance of those of order higher than the second.

Fig. 8.1*c* gives the waveform across the input of a network in the form of a large number of nominally identical sections. The input has been deliberately mismatched; the train of echoes from slight mismatches between the component sections of the network is very evident, the total echo delay in this instance being 4 μs. The earlier echoes corresponding to the leading edge of the input pulse fall within its duration, and consequently appear as waveform distortion. When the output end of the network was open-circuited, the waveform of Fig. 8.1*d* was obtained. As is predicted by the expression for the reflection coefficient, the echo has the same polarity as the input signal. If the output termination had been smaller than the network impedance, the echo would have been inverted, corresponding to a negative reflection coefficient. The loss in the network due to internal dissipation was not particularly low, as is indicated by the rather small amplitude of the reflection from the open-circuited end.

In view of the importance of controlling the amplitudes of echoes generated at terminations, it has become standard practice to allocate tolerances to these in terms of the amplitude of the echo produced. The quantity used generally in communication engineering for specifying the quality of a termination is the return loss, which is the reciprocal of the modulus of the reflection coefficient expressed as a voltage ratio in decibels, i.e.

$$20 \log_{10}|(Z_T+Z_0)/(Z_T-Z_0)|$$

From the definition above of the reflection coefficient, this is also equal to 20 $\log_{10}$ of the modulus of the ratio of the incident voltage to the voltage of the reflected component, which suggests an alternative formulation of return loss more appropriate to television, as is explained below.

8.2 Specification

The above conventional definition of return loss has been framed in terms of the reflection coefficient at a given frequency, which does not lend itself very well to the specification of limits for the videoband. The most common systems used either lay down maximum figures for the various parts of the band or, to simplify the specification, state the worst allowable figure for any part of the band.

A single-figure specification proposed by the author has been standardised by the BBC for a good many years*, and has more recently been adopted internationally (CCIR, 1966). This consists in recognising that we are not concerned with reflections at various frequencies but with the reflections experienced by video waveforms, and that it would consequently be much more logical to specify return loss in terms of the maximum peak-to-peak amplitude at the point of measurement of the reflection of a standard test waveform. In the original proposal, the standard waveform was to be a 1 T sine-squared pulse which had been band-limited by means of a linear-phase lowpass filter cutting off at the upper limit of the videoband. It is now suggested that this should be replaced by either the chrominance–luminance pulse-and-bar waveform or the augmented pulse-and-bar waveform (Chapter 6), which are suitable for both monochrome and colour, together with the standard 50 Hz square wave employed for K rating purposes.

The formal definition is then: if the peak-to-peak amplitude of the incident waveform is E_i and the peak-to-peak amplitude of the largest component of the reflected waveform is E_r, the waveform return loss is defined as $20 \log_{10} E_i/E_r$. This provides a single figure of merit for a termination which has a simple physical significance. If desired, it would be perfectly simple to derive from either of the proposed test waveforms individual figures for both the luminance and chrominance regions; but at present this does not seem worthwhile.

* WEAVER, L. E.: 'The measurement of return loss', BBC Designs Department Technical Memorandum 9.10(60) (unpublished)

8.3 Impedance method

The least sophisticated method of measuring return loss is to measure the impedance directly with an impedance bridge at a number of frequencies. If the impedance is found to be R_1+jX_1 at a frequency f_1, the return loss against 75 Ω is given by

$$(\text{return loss})_{f_1} = 10 \log \left| \frac{(R_1+75)^2+X_1^2}{(R_1-75)^2+X_1^2} \right|$$

The shortest possible leads must be used between the bridge and the termination. If the termination forms the input or output of equipment containing active devices, the equipment must be in an operative condition, except that, when measuring an output termination, some means must be found for preventing the output voltage from falsifying the measurement while ensuring that operating conditions are otherwise normal. This means that, if the output impedance of a signal generator is to be measured, it may be necessary to make a temporary modification to the circuit to reduce, or remove, the output signal.

8.4 Return-loss bridge

The basic circuit, given in Fig. 8.2, is immediately recognisable as a Wheatstone bridge in which the impedance to be measured is Z_x and the comparison standard is R_0. The test voltage is applied across the diagonal AB, and the reflection term is generated across CD. If it is assumed that the output voltage E_0 is measured with a high-impedance device, the proof is simple. The p.d. between C and B is evidently $\frac{1}{2}E_i$, and likewise the p.d. between D and B is $\{Z_x/(R_0+Z_x)\}\, E_i$. Hence

$$E_0 = V_{CB} = E_i \left[\frac{1}{2} - \frac{Z_x}{R_0+Z_x}\right]$$

$$= \tfrac{1}{2}E_i \left[\frac{R_0-Z_x}{R_0+Z_x}\right]$$

The return loss is then equal to $20 \log_{10} |E_i/2E_0|$.

The operation of the bridge is best visualised physically by means of the compensation theorem. If one considers the unknown Z_x as the sum of R_0 and an error term ΔR_0, the bridge is balanced as far as the first of these component terms is concerned. The output voltage is consequently a function only of the e.m.f. corresponding to the p.d. across the error term ΔR_0, i.e. a function only of the amount by which the unknown Z_x differs from the correct value R_0, as in the derivation of the expression for the return loss in Section 8.1.

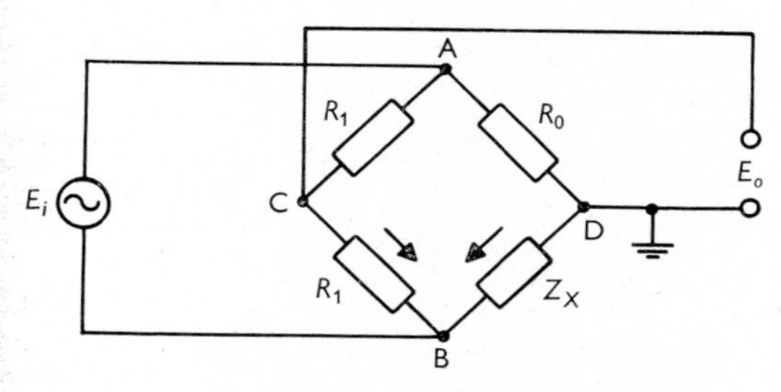

Fig. 8.2 Basic circuit of return-loss bridge

However, the basic Wheatstone-bridge circuit is not convenient in practice since either the output indicator or the signal source must be balanced to earth. A classic alternative replaces the bridge arms AC and CD of Fig. 8.2 by the centre-tapped secondary winding of a balanced-to-unbalanced transformer, giving the so-called 'hybrid coil' or 'hybrid transformer' circuit .The test waveform is applied to the unbalanced primary winding of the transformer, and consequently an effective electrostatic screen between the two windings, connected to the earth of the primary side, is essential to prevent the current flowing in the secondary winding from being modified by components injected from the primary via the interwinding capacitances.

While it is possible to construct transformers of this type with an extremely high degree of balance over a wide frequency band, it is not surprising, in view of the difficulties involved in their manufacture, to find that they are very expensive. A widely used alternative leaves intact the arms of the original bridge and utilises the transformer solely to convert an unbalanced voltage to balanced form, or vice versa. The removal of the requirement for the precise equality of the voltages across the two halves of the secondary winding makes the construction of the transformer very much simpler. Fig. 8.3 shows a modified form of an instrument which has given yeoman service for many years. A further version is discussed by Thiele (1964).

The input test waveform is applied, through a 75 Ω constant-resistance splitting pad, both to the primary of the transformer and to the input of a calibrated variable attenuator. The output voltage

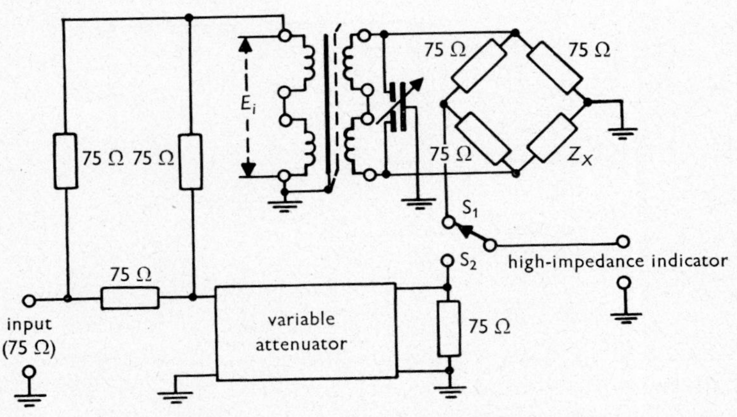

Fig. 8.3 Circuit of transformer return-loss bridge

from the bridge, which, of course, is proportional to the reflection coefficient, is derived from the junction of a pair of closely matched 75 Ω resistors forming one set of bridge arms. Care is taken to keep the wiring and the location of the components as symmetrical as possible with respect to earth. In spite of this, some residual capacitive unbalance is inevitable; this is compensated by means of the differential capacitor shown.

The return loss is measured by switching the output indicator, likely to be a high-impedance input of a waveform monitor, between the output of the bridge and the output of the attenuator until equality of reading is obtained. Preferably, the attenuator contains a fixed 6 dB pad, when the reading of the variable portion gives the return loss immediately. Otherwise, 6 dB will have to be subtracted from the reading.

8.4.1 Wide-range bridge

Whereas the performance of a transformer in a bridge such as that shown in Fig. 8.3 is acceptable when sinusoidal measurements are made, it is far less suitable for waveform measurements, owing to the greater bandwidth required. In particular, the very appreciably greater bandwidth required at the lower frequencies to avoid severe line tilt makes the achievement of an adequate high-frequency response more difficult and also considerably increases the cost.

However, two methods are available for improving the waveform response of the bridge with a saving in cost. The first replaces the transformer of Fig. 8.3 by a push–pull amplifier. Provided that a sufficiently good balance over the frequency range can be achieved, which will require some care, excellent results can be obtained.

The preferred method, however, recognises that the elementary bridge of Fig. 8.2 will still function when the test waveform is applied between points C and D, provided that the difference between the potential drops across AD and BD is used as the output. This is easy to manage by the use of a difference amplifier (often incorrectly called a 'differential amplifier'), which can be made without undue difficulty to provide a very high and stable balance over a wide frequency band. Only a single stage is required, since extra amplifications can be provided in unbalanced form after the difference amplifier itself.

A bridge has been constructed on this principle which will measure return losses of 15–65 dB over the frequency range 50–10 MHz with negligible waveform distortion. This is considerably better than can be achieved with a transformer. The accuracy has been optimised around 30 dB, where the greatest precision is usually demanded, and the error is then only $\pm 0{\cdot}2$ dB. With numerically increasing values of return loss, the error rises to $\pm 1{\cdot}5$ dB at 65 dB, which is very adequate for practical purposes.

The test signal (Fig. 8.4) is applied by means of a switch either to the input of an attenuator which feeds the centre point of the bridge, or to the input of a built-in calibrated attenuator which serves as the comparison standard. The values of the bridge resistors are chosen so as to terminate the input attenuator in 75 Ω. The output terminals of the bridge are applied to the two inputs of a high-grade difference amplifier with a very good and stabilised balance, which, in effect, acts as the equivalent of a very wideband, balanced/unbalanced transformer with a bandwidth from a few hertz to above 10 MHz.

The gain provided by the difference amplifier itself is relatively low; so its output is further amplified by a single-ended amplifier which, in the practical instrument, is provided with two useful facilities:

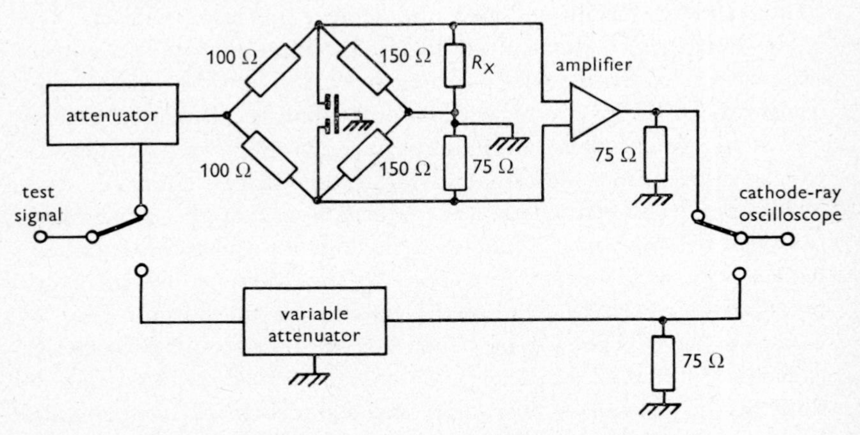

Fig. 8.4 Widerange transformerless bridge (simplified)

(*a*) a polarity-reversal switch which permits a positive-going output waveform to be obtained under all conditions, with an indication, at the same time, of the sense of the reflection; and (*b*) an auxiliary amplifier which is switched into circuit when the return loss is numerically greater than 30 dB, thus avoiding difficulties due to amplifier overloading.

The method of operation is to switch between the bridge, to which the unknown impedance is connected, and the measuring attenuator, until the two output signals, as observed on a waveform monitor, are of equal amplitude. The attenuator reading then gives the return loss directly. Calibration, when required, may be carried out using a standard resistor as the unknown impedance.

A useful feature of this type of bridge is the ease with which it can be improvised by making use of the difference-amplifier facility of a high-grade waveform monitor to replace the balanced amplifier. This arrangement cannot measure very high values of return loss, for which careful screening and balancing are required, but can easily cover the range most commonly met with.

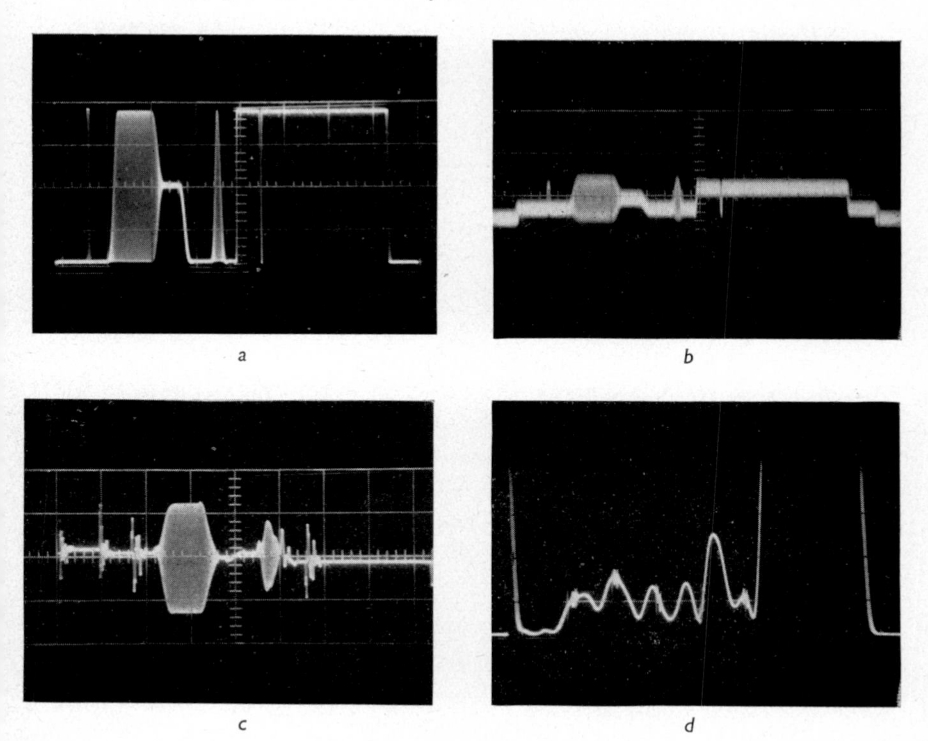

Fig. 8.5 Use of wide-range bridge

a Input luminance–chrominance test waveform
b Reflection from precision resistor: return loss = 57 dB
c Reflection from input of network: return loss at subcarrier = 18 dB
d Frequency sweep corresponding to (*c*): rectified

The method of operation of this bridge, and of the transformer type if its waveform distortion is sufficiently low, may be clarified with an example. Let the input test waveform be the chrominance–luminance pulse-and-bar signal of Fig. 8.5*a*, preferred since it provides an adequate test of the chrominance as well as luminance regions; even for monochrome operation, the behaviour of the upper part of the videoband can be examined without the spectrum of the test signal exceeding the nominal width of the videoband. If the impedance whose return loss is to be measured were a pure resistance, the displayed echo would have the same waveform as the test signal, but, of course, an amplitude reduced in proportion to the closeness of the resistance to 75 Ω.

Fig. 8.5*b* shows the waveform of the echo from an especially selected and mounted resistor. The measured return loss was 57 dB, and the amount of amplification which had to be introduced can be judged from the noisy appearance of the echo signal. The impedance of the resistor is not quite flat over the videoband, as is evident from the slight change in the relative amplitude of the chrominance component, but the error is commendably small.

An example in which the change in the impedance over the band is quite considerable is given in Fig. 8.5*c*, in which the test object was the input impedance of a correctly terminated network proposed as a signal delay for a video bandwidth of 5·5 MHz. The return loss is obviously at its poorest in the region of subcarrier frequency where the measured value was 18 dB, taken as the ratio of the peak-to-peak chrominance amplitudes in the input and echo signals. At the video frequencies corresponding to the upper spectral region of the $2T$ pulse, the return loss measured as the ratio of the input-bar amplitude to the peak-to-peak amplitude of the echo-signal pulse is about 3 dB better, and the echo-bar amplitude (the small pedestal on the right-hand side) gives the low-frequency return loss as 42 dB. It is suggested that situations of this kind should be dealt with by taking the worst value, in this case 18 dB, as the waveform return loss. Possibly, further experience may suggest that, for colour working, a weighting factor should be used for the chrominance region, to deal with a conceivable difference in the importance of the echoes in the luminance and chrominance regions; but experience to date seems to indicate that this weighting factor, if any, cannot be very great.

8.4.2 Frequency-sweep measurement

The return-loss bridges described above may also be employed when the test signal is a frequency sweep or a multiburst signal of appropriate form. Such a display of the return loss in terms of frequency, rather than of waveform, is sometimes useful in laboratory and investigational work. To provide a direct comparison, the measurement of Fig. 8.5*c* was repeated using a frequency sweep, and the result is given in Fig. 8.5*d*. It is evident that the rather poor return loss in the chrominance region is due to an impedance deviation which unfortunately happens to coincide with that important part of the spectrum; the large peak on the right-hand side can be seen to be located not quite symmetrically with respect to the 4 MHz and 5 MHz markers.

The very much improved return loss of the network at the lowest video frequencies, derived from the amplitude of the luminance bar in the corresponding echo waveform, is shown by the almost flat region of low amplitude at the extreme left-hand end of the trace; the large initial transient is due to the sweep generator, and is to be ignored. Finally, the somewhat irregular area of the trace between the 1 MHz and 4 MHz markers corresponds more or less to the distorted echo of the 2*T* sine-squared pulse in Fig. 8.5*c*.

It is tempting to compare the return-loss measurements by the sweep method with those by the waveform method by measuring the echo amplitudes in the various regions of Fig. 8.5*d*, but it should always be borne in mind that the amplitude scale in a detected sweep signal is usually very nonlinear as a result of the change in rectification efficiency with a change in level at the low voltages applied to the detector diodes. A detected sweep is often preferred because it is clearer and because the signal occupies such a small frequency band that any deviations in the frequency response of the oscilloscope are not impressed on the sweep waveform; if there is any trouble, it arises from the low-frequency behaviour of the instrument. Consequently, measurements on a detected sweep should always be made by comparison against the undistorted sweep which is detected in the same manner and at the same level.

Alternatively, if the waveform monitor has a flat frequency response over the required range, the unrectified sweep signal may be displayed directly as in Fig. 8.6, which gives the return loss as a function of frequency for a lowpass filter intended for use in the signal path. Nevertheless, care still has to be taken because, if the displayed signal is 'windowed' for more accurate measurement by increasing the gain, overloading effects may change the ratio of the higher to lower frequency gains. This is usually betrayed by a change in the slope of the

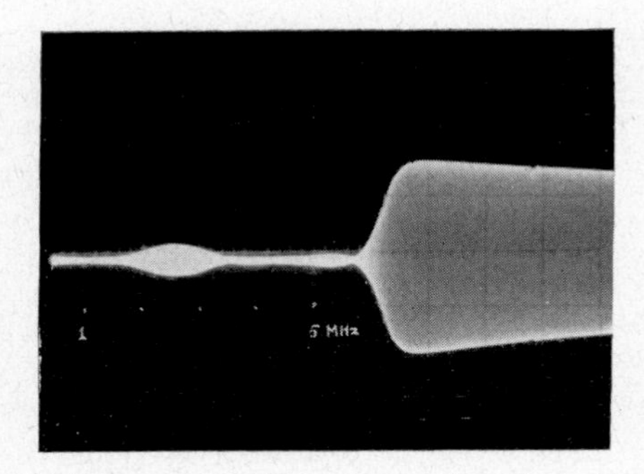

Fig. 8.6 Reflection corresponding to input impedance of lowpass filter

display as it is moved vertically over its maximum range by means of the vertical shift. In addition, rectification may introduce an asymmetrical type of distortion into the display.

Incidentally, the waveforms of Fig. 8.5*c* and *d* bring out very well the practical usefulness of a test signal such as the chrominance–luminance pulse-and-bar signal whose spectrum is restricted to the width of the videoband. The large transient at the right-hand side of Fig. 8.5*d* represents the expected massive echo at frequencies falling into the stopband of the network, which has a lowpass configuration. A very similar effect is also visible at the right-hand side of Fig. 8.6. If the spectrum of the test signal had extended into this region, the echo waveform would necessarily have included irrelevant information corresponding to the frequencies reflected in the stopband, which might very well have modified the measured value of video return loss.

The swept-frequency technique can also be used to furnish an accurate direct record of the return-loss/frequency characteristic. For this purpose the sweep generator is fed to the wide-range bridge described above, which is connected to the termination under examination, and the output from it is taken back to the sweep generator in the usual manner. The outputs from the sweep generator intended for the x and y inputs of the viewing oscilloscope are connected to the corresponding inputs of an x–y plotter. The desired frequency range is traversed either at a very slow rate, say 10 s per sweep, or manually.

By itself, this record would be of little use since the frequency and return-loss scales are not necessarily linear, but a simple extension of the technique permits the scales to be added. For the return loss, this is achieved by changing over from the 'measure' to the 'calibrate' function of the bridge, in which the echo attenuation is simulated by inserting a switchable attenuator in series with the frequency sweep. The run is then repeated with a number of appropriate settings of the attenuator, each of which traces a line of known constant return loss. The frequency calibration can be added by superimposing the frequency markers on a suitable member of this family of lines. At a later date, vertical lines may be drawn through these to form a frequency scale.

The author has found this device particularly valuable for the design of video networks. Fig. 8.7 shows the measured return loss of an experimental video delay line, and it is evident that the important values can be derived from this to a very adequate degree of accuracy for the purpose.

8.4.3 Comparison terminations

Bridge methods compare the unknown termination with a standard provided as a part of the instrument, and the accuracy therefore depends fundamentally on the accuracy of this standard. In bridges of this type designed in the author's laboratory, an axially cut metal-film resistor is used which has a resistance of $75\ \Omega \pm 0{\cdot}01\%$ at 20 °C. It is coaxially mounted in a socket identical with that used for the connection to the unknown so as to balance the impedance characteristic of the socket, and very great care is taken with the bridge layout.

8.5 Delay-cable method

The essential problem in the measurement of return loss lies in the separation of the reflected wave from the incident wave, so that the echo amplitude can be measured. A very simple method of bringing

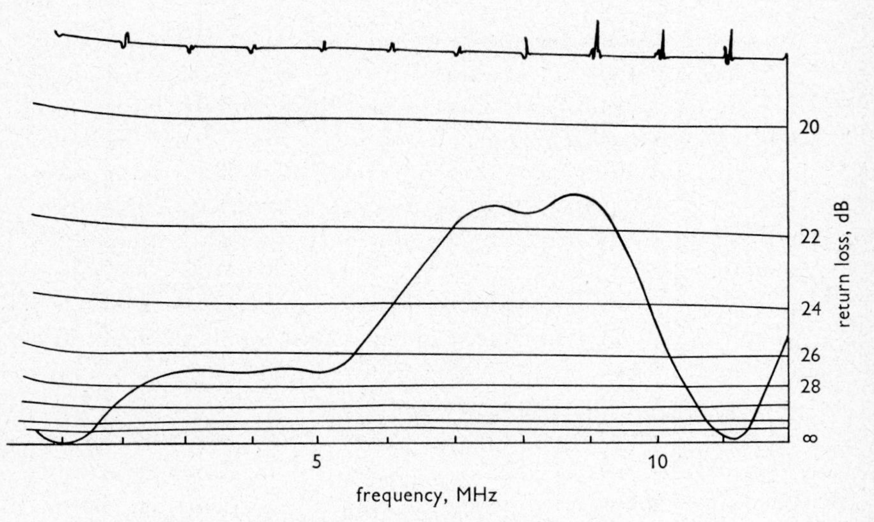

Fig. 8.7 Automatic plot of return loss of network

this about is the use of a cable to connect the test-signal generator to the unknown impedance which is electrically long enough to delay the echo by a time sufficient to provide a clear separation between the test signal and the echo.

Assume that a 1*T* pulse is used as the test signal. The overall width at the base for $T = 100$ ns is 200 ns, so that an absolute minimum delay of 200 ns is required. The delay between signal and echo is twice the signal delay of the cable, so that a length of cable equivalent to rather more than 100 ns is required, i.e. at least 70 ft (21 m) of good-quality flexible coaxial cable. A $2T$ pulse would require twice this length.

There are several objections to this method. First, the impedance presented by the far end of the cable to the unknown impedance is not precisely 75 Ω as it should be. Only a pulse type of test signal which has the minimum duration for the frequency spectrum concerned is of use, and, in particular, the use of a chrominance pulse is excluded since the delay required becomes impracticably long. Furthermore, the echo is distorted by its double traverse of the cable. Finally, large values of return loss are in any case difficult to deal with, since it is then necessary to measure a very small waveform in the close proximity of a waveform many times larger. The waveform monitor must consequently be capable of a considerable degree of 'windowing' without distortion. The method cannot therefore be recommended for general use.

8.6 References

ALLNATT, J. W., and PROSSER, R. D. (1965): 'Subjective quality of television pictures impaired by long-delayed echoes', *Proc. IEE*, **112**, (3), pp. 487–492

CCIR (1966): Documents of the XIth Plenary Assembly, Oslo, Recommendation 451

SHEA, T. E. (1929): 'Transmission networks and wave filters' (Van Nostrand), pp. 56–58

THIELE, A. N. (1964): 'Return loss bridge for video frequencies', Australian Broadcasting Commission Report 17

9 MISCELLANEOUS MEASUREMENTS

9.1 Time measurement

The principal time measurements required are

(*a*) risetimes of edges
(*b*) timings of waveforms including durations of pulses
(*c*) transmission time through circuits and apparatus

During such measurements, certain conventions must be observed so that the results are obtained in a standard form. For example, the risetime of an edge is defined in terms of the response to an ideal step waveform. The risetime is then defined as the interval between the points at which the output waveform reaches 10% and 90% of the difference between the original and final levels. This is a purely arbitrary definition, and has no theoretical significance.

The timings of waveforms and the durations of pulses are always defined in terms of the points at which the amplitude reaches 50% of the final amplitude, except that, with narrow pulses, the timing is taken from the peak of the pulse. For example, in the sine-squared-pulse and bar waveform, the timings of the component waveforms are measured with respect to the half-amplitude point of the leading edge of the line synchronising pulse, which is the normal reference point for a line waveform. The timing of the bar of the sine-squared-pulse and bar waveform is then measured between the timing reference and the half-amplitude point on the leading edge of the bar, and the duration of the bar is measured between the half-amplitude points on the leading and trailing edges. On the other hand, the timing of the sine-squared pulse is measured from the peak of the pulse, but the duration of the pulse is taken between the two half-amplitude points.

9.1.1 Measurement of risetimes

The method most frequently employed utilises the calibrated time-base of a waveform monitor, which must possess sweep rates fast enough for the duration of the transition to correspond to at least 1 cm of the screen width if the accuracy of the measurement is not to be impaired seriously. The spot size should be small enough to give a clearly defined trace and the sweep jitter should be negligible. This small spot size should preferably be combined with a brightness high enough to avoid the use of face-fitting hoods or any other device which acts as a constraint on the operator. With a high-grade waveform monitor, the accuracy of sweep calibration is not a great problem, since this particular measurement is not usually required to better than a few per cent. However, one must note that, at these high sweep rates, the calibration is most likely to be in error, particularly when the initial part of the sweep is in use and when the so-called 'magnifier' is brought into action.

Except in special cases, risetime measurements are not possible without a delaying sweep of some kind, since, with the high sweep rates needed for television waveforms, a trigger pulse occurring only a short interval before the beginning of the transition is indispensable. This may be derived either from a gated type of delaying sweep such as is fitted to many cathode-ray oscilloscopes, or from a television line selector, which may further be provided with special trigger pulses for waveforms such as the sine-squared pulse and bar (Chapter 6.2). In the former case, the oscilloscope must be provided with field trigger pulses, whether obtained from local sources or separated from the waveform under examination.

A delaying sweep or delayed trigger system, has the incidental advantage of making it possible to position the transition under examination in the centre of the screen, where trace linearity is likely to be optimum under all conditions. A great many errors in the measurement can also be eliminated, or at least substantially reduced, when this auxiliary sweep is calibrated, since one then only needs to use it to shift the transition between the 10% and 90% amplitude points and to note the change in delay; this is facilitated if the graticule has a central vertical line which is engraved with these points. Delayed trigger times can be made very closely proportional to setting of a variable resistor, and if a multiturn potentiometer or similar is used, the delay time can be considered as linear with rotation, which makes the measurement convenient and accurate

Such calibrated trigger delays or their equivalent are sometimes fitted as an integral part of some waveform monitors; if not, adequate examples are not unduly difficult to construct as an auxiliary for all time-interval measurements on television waveforms. The range of delay should vary from a few microseconds up to about one and a quarter line periods, assuming that the trigger is normally actuated by the leading edge of a line synchronising pulse, since this ensures that the whole of the line-blanking interval can be displayed.

For risetime measurement, the delay needs only to be calibrated over a very small range. This is most conveniently and accurately achieved if it is possible to insert, in series with it, a calibrated variable delay line with which the actual measurement is made. Such lines are discussed further in Section 9.1.3, and the method of calibration in Section 9.1.4.

Alternatively, an additional delayed trigger, which has been calibrated against a known delay line or a known pulse train, may be employed as the means of measurement.

Use is sometimes made of 'brightup' or '*Z*-axis' markers in the form of bright dots on the waveform monitor trace. These are not particularly convenient to apply to existing waveform monitors, since, to achieve sufficient accuracy with risetimes down to 100 ns, as is required in television, the cathode or grid of the cathode-ray tube must be driven with high-amplitude pulses which are short enough to locate points on the transition under examination with reasonable precision. However, this system has been used very successfully in commercial oscilloscopes where the pulse generation and application circuitry are designed as an integral part of the instrument. Instances are available in which a pair of these markers locate points at will on a displayed waveform, and the corresponding time interval is then given directly by a scale on the instrument, or even by a digital readout.

More sophisticated methods, principally for automatic monitoring of signals, are described by Steinmetz (1967). In each case, these depend basically on amplitude-discriminator circuits which generate short pulses when the amplitude reaches 10% and 90%, respectively, and can only be applied to repetitive waveforms.

One of these employs the 'counting down' facility, used in certain sampling oscilloscopes. Auxiliary waveforms are generated and locked to the input waveform, from which a very narrow sampling pulse is derived which either advances along the sampled waveform by a fixed amount in each successive cycle, or may also advance and at the same time skip an integral number of cycles.

The result is that the ensemble of samples of the waveform is only complete, and only represents one whole period, when, say, n periods of the input waveform have been counted. The period of the input has then apparently been stretched by a factor of n. It can further be arranged that samples occur at the 10% and 90% points to actuate amplitude discriminators and that a pulse is generated at each point. The time between the two pulses can then be measured with a counter, taking into account the 'stretch factor' of n. With the time stretch, a measurement is possible even for very short risetimes. Another method proposed by Steinmetz also utilises the 10% and 90% amplitude discriminators, but in this instance they are used to switch a bistable circuit so as to generate a rectangular pulse whose width is proportional to risetime. A train of such pulses can then be integrated and made to yield a voltage or current linearly proportional to risetime. In an example quoted, a risetime of 100 ns gave an output of 300 mV.

In every instance above, the measurement accuracy assumes the availability of a quasi-ideal signal with low distortion and low noise. In practice, of course, the signal is sometimes distorted so that it is difficult to decide precisely how the 10% and 90% points are to be derived, and it is also degraded by noise. In such situations, fully automatic methods tend to fail because it is not possible to build into them all the criteria which a human observer would use to obtain the best possible reading under the circumstances; the accuracy of measurement must inevitably be decreased.

9.1.2 Waveform timings

In general, these measurements follow the lines laid down in the previous Section, except that one no longer is concerned with the 10% and 90% points; the 50% amplitude points of transitions and the peaks of any pulses present are now the criteria.

For occasional measurements, a direct measurement from the screen of the waveform monitor is usually satisfactory, the accuracy of the sweep rate being checked if required. It is important to have a delaying-sweep facility available for this type of measurement, otherwise it may well be found that the area of the waveform concerned cannot be expanded sufficiently for any reasonable accuracy in the location of the 50% or peak-amplitude points. A calibration of the delaying sweep or delayed trigger is again advisable, but, unlike the measurement of risetimes, this must hold over a wide range, to accommodate time intervals as disparate as the front porch and the field-blanking region. It must also be possible to select points of interest in either field at will. These longer delays are very conveniently checked with a digital counter having a fast clock pulse, say, 10 MHz. If a sufficient number of successive readings are integrated by the instrument, the uncertainty can be less than 100 ns. With a suitable instrument and an integration time of a second or longer, the resolution of the measurement can be reduced to as little as a very few nanoseconds, but it is unwise to rely on such readings unless one is aware of how the counter functions with respect to the transitions of the triggering waveforms. A suitable interface unit could be designed if thought worth while; with this, an extremely large range of delays could be measured very accurately.

The accurate measurement of waveform timings is of no little importance for the standardisation of synchronising waveforms in studio centres and technical areas. Where such a measurement is carried out regularly, it may be worth while to ensure that the required accuracy is obtained reliably and with the minimum of time and effort. A specialised instrument of this type has been designed in the author's laboratory. It has the advantages of a digital readout and a simple and accurate method of locating the 50% amplitude points on the transitions of the waveform under measurement.

The latter is carried out by the linear addition to the waveform under test of an auxiliary rectangular pulse which must be perfectly symmetrical and must also have a risetime much shorter than that of any of the waveforms to be measured; with television waveforms, these conditions are by no means difficult to meet. The edges of this pulse can be shifted independently in time, and its amplitude is made variable at will.

A typical condition which may occur before the start of a measurement is shown in Fig. 9.1*a*. The auxiliary pulse happens to be occurring appreciably later than the synchronising pulse, the duration of which is to be measured. By the use of a variable control, the leading edge of the auxiliary pulse can be placed centrally over the leading edge of the synchronising pulse, which results in the double-peaked pulse seen on the left-hand side of Fig. 9.1*b*, and in more detail in Fig. 9.1*c*. The two trailing edges may also be superimposed with a similar result. In general, a step will be seen between the two double-peaked pulses, which can be removed by adjusting the amplitude of the auxiliary pulse to equal that of the pulse under measurement. The result will be similar to Fig. 9.1*b*.

It can readily be shown that the amplitudes of the positive and negative peaks of the double-peaked pulse are equal only when the 50% amplitude points of the two pulses are in coincidence, a fact which provides a sensitive criterion for the adjustment of the duration of the auxiliary pulse to precise coincidence with that of the pulse under measurement. For this, two conditions have to be fulfilled: the auxiliary pulse must be perfectly symmetrical, and the portion of the measured transition intercepted by it around the 50% point must be skew-symmetrical. The first condition is achieved in the design, and the second is met with sufficient accuracy in practice by ensuring that the ratio of the transitions is sufficiently high, say, 10:1 or more. This additionally confers a good sensitivity on the adjustment.

The superposition of the leading edges is carried out by means of a continuously variable control, and that of the trailing edges in steps of 1 μs together with a vernier providing 25 ns steps, the information from these being utilised to operate a digital display which then gives the duration directly in microseconds to the nearest 50 ns. Durations of either polarity up to almost a complete line period can be measured quickly and reliably.

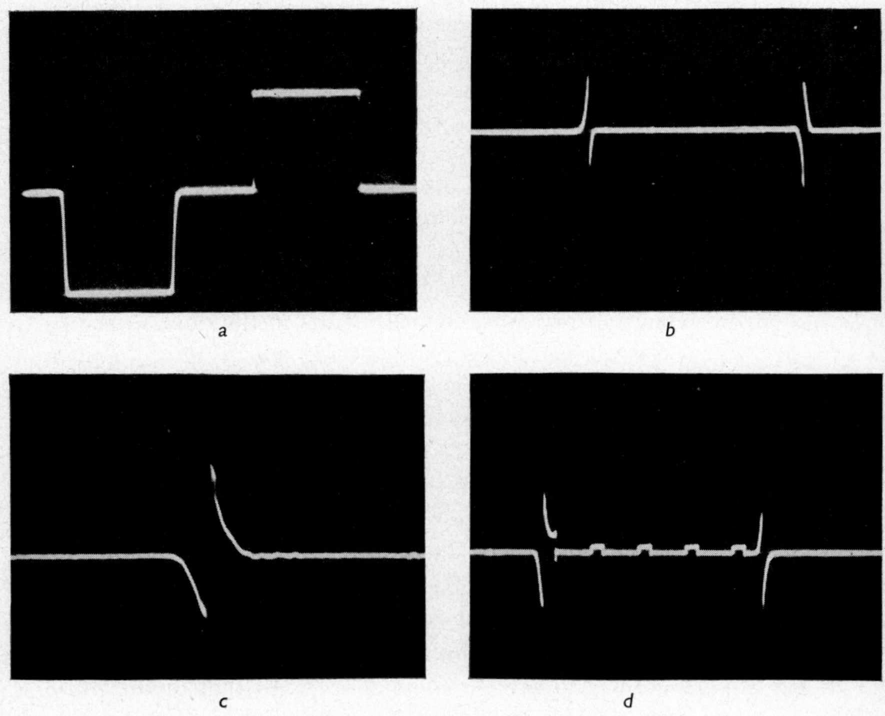

Fig. 9.1 BBC method of measuring pulse durations

a Synchronising pulse with added auxiliary pulse
b Adjustment of duration of auxiliary pulse
c Detail of double-peaked transition waveform
d (*b*) with 1 MHz clock pulses added

The 1 MHz clock frequency is generated by multiplying the line-repetition frequency by 64, thus ensuring that the two are constantly locked. This also affords a simple means of shifting the leading transition of the auxiliary pulse without the necessity for re-establishing the counter zero. In Fig. 9.1*d*, the 1 μs pulses have been added to the display for purposes of illustration. In this example, four pulses are visible between the two double-peaked pulses, the synchronising pulse duration as given by the readout being 4·70 μs. Since the vernier control has only 25 ns steps, some lack of symmetry in the second double-peaked pulse is to be expected; in this instance, it corresponds approximately to 10 ns.

A display of the type of Fig. 9.1*d* could replace the digital readout if simplicity were important, but means would have to be provided to reduce errors with long durations, such as marking every tenth pulse.

9.1.3 Transmission times

If the aim is to ensure synchronism between video waveforms arriving at the same point, it is advisable to utilise measuring waveforms having a standardised form and the same bandwidth as normal video signals. The reason is that, if the bandwidth is less than the nominal bandwidth of the video channel, precision is lost because of the increased time of rise and fall of the transitions, whereas, if it is wider than necessary, the distortion introduced by any frequency limitation occurring in the channel may well cause an apparent shift in the transition from which the time of arrival at the output is measured.

Very suitable waveforms for the purpose are the various sine-squared-pulse and bar waveforms described in Chapter 6. In particular for colour purposes, the use of one of the combined chrominance and luminance waveforms makes it possible to investigate any difference between the transmission delays of the luminance and chrominance channels.

When the two ends of a circuit are simultaneously available, the measurement of the transmission time does not present any notable difficulty. The simplest method employs a waveform monitor to measure the displacement between the input and output waveforms, the accuracy being improved somewhat by the use of an accurate calibrating waveform to check the monitor sweep.

A somewhat more sophisticated technique of this type utilises a counter for the accurate measurement of the time interval, the oscilloscope merely indicating coincidence. It is more suitable for delays greater than, say, 500 ns, since, with very short delays, the inaccuracies of setting become an increasing percentage of the time measured.

The basis is a generator furnishing either very narrow pulses or pulses with very fast edges, whose period can be measured with an electronic counter; this is the waveform shown in Fig. 9.2*a*. It is displayed on one of the two sweeps of a dual-trace oscilloscope.

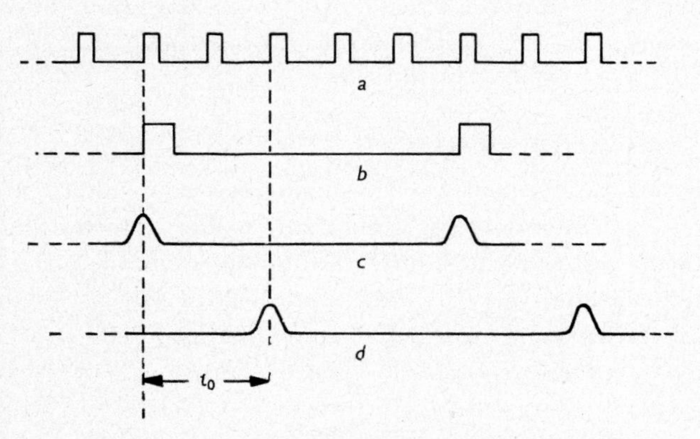

Fig. 9.2 Measurement of transmission times

a Initial pulse train
b Frequency-divided waveform
Divided by $n = 5$
c Input pulses
d Output pulses

This pulse train is further employed as the input to a divider which yields a similar pulse train having a period multiplied by a suitable integer; from the latter is derived a series of sine-squared or similar well shaped pulses whose bandwidth is approximately equal to the nominal video bandwidth, assuming that the circuit under test is to be used for the transmission of video signals. During the initial adjustments, these pulses are displayed on the second trace of the oscilloscope. A reference time relationship is then set up between the two sets of pulses, e.g. coincidence of the peak of a sine-squared pulse with the leading edge of one of the original train of pulses (Figs. 9.2*a* and *b*). To achieve this, it will be necessary to insert a small variable delay in series with one of the two displayed signals. The trigger for the oscilloscope sweep is taken from the original set of pulses.

The sine-squared pulses are now applied to the input of the circuit under test, and the output signal is displayed on the second sweep of the oscilloscope, the trigger being taken as before from the original pulses. There will now be a time displacement between the two sets of pulses; but, by suitably varying the repetition frequency, the original condition may be restored except for a shift of an integral number of periods of the original pulses. In Fig. 9.2, this has been taken as 5 for simplicity, but in practice it will usually be considerably larger. A measurement of the frequency or period of these latter pulses with a digital counter will then give the transmission time with a good accuracy. A variant of this technique has been used by Royle (1969) for the measurement of pulse durations.

This method is very suitable for circuits which do not require the test signal to be provided with synchronising signals, and although the latter may be added to the trains of pulses if desired, the complication may be thought not worth while. A completely versatile technique which may be employed instead involves the use of a calibrated delay box, so arranged that any given delay up to the maximum provided may be switched in as required. This, incidentally, is a most valuable device, and can be utilised in a wide variety of ways. Delay lines very suitable for this purpose are described by Poullis, Weaver and Cooke (1970). With these very high-quality delays, up to 4 or 5 μs can be realised, switchable in steps of 5 ns; and, with a slight degradation in performance, considerably longer delays can be achieved.

In general, the following is a useful guide to the construction of a calibrated delay box (methods of calibration are given in Section 9.1.4):

(*a*) The quality of the component lines should be such as to give low distortion on a sine-squared-pulse and bar waveform. In particular, marked overshoots on one side only of the pulse are to be avoided as potential sources of error. Constant-resistance sections can be employed to equalise these lines; if so, placing them between the lines may improve the matching.

(*b*) The return losses of the individual sections should be adequate to ensure that any additional waveform distortion due to reflection effects is negligibly small for the purpose under any conditions of interconnection.

(*c*) It is essential to provide an internal reference path with respect to which the box is calibrated. If any of the delay lines have appreciable attenuation, it is useful to balance this by inserting an equal loss into the reference path.

(*d*) An input socket for the test waveform should be provided which feeds both the reference path and the calibrated delay path either from a constant-resistance splitting pad or from a high-quality distribution amplifier.

To make a measurement, the circuit under test is connected to the reference path, and the calibrated delay is adjusted until the two sets of output pulses are exactly in time coincidence, which experience has shown is best judged by observing on a waveform monitor the point at which the sum of the two output waveforms is a precise maximum. This addition must be carried out by means of a pad or such other means as will still correctly terminate the two outputs. Subtraction of the two waveforms in a difference amplifier will give rise to an error if their amplitudes are not exactly equal. Due allowance must be made for the delays of the connecting leads which, provided that they are short, may be done by measuring their lengths. The velocity ratio of flexible coaxial cables is usually very close to 0·66, and the transmission time of an electromagnetic wave over 1 ft (30 cm) of free space is 1 ns. Hence one can usually assume that each foot of connecting cable contributes 1·5 ns of delay.

A very special problem occurs when both ends of the circuit to be measured are not simultaneously available. If only one such circuit exists, the information required cannot be obtained unless it happens to be bilateral, e.g. a cable, when the delay between the input waveform and the 1st-order echo from an open- or short-circuited far end provides a figure for twice the transmission time.

On the other hand, if at least one other circuit exists between the two points, the transmission time of each can be measured individually, provided that (*a*) one of them contains no one-way devices, or (*b*) such one-way devices as may be present can be reversed in one of the circuits without materially altering its transmission time. For this, two of the circuits are connected in series and the total transmission time, say $t_1 + t_2$, is measured. The two circuits are then fed in parallel with the test waveform, and a measurement is made at the receiving end of the difference in times of arrival of the waveforms at the two ends of the circuits. This will be $t_1 - t_2$, so that two equations are available from which the two unknowns can be found. If there are a number of such circuits, they may be measured in this way in succession against the same reference circuit.

9.1.4 Special methods

The techniques described here for the measurement of transmission delays are more applicable to the standardisation and calibration of networks and equipment than to operational practice, although their use is not excluded elsewhere, and, for that reason, they are grouped separately.

The first of these differs radically from the others in that it is not an analogue method. The latter type of measurement, in which the video signal is simulated by a standardised, stylised signal such as the sine-squared pulse, has considerable advantages since it makes a minimum of assumptions, provided that the signal is suitably chosen. Nevertheless, a measurement in terms of the insertion phase characteristics of the network or circuit under test has advantages for certain applications.

The quantity usually taken to decide the time of transmission of a complex waveform through a network is the group delay, often also termed envelope delay; but, since this is always to some extent a function of frequency, it is necessary to look more closely at what is meant by group delay in the present context.

The first difficulty encountered is the need to specify what is meant by the position of a test pulse, such as the sine-squared pulse. In the analogue methods, the peak of the pulse is always used, since it is the portion of the waveform which is most readily identifiable and which can be located within quite close limits; moreover, it corresponds closely to the position of such a pulse as located visually on the screen of a receiver or monitor. Unfortunately, this is difficult to correlate with the group-delay characteristic.

The most acceptable alternative is the time co-ordinate of the centroid of the pulse, the position of which can be shown to be determined by the value of the group delay at zero frequency. The writer is indebted to G. G. Gouriet of the BBC Research Department for the following very brief and elegant proof, given in a private communication.

In a linear system with transfer function $f(p)$, p having the usual meaning of $j\omega$, the group delay is given by Gouriet (1958):

$$\tau(\omega) = \mathrm{Re}\frac{f'(p)}{f(p)}$$

in which $f'(p)$ is the first derivative of $f(p)$.

If, now, $f(t)$ is the response of the system to a unit impulse,

$$f(p) = f(t)\,e^{-pt}dt$$

and, accordingly,
$$\tau(\omega) = \frac{\int_{-\infty}^{\infty} tf(t)\,e^{-pt}\,dt}{\int_{-\infty}^{\infty} f(t)\,e^{-pt}\,dt}$$

from which
$$\tau(0) = \frac{\int_{-\infty}^{\infty} tf(t)\,dt}{\int_{-\infty}^{\infty} f(t)\,dt}$$

The right-hand side of this expression is clearly the time co-ordinate of the centroid of $f(t)$, which may be taken to be the test pulse. In other words, the group delay at frequencies close to zero gives the shift of the centroid of the pulse, which for symmetrical pulses is also the displacement of its peak. The two, however, will no longer be identical if the system causing the time delay introduces asymmetrical distortion, although, in high-quality video systems, such distortion should be negligibly small.

The most obvious application of this principle is the direct measurement of the network or circuit with a group-delay measuring set; but such a specialised equipment may not be available.

An alternative, valid if the delay is relatively great and at the same time if the group-delay characteristic is flat at the lower video frequencies, is attractive because it requires much less specialised apparatus and the precision of the measurement is also quite high. The technique consists in the determination of the frequency at which the insertion phase shift through the network or circuit becomes precisely 180°, indicated by the lowest frequency at which the Lissajous figure degenerates into a straight line, apart, of course, from the trivial case of zero frequency (Section 9.4).

The apparatus required consists of an oscilloscope having adequate and approximately equal gains on both the x and y channels, an oscillator and a frequency counter. It is very desirable for the output of the oscillator to be provided with a lowpass filter with a suitable cutoff frequency to improve the waveform purity.

The output of the oscillator is split into two 75 Ω paths by means of a constant-resistance splitting pad or a distribution amplifier, which initially are connected directly to the two terminated inputs of the oscilloscope. By inserting a small delay element in one of the paths, or by adjusting the lengths of the leads, the trace is made a straight line, and should remain so over the frequency range concerned. Because of the harmonics present, omission of the lowpass filter is likely to give a trace of a 'loop' or 'butterfly' shape, which is useless for adjustment purposes.

The network or circuit to be measured is then inserted into one of the paths, and the first 180° point is found; it is assumed that, in so doing, the test object is correctly terminated. The counter is then used to measure the frequency, say, f_1. Then the phase delay, equal to the group delay if the phase characteristic is linear over the frequency range concerned, is given by $t_g = 1/2f_1$.

The reason for the restriction of the method to long delays is now clear: even at $t_g = 1\ \mu s$, f_1 is as high as 500 kHz, which evidently reduces the likelihood of the phase characteristic remaining flat. More correctly, the requirement of the measurement to be valid is that the slope of the chord drawn from the origin to the 180° point on the insertion phase/frequency curve should be equal to the slope at the origin; evidently, this is more likely to be true the higher the delay and the lower the value of f_1, although, of course, not necessarily so.

Such errors can be lessened, and at the same time the measurement of shorter delays may be undertaken, if the unknown is connected in series with an auxiliary network whose delay is known to be relatively high and also uniform from zero to some sufficiently high frequency. After the measurement of the two in series, the auxiliary network is measured by itself, and the delay of the unknown is obtained by subtraction. Care must be taken that all networks are correctly terminated at all times, and it is wise to insert an amplifier with impeccable input and output impedances between the auxiliary network and the unknown, this amplifier being considered as part of the auxiliary network and measured with it.

There appears to be an advantage in providing an auxiliary network with as high a delay as possible, to reduce the value of f_1 to a minimum; but, on the other hand, the likely error in setting the Lissajous figure to the straight-line condition, which is a constant phase angle for a given display and amplifier gains, represents a larger error in delay, the lower f_1 becomes. For example, the author has constructed an auxiliary network with a delay of 5 μs, maximally flat to a little over 100 kHz, the highest frequency at which it could be used for measurement. Networks of this type with a flat delay over a quite restricted frequency range are not unduly difficult to design. It is shown in Section 9.4 that, with moderate amplifier gains, a setting accuracy of about 0.1° is likely to be achieved, which, at 100 kHz, represents a delay error of 3 ns in round numbers. Hence, since two separate measurements are involved, a maximum error of 6 ns is possible, which, for the measurement of very short delays, might be considered excessive. On the other hand, as, in fact, in the example cited, the amplifier gains could well be considerably higher than the assumed values, which are quite conservative, with a corresponding improvement in the accuracy.

An especially powerful application of the Lissajous-figure technique is the adjustment of the delays of sets of networks to equality, where it has the advantages of simplicity, sensitivity and reliability. It has been found practicable to employ it for the manufacturing alignment of groups of matched filters for colour cameras, for example, where it was required to equalise the delays to ± 1 ns.

The second of these special techniques is particularly suitable for the measurement of video delay lines and similar networks, and can cope with a wide range of delay times. It was originally devised by the author (Weaver, 1968) for the production testing of preset video delay lines for source synchronisation in television studios, but it has also found several other applications. Under optimum conditions, the measurement is repeatable to 0·1% of the measured value, but the accuracy is to some extent dependent on the bandwidth and the delay of the networks measured.

It is an analogue method, which has the advantage that it makes no assumptions about the shape of the phase/frequency curve, but with the unusual feature that the pulse used is transmitted repeatedly through the network under test, for which reason it is known as the 're-circulated pulse' method. The basic circuit is quite simple, and can probably be put together almost on the spur of the moment in most laboratories and test rooms dealing with television equipment. In the form described it gives excellent results, but fairly obvious refinements may be added if desired.

The heart of the circuit is the closed loop on the right-hand side of Fig. 9.3, consisting of a pair of 75 Ω constant-resistance splitting pads S_1 and S_2, a variable attenuator, a high-grade video amplifier and two delay networks of which one is the unknown. These are so disposed that, provided that the amplifier has input and output impedances of high return loss against 75 Ω, both networks are properly terminated.

The variable-frequency pulse generator on the left of Fig. 9.3, which may be a triggered pulse generator controlled from a video oscillator,, supplies a train of narrow initiating pulses via an attenuator to the pad S_1, at which point they enter the loop. These pulses must be short in duration compared with the delay to be measured, and well-shaped; in most instances, it is advisable to start with a very narrow pulse and to shape it with a Gaussian filter (Dishal, 1959). The frequency or period of the pulse train is measured with an electronic counter while its waveform is displayed on the first trace of a dual-trace oscilloscope.

The attenuation within the loop is adjusted so that the net loop gain is a little less than unity; the value is not particularly critical, and it is not advisable to try to approach the unity-gain condition too closely. As a result, each initiating pulse gives rise to a slightly damped train of pulse within the loop having a pseudoperiod equal to the total transmission time around the loop (Fig. 9.4).

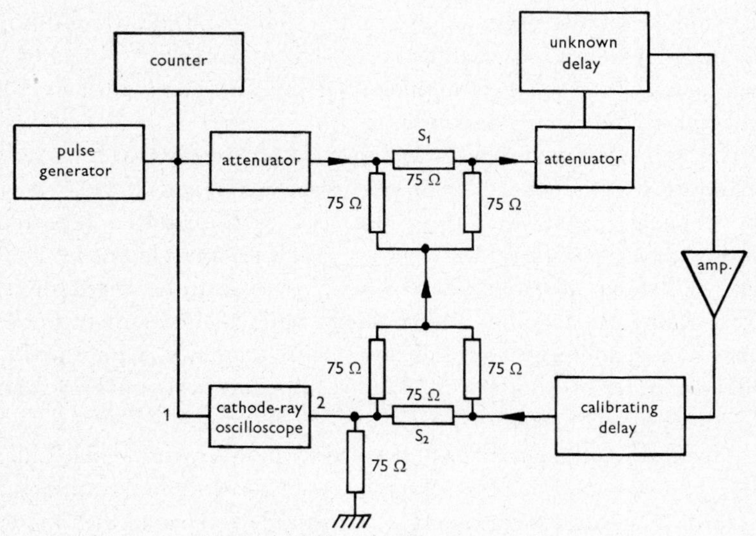

Fig. 9.3 Video-delay measurement

By adjusting the pulse-repetition frequency, it is possible to cause each successive initiating pulse to coincide precisely with one of the recirculated pulses, the interval between them being an integral number, say n, of the recirculated pulses. This is indicated by each nth pulse attaining its maximum amplitude, a condition shown in Fig. 9.4 for $n = 20$. The sum of the initiating-pulse train and the recirculated-pulse train is displayed on the second of the oscilloscope traces, so that a comparison of the two, as illustrated, enables the value of n to be found. It is not critical, and need not be unduly large; the number shown in Fig. 9.4 was deliberately made fairly large to illustrate the shape of the recirculated-pulse train. Then, if the period of the initiating-pulse train is indicated by the counter as t_p, the time of transmission around the loop is clearly t_p/n.

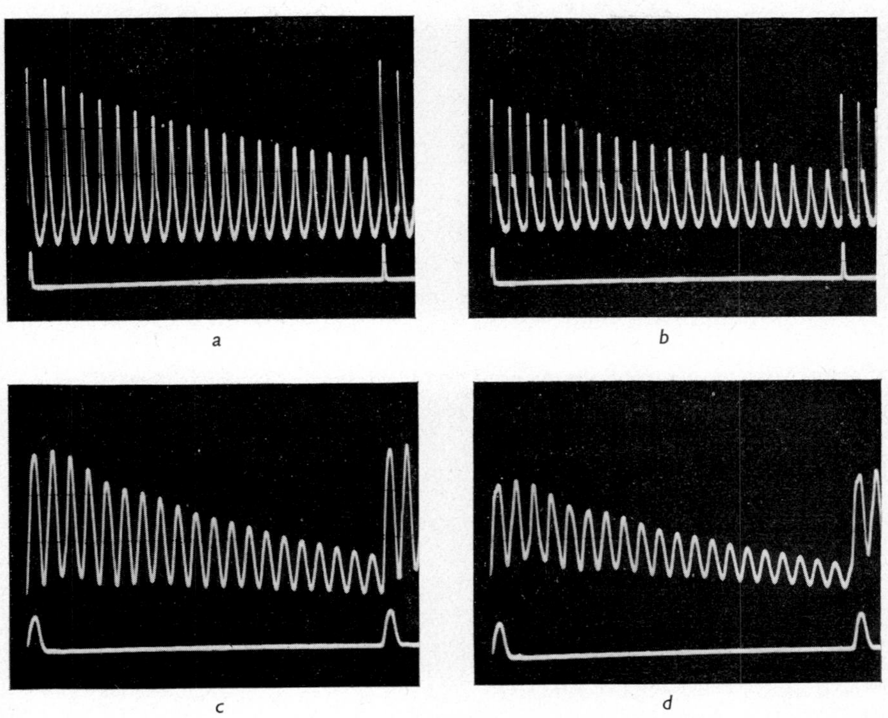

Fig. 9.4 Waveforms of video-delay measurement

a Correct adjustment ($n = 20$)
b Frequency shifted from the condition of (*a*) by 1 %
c Slight interference between pulses
d Severe interference between pulses

The first measurement, of course, does not give the time of transmission through the unknown but the sum of the delays of the unknown and calibrating networks together with the contributions from the leads, the amplifier and the various residuals, all of which are too small to be capable of accurate measurement without the addition of the calibrating network. A second measurement is accordingly made with the unknown removed, and the difference of the two readings is taken as the time of transmission through the unknown.

The accuracy and sensitivity of this technique are determined very largely by the product of bandwidth and delay for the networks concerned, which means that it is eminently suitable for video-delay lines with low distortion. The order of sensitivity to be expected under suitable conditions can be assessed from a comparison of Figs. 9.4*a* and *b*, in which the former shows the correct adjustment and the latter a deliberate mistuning by slightly less than 1%. The decrease in the combined amplitudes of the superimposed pulses and the change in the shape of the remaining pulses are very evident.

The most likely cause of error is interference between successive recirculated pulses as a result of one or more of the following:

(*a*) an initiating pulse with too great a width
(*b*) too low a delay around the loop
(*c*) too low a bandwidth in the networks or amplifier causing an undue increase in the recirculated-pulse width.

The presence of such interference is quickly betrayed by a lack of flatness of the baseline of the pulses; Figs. 9.4*c* and *d* show slight and more severe forms of this distortion, respectively, which have been deliberately brought about by an increase in the width of the initiating pulse. The most general remedy is an increase in the delay of the calibrating network, assuming adequate bandwidth, which increases the pseudoperiod of the recirculated pulses; it consequently plays a dual role in the method.

The above technique is the most basic, but it has nevertheless been found to operate extremely satisfactorily. Fairly obvious modifications and improvements may be made, taking care not to impair the accuracy; e.g. the counter may be connected to a point within the loop to read the time of transmission around the loop directly, provided always that it is not affected by the change in the shape from pulse to pulse. In the circuit of Fig. 9.4, the pulses applied to the counter have a standardised shape.

9.2 Jitter

The measurement of jitter is a particular case of time measurement, but the techniques used are very different from those described above.

The most general definition of jitter is an unwanted and variable difference between the expected and actual instants of occurrence of a repetitive event. It may be quasiperiodic or random in nature; most usually, it is a combination of the two.

If the jitter is due to purely random causes, the distribution of the variations in time will be Gaussian, and, for that reason, Mantel (1966) has proposed that it should be stated in terms of the standard deviation of the error. However, not only is this difficult to measure, but it is by no means certain in general television work that such an assumption is justifiable. It is more usual, therefore, to quantify jitter in terms of the peak-to-peak or quasi peak-to-peak variation of the event under observation. Admittedly, when the jitter is mainly or entirely random in character, the quantity measured is not unambiguously defined; but then, in television practice, it is very unusual to require the amount of jitter to be known very precisely.

9.2.1 Direct measurement

The measurement of jitter by triggering a waveform monitor from each successive edge of the waveform under observation is only likely to be successful, assuming that a delaying sweep is used to obtain the display, if the time between the actual triggering edge and the transition under observation is very short; otherwise, the delaying sweep will add its quota of jitter to that already existing on the transition. The importance of this will depend on the type and quality of the delaying-sweep circuitry and on the desired accuracy. Furthermore, one has to be assured that the waveform-monitor trigger circuits are accurately following rapid jitter.

A method for measuring the total peak-to-peak jitter which makes rather fewer assumptions utilises the so-called 'flywheel' technique to regenerate from the input signal a train of line trigger pulses which have been freed from the superimposed jitter, and employs these to trigger the waveform monitor which is displaying the transition to be measured. In addition, a 'brightness strobe' replaces the delaying sweep. The apparatus is naturally much more complicated; but, if only very occasional measurements are needed, it may be possible to improvise a large proportion of the circuitry from equipment already available. Fig. 9.5 shows one version.

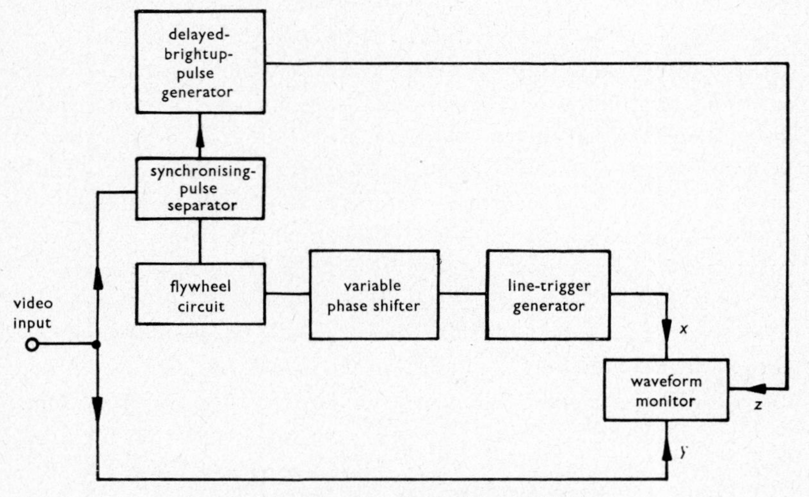

Fig. 9.5 Flywheel method of measuring jitter

The input signal, either a video signal or one of the outputs of a synchronising-pulse generator, is split into two paths. In the upper path, the synchronising information is separated from any picture information present, and passed through a narrow-bandpass filter which extracts the sinusoidal fundamental line-frequency component, at the same time removing the phase-modulation sidebands corresponding to the jitter. Alternatively, it may be more convenient to employ the separated synchronising pulses to drive a more conventional type of flywheel circuit, such as a blocking oscillator or locked oscillator, which then requires much less efficient filtration to yield the line-frequency fundamental component.

The separated line-frequency fundamental component is passed through a variable phase shifter, and is finally utilised to regenerate line-frequency trigger pulses, now completely free from jitter, which trigger the waveform-monitor sweep. The phase shifter positions the displayed waveform, since, for the measurement of jitter, the fastest sweep rate will have to be used and any shift control must itself be inherently free from jitter.

The display thus produced will be composed of the superposition of all leading edges of the line synchronising pulses, say, in a given field. However, it may be required to examine in detail the jitter of a

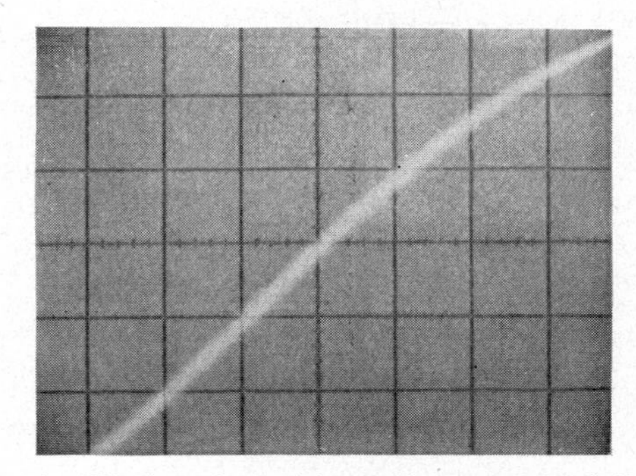

Fig. 9.6 Display of transition with 1·5 ns pk–pk jitter

given edge or series of edges. This may be achieved, again without adding to the jitter, by the use of a 'brightness strobe', i.e. a pulse of variable width whose delay from a given field pulse may be chosen at will, which is applied to the cathode of the cathode-ray tube and which brings the trace up to full brightness only for the duration of the pulse. This has been very much the same effect as a delaying sweep, with the exception that any positional errors of the brightup pulse will not add to the jitter on the displayed edge since it only shifts the displayed area slightly.

At the high sweep rates necessitated by the measurement of small amounts of jitter, the visibility of the trace is naturally rather low, and the outer limits of the random positional movement of the trace are not easy to determine by inspection; so it is usually more satisfactory to photograph the trace with a superimposed graticule for calibration purposes. Fig. 9.6 gives an example, in which the sweep rate was effectively 10 ns/cm, the highest sweep rate of 100 ns/cm being supplemented by a × 10 magnifier. The intercepts made by the trace on the horizontal and vertical graticule axes are, fortuitously, both equal to about 2·5 mm. The horizontal component is therefore 1·8 mm, which could reasonably be rounded off to the next lower 0·5 mm, giving an estimated jitter of 1·5 ns.

The exposure in this instance was 4 s with a film of speed 3000 ASA. The boundaries of the trace movement would be better defined with a longer exposure, but the use of the magnifier at the maximum brightness afforded by the cathode-ray tube, somewhat aggravated by the brightness strobe, tends to cause background illumination of the screen, which reduces the contrast as the exposure is increased, so that the optimum exposure may have to be found by experiment.

9.2.2 Discriminator measurement

Jitter in a waveform transition corresponds to a random positional modulation of the transition, so it should be possible to separate the random from the nonrandom components with a phase discriminator. Such a device has been described by Grünewald*; it is intended for the direct display of the jitter of a synchronising-pulse generator either on an oscilloscope as a waveform or alternatively as a reading on a meter.

As in the method described above, a 'clean' version of the input line-frequency synchronising information must be derived from the signal, but the problem is rather more severe here since no irregularities at all are permissible in these reference pulses, whereas the previous method required clean triggering edges but was otherwise relatively tolerant, particularly to vestiges of the halfline and equalising pulses. In the discriminator method, these would contribute to the reading.

The synchronising-pulse generator (Fig. 9.7) is consequently more elaborate than usual, since it also contains circuitry for removing the halfline edges. This is achieved first by extracting, from the input

* GRÜNEWALD, A. (1962): 'Gerät zur Messung horizontaler Bildschwankungen beim Fernsehen', Bericht 63, Institut für Rundfunktechnik, Munich (unpublished)

pulses, the fundamental line-frequency component by rather less stringent filtering than was required above. The synchronising pulses are differentiated and then added to this line-frequency sine wave with a phasing such that the line-frequency spikes are situated on the peaks of the sine wave. This implies that any halfline frequency information must be located in the troughs, so that, if the line-

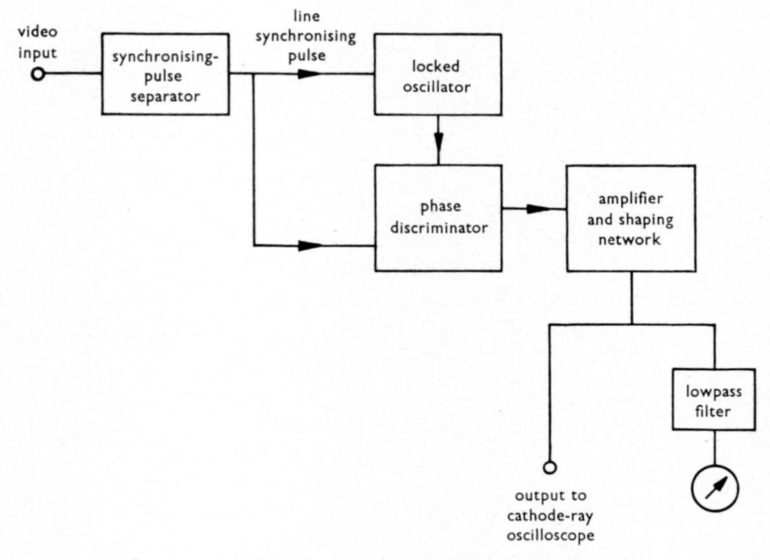

Fig. 9.7 Discriminator method of measuring jitter

frequency spikes are clipped off for conversion into synchronising pulses, the halfline edges are completely removed.

The purely line-frequency pulses thus derived still contain the jitter, and can consequently be employed as one of the inputs to the phase discriminator. They are also utilised to lock a flywheel oscillator of very high inertia, the output of which is a sinusoid free from the jitter components and consequently suitable for use as the reference input to the phase discriminator.

The output reading is obtained from the demodulated jitter in either of two ways. The first is a direct display of the waveform on an oscilloscope with the frequency weighted by a slope of 6 dB per octave to convert the peak-to-peak amplitude of the display into peak-to-peak jitter. In the second, the waveform is detected and the resulting current is employed to drive a meter. The calibration is then in hertz per second, i.e. in terms of the time rate of variation of the line-frequency synchronising pulses, which, according to Grünewald (loc. cit.), is a measure of the visual impairment of the resulting picture. One precaution has to be taken, however; i.e. to pass the demodulated waveform through a 35 Hz lowpass filter before detection, to remove any 50 Hz components resulting from the field information in the line pulses. This is unnecessary with the visual display since the eye is able to discriminate between the demodulated jitter and the 50 Hz 'spikes'.

An instrument of this type is rather specialised and inflexible, in that only the total jitter of the line synchronising pulses is indicated, and it is not possible to investigate the performance of any selected edge. On the other hand, it is very useful, both as a speedy method of checking synchronising-pulse generators in a standardised manner and, since a current which is a function of the jitter is available at the output, as a means, for example, of driving a pen recorder or similar instrument for the investigation of a generator during acceptance testing.

9.3 Frequency measurement

By far the most important measurement of this type in modern video practice is the measurement of colour-subcarrier frequency. The working tolerance on the nominal frequency is not everywhere the same, but, at the time of writing, the standard adopted in the UK is 4 433 618·75 Hz $\pm$ 1 Hz, say 2 parts in 10^7 in round numbers. An accurate means of checking this frequency to the standard required is desirable at all points where signals are originated.

The most obvious method is the use of an electronic counter for a direct measurement of the frequency, but this requires the counter to possess an internal reference oscillator with an uncertainty, at the time of measurement, of a few parts in 10^8; and, since a large part of the cost of a counter of high accuracy stems from the expense of providing the reference oscillator together with a display with an appropriately large number of digits, such counters can hardly be inexpensive.

However, experience has shown that it is perfectly possible to make extremely accurate measurements of subcarrier frequency by an adaptation of a counter of quite modest performance and cost. The conversion depends on two facts, often overlooked. First, a frequency standard some orders of magnitude better than is required for the desired accuracy is available in most countries in the form of the carrier frequency of a standard-frequency radio transmitter. The UK is fortunate in possessing the high-powered Droitwich transmitter with the remarkably convenient carrier frequency of 200 kHz, which maintains its frequency on a long-term basis within ± 2 parts in 10^{11}, and ± 1 part in 10^{11} daily. Secondly, many counters will 'over-range' or 'overspill'; i.e. if the number it is called on to display has more digits than can be accommodated, the surplus will be dropped without detriment to the displayed number, apart, of course, from the loss of the digits. Hence it is possible to utilise a counter with a small number of digits in the display to indicate, not the initial part of the number, which is redundant in any reasonable operation, but the significant later portion which shows the count of hertz and tenths of a hertz, with a corresponding saving in expense.

Consider how the Droitwich carrier may be utilised to improve the accuracy of an inexpensive counter with the common reference oscillator frequency of 1 MHz. One method employs a straight receiver, i.e. one which involves no frequency changing, to amplify the signal from an aerial to a level suitable for clipping to remove as much as possible of the modulation envelope. The waveform is then distorted by a well known technique to yield a pulse-type waveform rich in harmonics, and the fifth harmonic is selected by means of a bandpass filter. Since any negative-going components of high modulation depth are not affected by clipping, the filter should take the form of a very narrowband crystal filter or a locked crystal oscillator. The resulting sinusoid is applied to the 'external reference' input found on most counters, when it replaces the normal internal reference.

An alternative method, the practicability of which depends on whether it is regarded as permissible to modify the circuit of the counter to a not very large extent, makes use of the internal oscillator for filtration. A very simple method which has been found to operate most reliably merely injects a suitable level of the fairly crudely filtered 1 MHz in series with the internal crystal, so that the internal oscillator circuit acts both as a filter and as the drive for the counter. Since the aging rate of the internal crystal is likely to be fairly high, it may be necessary to reset the crystal-oscillator frequency from time to time to the centre of its locking range; but, as the latter may readily be made fairly wide, the intervals need not be unduly frequent. In addition to minimal changes to the counter circuitry, this device has the advantage that misleading side-locking effects are absent. Furthermore, commercial receivers are available which can be used directly to drive the internal oscillator.

It cannot be denied that the problem is made much simpler when the standard broadcast frequency is a number as convenient as 200 kHz; if the frequency cannot easily be converted to the internal counter reference frequency, the method is inapplicable unless one is prepared to put up with a number on the counter display which is proportional, instead of equal, to the correct frequency.

A counter of medium quality may also be employed in a somewhat simpler circuit if it can be assumed that another subcarrier source known to be within the tolerance is always likely to be available, e.g. 'off air' or from another programme source, since it can then be used to measure the difference in frequency between the two. This device is evidently in no way as satisfactory as the direct measurement, unless an exact measurement of the comparison frequency can be obtained, but it may nevertheless be very acceptable to know that the difference between the local and remote subcarriers is small enough for the latter still to be usable.

The conventional method for obtaining the difference frequency in such cases is to beat the two frequencies together in a modulator or some suitable nonlinear device and to select the lower sideband, which can be effected very simply, since the frequency difference is so large. The separated difference frequency is then applied directly to a counter. However, the great simplicity of this arrangement is offset to some extent by possible errors from hum voltages and from any field or picture components present in either or both subcarriers if they have been regenerated from the burst. Furthermore, which may well be very important, no indication is given of the polarity of the frequency error.

A little-known alternative method, believed to have been originated jointly by the author and J. Lewis of the BBC Designs Department, has been used operationally. It has the advantage of being insensitive to low-frequency components and provides a clear indication of the direction of the error. It is slightly more elaborate than the former method; but, with modern circuitry, the extra complication and expense are by no means forbidding, particularly in view of its very satisfactory behaviour.

The measurement requires a counter capable of indicating a frequency ratio. Assume that the higher of the two frequencies is applied to the normal or A input of the counter and the lower is applied to the gate or B input, but only after having been divided by 44. Then, even if the counter has no more than six digits, as is likely to be the case since a counter of medium quality has been assumed, a count can still be selected in such a way that the number displayed is the error in hertz and tenths of a hertz. On the other hand, if the two inputs are reversed, the display will show a row of 9s. Hence the higher of the two frequencies can be indicated unambiguously.

The derivation of the algorithm used is as follows. Let the lower of the two frequencies, i.e. the one which is to be divided by a factor, say n, be F. Also let the higher frequency be $F+f$. The counter will then indicate the ratio $n(F+f)/F = n+(n/F)f$, i.e. the number corresponding to the division ratio plus a constant times the frequency difference f. If it is assumed that $F = 4{\cdot}43361875$ MHz and $n = 44$, the counter reading will be $44+10f$, apart from an error less than 1% of f and consequently completely negligible for the purpose. If the display consists of no more than six digits, the initial 44 will disappear and the counter will appear to read the difference frequency directly in hertz and tenths of a hertz.

For many purposes, it may be possible to accept an error of 10% of f, in which case the division ratio need only be 4 instead of 44.

9.3.1 Ratio of subcarrier frequency to field frequency

In the Pal and NTSC systems, there is a precise relationship between the frequencies of the colour subcarrier and the field, namely 709 379 complete periods of the subcarrier in eight fields with Pal and 286 366 in four fields with 525–line NTSC. Secam is exceptional, on account of the frequency modulation of the subcarrier.

This relationship is at present exploited in the equipment which automatically corrects in colour videotape recorders for the random-phase errors introduced by the mechanical systems involved; if it is not correct, it will not be possible to replay a tape without distortion. The subcarrier-generating equipment is designed for the greatest possible reliability, and faults occur extremely rarely; nevertheless, one must consider the fact that the loss of the colour information in a programme can be highly embarrassing, particularly with an outside broadcast which cannot be repeated. An unfortunate feature of such a fault is that, if no warning is given in advance, it may not be discovered until much later when a replay is attempted.

It may consequently be thought worth while to provide a means for monitoring this frequency ratio, all the more since it has been found practicable to include it as a facility of a versatile subcarrier-monitoring equipment which also utilises the techniques for precise subcarrier measurement and comparison described in Section 9.3.

The basis is a counter of medium quality (as described in Section 9.3), which has provision for ratio measurement. The synchronising information is stripped from the incoming video signal and employed to generate a rectangular wave with a period of precisely eight fields, care being taken to ensure that its edges are completely unaffected by picture content. This is used as the B input to the counter. The corresponding A input may consist of either a feed of a local subcarrier or a subcarrier which has been regenerated from the burst.

The display then reads the ratio A/B, with the proviso that the count is so selected that the first decimal place is displayed; otherwise the last digit corresponding to the number of units in the ratio will not be trustworthy. However, the visual display is much less important than a fault alarm, which can be derived from the binary-coded decimal output of the coder. In an example in operational use, this actuates an audible alarm, but extra confidence is built into the system by logic circuitry which rejects a miscount until it has been repeated four times in sequence, thus insuring against false alarms resulting from noise and switching transients.

9.4 Phase measurements

9.4.1 General

Phase shift, as distinct from group delay, has little significance in television except at colour subcarrier frequency where its measurement is of considerable importance in all areas associated with the generation and handling of colour signals. Such measurements have to be carried out to a standard of accuracy which, even with routine operational work, may well be within $\pm 1°$, and for some purposes $\pm 0{\cdot}1°$ or even better. These are quite stringent working tolerances, especially at the frequencies involved.

Since the majority of the available equipment for phase measurement is relatively expensive, and in many instances not sufficiently accurate for television purposes, it seems desirable to deal with the basic principles in fair detail and to show that it is possible to make precision measurements with fairly simple equipment. The provision of standards is also important, even when specialised apparatus is at hand, since these make it possible to resolve uncertainties with minimum loss of time.

The most widely available equipment in working areas is the vector monitor (Section 11.5), which in modern forms provides for the measurement of phase angle over 360° by means of a polar display, which is supplemented by a vernier phase shifter of relatively restricted range operating on a line display. In view of the large number of potential sources of error in a polar display, it is far from simple to guarantee an accuracy over long periods even within $\pm 2°$ at all points over the circle, although the phase shifter used in conjunction with the line display is capable of considerably higher accuracy. Instruments of much improved intrinsic accuracy are feasible using digital techniques, but it is likely to be some time before they come into widespread use. A much simplified version of one such sophisticated design is described below.

9.4.2 Lissajous figure

There are two very basic methods for measuring phase angle, the wattmeter method and the Lissajous figure. The former is of little use at subcarrier frequency, since the errors are likely to be large, but the latter is of great use if correctly applied, with the additional advantage that the apparatus needed is simple and readily available.

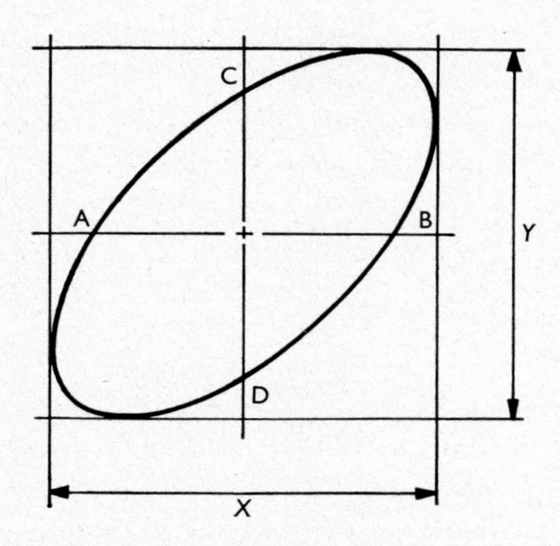

Fig. 9.8 Measurement of phase angle using Lissajous figure

$$\sin \beta = \frac{AB}{X} = \frac{ab}{XY} = \frac{CD}{Y}$$

if a = major axis, b = minor axis

The Lissajous figure (Fig. 9.8) is the display obtained when the two frequencies, the phase shift between which is to be determined, are fed into the horizontal and vertical amplifiers of an oscilloscope, which, to be suitable, should have an adequate frequency response through both amplifiers and roughly equal gains; contrary to popular belief, it can easily be shown that there is little advantage to be gained in departing substantially from the equal-gain condition.

To make a measurement, the phase shifts through the two channels must first be equalised by applying a single subcarrier source to the two inputs paralleled, when any difference may be compensated for either by adjusting the lengths of the connecting leads or by inserting a variable phase shifter in the appropriate side. Two useful circuits for this purpose are given in Fig. 9.11; both are suitable for insertion in 75 Ω circuits. That of Fig. 9.11*a* can furnish a change of nearly 180° if the reactance X is made in the form of a variable capacitor, possibly with switched capacitances in parallel to increase the range. A further 180° is, of course, possible by the reversal of one of the transformer windings. The circuit of Fig. 9.11*b* may be made variable in steps by altering the number of sections, or continuously variable by making the shunt capacitors a ganged variable capacitor, provided that the cutoff frequency with maximum capacitance is not too low. Continuously variable delay lines are also available commercially, and can be utilised very effectively.

Unfortunately, the Lissajous-figure method does not provide sufficient accuracy for the direct measurement of phase shift from the display, since it is not possible to derive the requisite dimensions with adequate precision; also the sensitivity becomes poor as the angle approaches 90°, as is evident from the expression given in Fig. 9.8. On the other hand, it is extremely useful as a null detector, since, for $\beta = n180°$, n being zero or an integer, the trace becomes a straight line, independent of gain, except for the slope of the line. Hence, if one starts with the two channels equalised, an unknown phase shift between the two inputs may be found by inserting a calibrated phase shifter into the appropriate path to restore the straight-line condition, when the unknown phase shift becomes equal to the inserted known phase shift. There is a potential ambiguity of 180°, which can be resolved by noting whether or not the slope of the straight-line display is the same as in the zero condition.

An estimate of the sensitivity of the null adjustment is readily derived. When the phase angle β is small (Fig. 9.8), we can write $\beta = ab/XY$, where a and b are the minor and major axes, respectively, of the ellipse formed by the display and X and Y are the sides of its bounding rectangle; for fixed channel gains, X and Y are constants, independent of β. For small angles, b must closely approach the diagonal of the rectangle, so that we can write $\beta \simeq a\sqrt{(X^2+Y^2)}/XY$. The homogeneity of this last expression for X and Y demonstrates why no great advantage is to be gained by departing from the equal-gain condition. Consequently, X and Y may be assumed to be equal, giving $\beta \simeq \sqrt{2a/X} = AB/X$, where AB is the intercept made by the ellipse on the horizontal line through its centre.

For example, assume that X and Y both have the moderate value of 10 cm. Experience shows that a change is likely to be detectable when the display moves by a distance equal to one-half of the spot diameter; hence the minimum value of a is 0·25 mm, say. The minimum detectable angle is consequently $2 \times 2{\cdot}5$ mrad, equal to $0{\cdot}2°$. With suitable precautions to prevent overloading one can improve the sensitivity by increasing the amplifier gains, so as to provide a discrimination of $0{\cdot}1°$ or even better.

The most significant potential source of error in Lissajous-figure measurements is the presence of harmonics in the two input frequencies; the discrimination of the adjustment is spoilt by even very small percentages of harmonics, which turn the straight line into a loop or butterfly shape. It consequently becomes essential to insert a pair of identical lowpass filters into the two measuring paths; the requirements are not stringent.

9.4.3 Measurement at reduced frequency

The measurement of phase shift is in general easier at low than at high frequencies because of the lesser effect of the circuit strays; any given length of connecting lead introduces a phase shift proportional to frequency, as also do stray inductance and capacitance; also the metering system is usually easier to engineer.

It would therefore be advantageous in some instances if a simple method could be employed for transforming the inputs to some lower frequency. This cannot be achieved by division, since the phase shift is reduced in proportion, but it can be managed by making use of a modulation process to subtract a constant from the frequency of the inputs whose relative phase is to be measured.

The basic circuit is given in Fig. 9.9. The two input waveforms are applied to two identical modulators whose carrier is supplied from a common oscillator. The outputs of the two modulators are followed by two identical filters which select the lower sideband $F-f$ only; then, if the inherent phase shifts through the two paths have been equalised, the two output waveforms at a frequency $F-f$ have precisely the same phase relationship as the two input waveforms at a frequency F.

The proof is simple. Any modulation process involves implicitly or explicitly the multiplication of the two waveforms concerned. If these are, say, $\cos pt$ and $\cos(qt+\theta)$, the 1st-order sideband terms are given by $\cos pt \cos(qt+\theta) = \cos\{(p+q)t+\theta\}+\cos\{(p-q)t-\theta\}$. Hence, for any angular frequency q, the phase shift between any pair of corresponding sidebands is equal to the phase shift between the

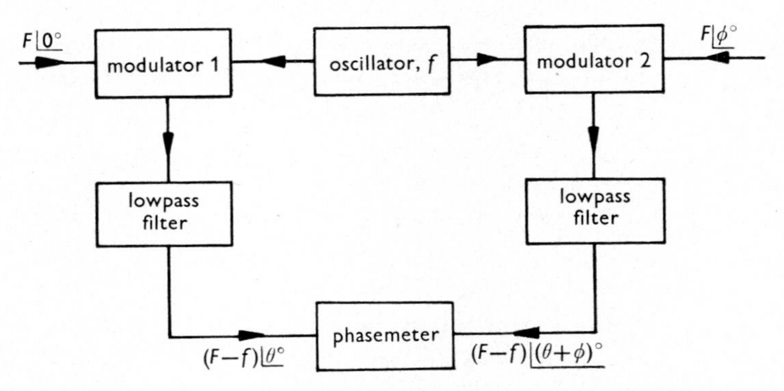

Fig. 9.9 Measurement of phase angle at reduced frequency

original frequencies from which they were derived. The change in sign between upper and lower sidebands is, of course, trivial.

Usually, little difficulty is experienced if care is taken to construct the two paths as symmetrically as possible to ensure that any changes in the phase shifts of the two channels due to aging or thermal effects are always equal. In particular, arrangements should be made to keep the two modulators always at identical temperatures.

Once the frequency has been sufficiently reduced, it becomes potentially feasible to increase the phase angle by frequency multiplication, which increases frequency and phase shift by the same factor. A basic source of instability is inherent in any multiplication process, since the multiplied waveform receives positional information from the original waveform only once in n periods for a multiplication factor n, whereas in division where the divided waveform is provided with timing information period by period. However, Augustin (1958) has described a multiplier operating at 10 kHz which is claimed to be capable of furnishing a resolution of 0·01°.

9.4.4 Digital methods

At any fixed frequency, a phase shift is equivalent to a time displacement, which suggests that the determination of the time interval between a pair of corresponding points in two waveforms will enable their relative phase to be measured with a digital counter, with the attendant advantages of a completely linear scale and a direct digital presentation of the measured quantity. This principle has been used, for example, for radio navigational purposes (Dunworth, 1966), but it is believed that it has not previously been applied to colour-subcarrier phase measurements. The obstacle seems to have been the very short times involved, since, at subcarrier frequency, one degree is equivalent only to 627·5 ns; but by subtracting a constant from the frequency by means of modulation with an auxiliary frequency (Section 9.4.3), the method becomes perfectly practicable.

In a simple application of the principle (Fig. 9.10), the two inputs are modulated with the output from a stable crystal oscillator whose frequency differs from that of the subcarrier by 2·778 kHz, and the lower sideband at 2·778 kHz is selected by means of filters. From these two sidebands are derived two pulses having, in each case, one transition coincident with the crossover point of positive slope, which are then utilised to generate a single rectangular pulse whose width t_0 is a measure of the phase shift between the original subcarrier inputs. The final operation is the stretching of this pulse by a factor of 100 in a divider type of pulse stretcher, which ensures the accuracy of the scaling factor, after which its duration is measured with a suitable counter.

The choice of constants has resulted in one degree at subcarrier frequency being equivalent to a final pulse duration of 100 μs, so that even an unpretentious counter will provide an accuracy within $\pm$ 0·1°, with, of course, a display which indicates phase angle directly (Spencer, 1963).

The overall accuracy of the measurement depends very largely on the ability of the local oscillator to maintain the difference frequency at precisely the correct value, assuming that the accuracy and stability of the subcarrier frequencies to be measured are significantly better

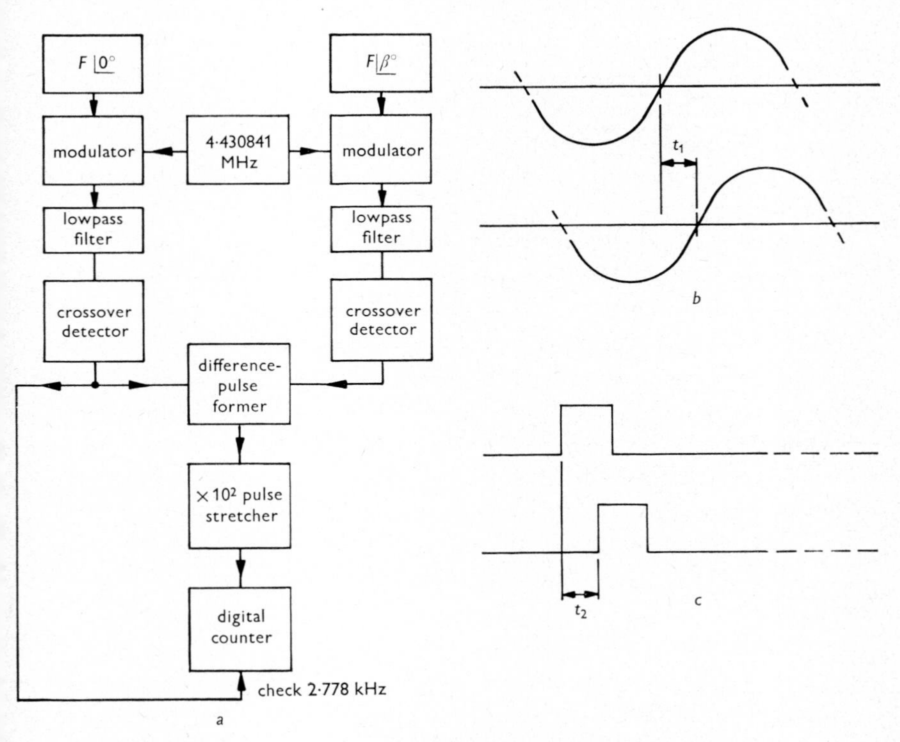

Fig. 9.10 Principle of simplified digital technique for measuring phase shift

a Basic circuit
b $1° \equiv 627{\cdot}5 \times 10^{-12}$ s
c $1° \equiv 10^{-4}$ s; $\beta \equiv t_2 \times 10^4$ degrees

than those of the local oscillator; to enable the local oscillator frequency to be monitored, a feed is taken to the counter. Since the oscillator may be pulled into the correct adjustment, the important factor becomes its short-term stability. A total error of 9 Hz changes the period of the difference frequency by 0·3%, say, in round numbers an error of 0·3° in 90°, so that the requirements are not impracticably severe, but evidently it would not have been reasonable to aim for a difference frequency of 277·8 Hz in an attempt to improve the resolution.

A considerably more sophisticated version of a digital phasemeter of this type has been designed for operational use in which the problem of the difference frequency has been overcome by generating it both by modulation and by direct division in two parallel paths, and then employing the error between them to control the frequency of the local oscillator by a phase-locking system. Hence no drift is possible.

A further improvement with important implications is the result of the choice of the difference frequency to be in the ratio of $1:360 \times 10^n$ to the input frequency (in the present example, $1:360 \times 10^2$), and the employment of the latter as the clock frequency for the counter. This is equivalent to a division of the period of the input waveform into 360×10^n equal units each of 10^{-n} degree, the counter then indicating the number of these in the fraction of the period corresponding to the phase difference. Other refinements are the elimination of the effect of phase jitter in the input frequencies and the provision of stability against aging and thermal effects. A detailed description of an instrument of this type is given in a paper by Weaver and Lewis (1970).

9.4.5 Phase standards

Standard phase shifters are invaluable tools in most areas where colour signals are generated, processed and distributed, apart from their use in design work and the manufacture of equipment.

Variable phase shifters may be constructed from networks of the type shown in Fig. 9.11*a*, from 4-phase capacitive phase shifters, from inductive goniometers and from variable delay lines. For most purposes, the last type has much to recommend it since commercial units are available with a very linear law of phase shift against rotation over 360°. The second two can also be made linear in principle, but satisfactory examples are harder to come by. The impedances of

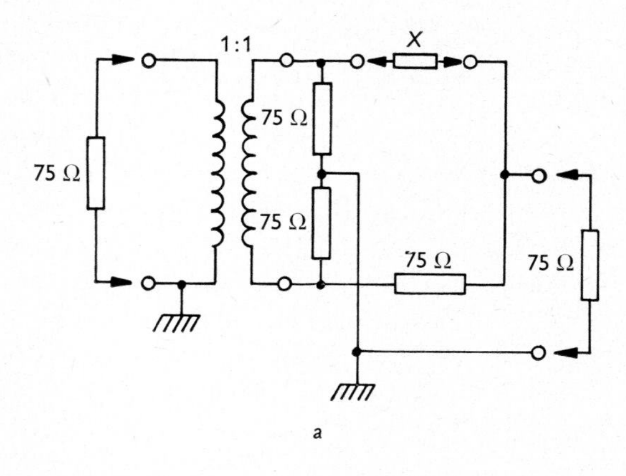

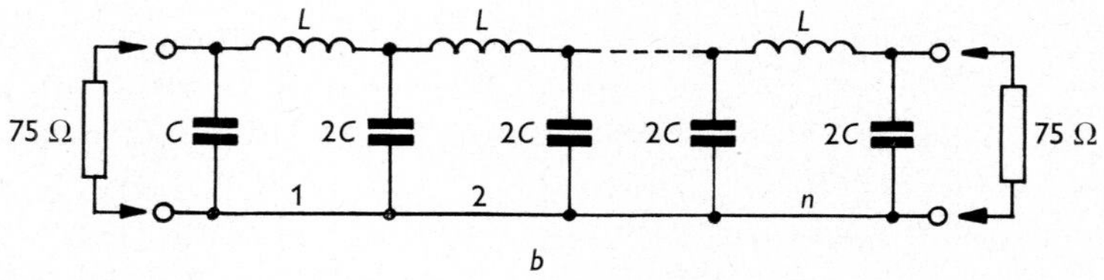

Fig. 9.11 Examples of phase shifters for use between 75 Ω terminations

a Bridge type

Insertion phase shift $\beta = 2 \tan^{-1} \dfrac{X}{75}$

Insertion loss = 12 dB

b Ladder type

for $f_c > 16$ MHz, say, and $f = 4{\cdot}433$ MHz,

$$\beta = n \sin^{-1} \frac{4{\cdot}433}{f_c} \quad (f_c,\ \text{MHz})$$

$$L = \frac{23{\cdot}8}{f_c} \text{ microhenrys}$$

$$C = \frac{2120}{f_c} \text{ picofarads}$$

suitable variable delay lines are rarely as low as 75 Ω, so all except the first will need to be provided with buffer amplifiers to maintain the correct internal and external impedances; the first type may also require an amplifier, but only to make up the loss in the network. This phase shifter has the drawback of a nonlinear and rather cramped scale over one end of the range; but, for zero-setting and similar purposes, it is very convenient, in which case it is useful to make provision for changing the sign of the phase shift by arranging for one of the transformer windings to be reversed when required.

Hybrid variable phase shifters may also be constructed from switchable phase networks or delay lines used in conjunction with a variable section with a swing sufficient to cover the gap between the fixed sections.

Unless fixed phase shifters can be purchased, or unless one is prepared to spend a good deal of time and effort on the design of networks with the requisite stability and low temperature coefficient, there is no alternative to the use of flexible coaxial cable. The velocity ratio of suitable high-quality coaxial cable is usually close to 0·67, which leads to the easily remembered rule of thumb for 625-line colour subcarrier of 5 in per degree of phase shift. The longer phase shifts consequently demand the use of considerable lengths of cable, e.g. 90° needs 37·5 ft, but experience has shown that it may be tightly coiled without detriment either to the cable or the accuracy of the calibration. Some high-grade miniature types are useful where space is at a premium.

An important precaution when constructing standard phase shifters is the provision of a reference path which is switched in when the preliminary zero adjustment is made during a measurement. The length of the standard is selected with respect to this reference path, so that, when it is switched in, the exact phase difference wanted is obtained. This device completely eliminates uncertainty about the conditions under which the standard phase shift is available.

A great difficulty with cable is cutting it to the correct length, first because the end has to be reterminated at each attempt, and secondly because the precise position of the electrical end of the cable is somewhat indeterminate. Some of the labour can be removed by cutting the cable very slightly shorter than required, and making up the deficit in length with a small network providing only a very few degrees and variable over a range. A section or halfsection of lowpass filter with a sufficiently high cutoff frequency is suitable, and need cause no mismatch. The variation is normally obtained by adjusting the series-arm inductor. Since the phase shift contributed by this auxiliary network is very small, its stability becomes correspondingly less important.

Cable standards have been found to possess excellent stability provided that adequate mechanical precautions are taken; e.g. the method of ending the outer conductor must ensure that the sheath cannot move in the course of time with respect to the inner conductor.

The 5 in per degree of phase shift given above is naturally only a rough guide, and the precise length must be found by experiment in each case. One method of determining the length required, although somewhat roundabout, has at least the merit of needing the minimum of specialised equipment.

The first part of the procedure consists in the measurement of the velocity of the cable selected as a function of frequency, since it is not altogether constant, depending on the type of cable. The largest variations occur towards the lower frequencies, say, below about 2 MHz, i.e. in a region which may well be required for the second part of the procedure. The measurement is carried out by feeding the cable, either short-circuited or open-circuited at its far end, from an oscillator via a series resistor. The voltage minima are observed with an amplifier–detector or oscilloscope across the sending end of the cable, and the frequencies of these minima are found by means of a counter. It will be recalled that the minima occur with an open-circuited line at odd numbers, and with a short-circuited line at even numbers, of quarter wavelengths. The input minima are chosen in preference to the maxima, since the indication is usually much sharper.

The electrical length of the cable is now known in terms of frequency; and, consequently, the variation of velocity with frequency is established. The end effect can be minimised by using another piece of cable as a reference path. If more than one length is utilised, it should preferably be from the same batch. This information makes it possible to predict not only a more accurate value for the length of the particular cable in use, but also the frequency at which a suitable quarter-wave resonance may be obtained to provide a sensitive indication for the final adjustment of the length. This will most often occur at an appreciably lower frequency than the subcarrier frequency, for which reason accurate values for the velocity are required over a fairly wide frequency range.

Experience shows that this procedure, once mastered, permits, the construction, if sufficient care is taken, of phase shifters which are correct to within a very few tenths of a degree of the nominal value. It is somewhat tedious, but makes a minimal demand on resources.

An even simpler technique, literally of the 'cut and try' variety, is available for the construction of 90° phase shifters, 90° being by far the most important single value to possess in areas dealing with colour signals, since it furnishes a considerable number of basic checks. A pair of nominally identical lengths of cable, equivalent to a very little more than 90°, are compared by the Lissajous-figure method (Section 9.4.2.); if they do not have identical phase shifts, the longer is adjusted to the shorter. They are then connected in series using the minimum amount of connecting lead, and the Lissajous figure is measured.

Since the cable lengths were deliberately made slightly too long, the phase shift should be a corresponding amount greater than 180°, which can be found by the approximate expression for small angles $\beta = ab/A$, in which a and b are the minor and major axes, respectively, and A is the area of the rectangle bounding the ellipse. Slightly less than half the length corresponding to the excess phase shift is then removed from each cable and the process recommenced. After a very few trials, one should finally be in possession of two identical 90° phase shifters. The process is made more convenient if one pads out the cable lengths with small variable phase shifters, as has already been suggested, not only because of the lesser amount of work entailed but also because the consequence of overestimating the amount by which the electrical length of the cables has to be reduced is no longer disastrous.

The Lissajous-figure technique is not the only procedure available for the above adjustment. It can also be carried out on a vector monitor, preferably using the line display, or even with a high-grade difference amplifier. However, in the present application, the Lissajous figure is to be preferred, since it is not only quite sensitive but also completely independent of amplitude effects which can give rise to errors in the other two methods.

For the calibration of variable and fixed phase-shifters with certain specified values, an excellent method is available which has the merit of being absolute, since it depends on the effective division of the period of the calibrating frequency into a number of aliquot parts as a result of a fixed relationship between it and an auxiliary frequency. It was originally devised by the author for the calibration of the

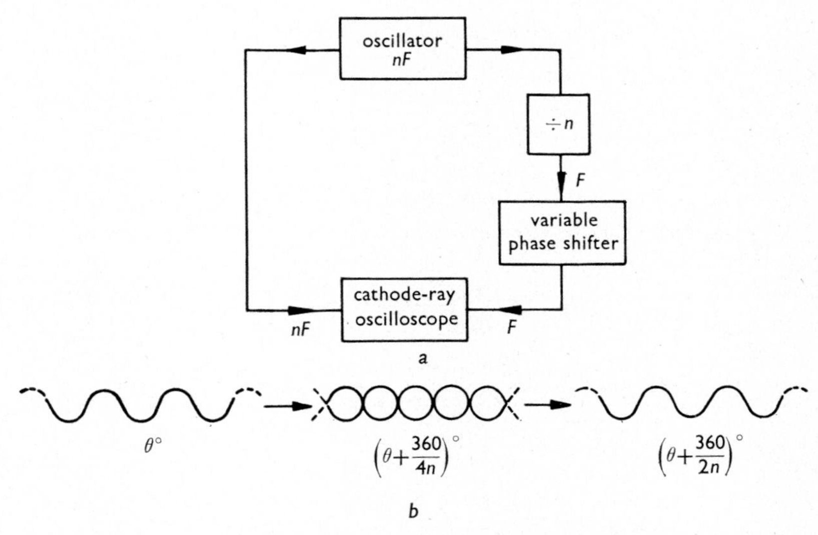

Fig. 9.12 Absolute method of phase-shifter calibration

ranging of a very early example of pulse-echo system, and has subsequently been employed as a standard method for setting up phase-shift networks. It has certain limitations at subcarrier frequency, but it is nevertheless to be recommended wherever practicable.

An oscillator (Fig. 9.12*a*) of frequency nF is divided by a factor n to yield the frequency F at which the calibration is to be performed. The output of the first oscillator is taken to the vertical input of an oscilloscope, and the output from the divider is taken to the horizontal input. For purposes of description, it will be assumed that the phase shifter is a variable unit and that adequate provision is made for terminating it.

The resulting display is a Lissajous figure with n horizontal loops, in which the forward and return traces corresponding to the two half-cycles of the frequency F are superimposed. Consequently, it is possible to adjust the phase shifter, which causes the trace to rotate until the forward and backward traces are in precise coincidence and the display has the appearance of a segment of a sinusoid (Fig. 9.12*b*).

If now the phase shifter is steadily rotated, the display passes through a position of complete symmetry of the trace to a point where coincidence of the forward and backward traces is again achieved. The change of angle is then precisely $360°/2n$, the presence of the factor 2 being accounted for by the fact that each of the two traces moves by an equal amount to meet the other. A variable phase shifter may therefore be calibrated in steps of an incremental phase shift $360°/2n$, which can be set with precision. Similarly, fixed-phase-shift networks can be set to the incremental phase shift.

While this procedure may be adopted at low frequencies for very small incremental phase shifts, this is no longer possible at subcarrier frequency since the frequency nF eventually reaches a value such that it is inconvenient to display it because of limitations in the oscilloscope. Nevertheless, with a suitable instrument, incremental phases of $5°$ are not out of the way. The method may be recognised as the general case of which the Lissajous-figure technique for null adjustment is the special case when $n = 1$. Like that, it has the virtue of being independent of amplitude changes except in so far as the sensitivity may be modified by any large variation.

Fig. 9.12*a* does not show two additions which are desirable to avoid degradation of the display and for the measurement of fixed phase shifters. The first is the inclusion, in each path, of a lowpass filter to make the harmonic content negligible; and the second is the insertion, in one path or the other, of a small variable phase shifter for zero-setting purposes. A useful refinement, if the method is frequently employed, is a very narrow pulse synchronised to a zero-crossing of the frequency F at the output of the divider and provided with a variable delay. Whether it is added to nF or utilised to modulate the brightness of the trace, it serves to identify one of the loops so that a record may be kept of the number of coincidences passed through.

The stability of the frequency nF need not be remarkably high, the tolerance depending on the rate of change of phase with frequency of the network or cable under test, as well as the desired accuracy. Usually a good signal generator is adequate, in which case a counter should be connected across the output of the divider so that F may be adjusted to subcarrier frequency precisely. It is feasible to apply the frequency-reduction technique of Section 9.4.3 to this type of measurement, except that a fair degree of complication is thereby introduced and the admirable simplicity of the method is lost.

9.5 References

AUGUSTIN, E. (1958): 'Phasenverfielfacher', *Hochfrequenztech. & Elektroakust.* **67**, (3), pp. 84–87

DISHAL, M. (1959): 'Gaussian-response filter design', *Elect. Commun.* **36**, (1), pp. 3–26

DUNWORTH, A. (1966): 'A digital phase meter for electronic navigational aids', *Proc. Instn. Radio Electron. Engrs. Aust.*, **27**, (8), pp. 213–221

GOURIET, G. G. (1958): 'Two theorems concerning group delay with practical applications to delay correction', *Proc. IEE*, **105**C, pp. 240–244

MANTEL, J. (1966): 'Definition of jitter and a fast method to derive it', *Electron. Lett.*, **2**, pp. 300–301

POULLIS, P. A., WEAVER, L. E., and COOKE, J. J. (1970): 'Video delay modules', *IEE Conf. Publ.* **69**, pp. 236–239

ROYLE, R. J. (1969): 'A method of measuring pulse duration', Post Office Engineering Department research report 73

SPENCER, A. J. (1963): 'The measurement of phase angle using a counter', *Marconi Instrum.*, **9**, (3), pp. 60–63

STEINMETZ, H. J. (1967): 'Verfahren zur Messung von Impulsflankenanstiegzeiten im Videokanal', *Tech. Mitt. RFZ*, **11**, pp. 49–54

WEAVER, L. E. (1968): 'Measuring delays accurately', *Wireless World*, Oct., pp. 355–356

WEAVER, L. E., and LEWIS, J. (1970): 'The digital measurement of colour subcarrier phase', *IERE Conference Proc.* 18, pp. 27–38

10 TEST PATTERNS

10.1 General

Test patterns, or test cards as they are often called, may be defined as test objects of standard form, either transparencies or opacities, intended principally for examining and checking the optoelectrical conversion inherent in the generation of a television signal, or alternatively the inverse process. The main uses of these patterns have been summarised by Hersee (1967) as follows:

(*a*) exploring the transfer characteristics of optoelectrical transforms, i.e. camera tubes and flying-spot-tube/photocell systems
(*b*) exploring the transfer characteristics of the reverse transform, i.e. receivers and picture monitors
(*c*) geometry tests. In this category come not only testing of the scan geometry of camera and display tubes, but also such uses as the registration of multitube colour cameras; the registration of multi-projector systems such as are used in caption equipments and telecines; and hop-and-weave tests on telecines and film cameras.

The word 'measurement' has been carefully avoided in the above, because only in a minority of instances can true measurements be made by using test patterns, chiefly on account of the difficulty of standardising the patterns, the method of illumination and the adjustment of the conversion device, and also certain difficulties inherent in the pattern itself. Nevertheless, they have many important uses.

The number of test patterns is very large indeed; moreover, the precise form is apt to change with time as modifications and improvements are introduced into the various techniques. Accordingly, only a small number of test patterns will be described here, more to bring out the basic points of their design and use than list the most important or the most widely used patterns.

10.2 Test card D

This pattern (Fig. 10.1) is the successor to the very successful test card C radiated by the BBC from 1947, shortly after the resumption of television broadcasting, until 1964. It was designed by a committee of representatives from the two television administrations and industry, the latter being very important since test card D is primarily intended for radiation during fixed periods of the day for the benefit

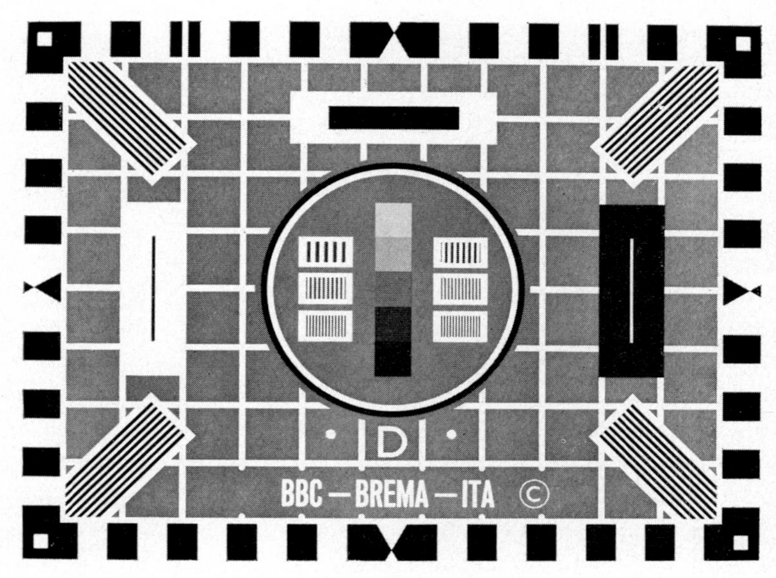

Fig. 10.1 Test card D

of television manufacturers, servicemen etc. Its design is consequently orientated towards providing the maximum of information for the correct adjustment of receivers and for showing up faults in picture reproduction.

A white grid on a grey background is framed by 'castellations', i.e. a border of alternate black and white rectangles. The rectangles allow the scan linearities to be checked over the whole picture area, and the castellations define this area so that the picture can be correctly framed with respect to the screen of the cathode-ray tube. The 'hour-glass' designs in the centres of the four sides provide precise references for this purpose. The left-hand vertical set of castellations also provides a good test of the synchronising-pulse separator. A minor point is that two of the vertical lines of the white grid are drawn through the black castellation blocks on the upper edge only. This enables one to identifiy the top of the picture quickly when the waveform is being examined by means of a line selector.

Just below the upper edge of the pattern is the 'letterbox', a long horizontal black bar in a white rectangle, the purpose of which is to show up faults at fairly low video frequencies which cause streaking effects.

The two vertical rectangles, on the other hand, contain only narrow vertical bars, to show up the presence of echoes in the displayed signal. The worst of these usually arise from multipath propagation or from a mismatch in the receiving-aerial system, but, in principle, they may originate in many places. Both a white bar on a black background and a black bar on a white background are provided, so that both positive and negative echoes can be identified. The duration of the bars is 0·3 μs for 405 lines and 0·2 μs, which makes it possible to see even very short-term echoes.

The central circle, which in itself is included principally as a check of the linearity of the display in the central region, contains two important patterns. The first is the set of resolution gratings, which is perhaps unique, in that it is an attempt to utilise, in a test card intended for reproduction in fairly large numbers, gratings designed to provide sinusoidal waveforms in the resulting video signal. This entails a number of technical difficulties; which are explained in Section 10.3. The points raised are of considerable significance in the design of test cards and resolution patterns.

The second pattern in the circle is the double set of five steps of luminance, the tone wedge. Here again, the easiest course from the point of view of the preparation of the master pattern is not the best for the ultimate purpose. The simplest would be to provide a set of five equal increments of voltage, but such a step wedge does not appear subjectively on the screen as five equal steps of brightness, which is what is required if the serviceman is to be able to set up the receiver correctly by eye.

The solution adopted was to design the master grey scale on the basis of data supplied by the BBC Research Department, so that, after γ correction, the reproduced grey scale has steps of equal subjective brightness, with an overall contrast range of 30:1. A useful refinement is the addition of small, slightly lighter circles in each of the two outer steps of the grey scale, which form a very sensitive indication of white or black crushing.

The only remaining feature of test card D is the set of four diagonal grids in the corners of the pattern. They are primarily intended for checking the focus in the corners of the tube, but are incidentally useful for revealing certain types of echo.

The pattern must also be considered as a whole. The existence of receivers using mean-level a.g.c. circuits makes it necessary to ensure that the contrast range of the receiver is set correctly for the average picture, which is achieved by selecting the component patterns and the background tone so that the corresponding average picture level is a little less than 50%. This, of course, only applies to trade test cards used in conjunction with domestic receivers.

10.3 Resolution patterns

Resolution patterns, as their name implies, were originally conceived for measuring the overall resolution of a television system; but, with improvements in techniques and equipment, television bandwidths are now better maintained, and this usage is now becoming less common. With signal sources and display devices at present, they are usually employed to measure the spatial-frequency characteristic, i.e. the equivalent amplitude/frequency characteristic of the device considered as an electro-optical convertor; associated amplifiers or other equipment may be included (Section 12.5).

In their most common form, these consist of groups of patterns in which uniform, rectangular black and white bars of equal widths are alternated. When they are located perpendicular to the direction of scanning, the output voltage from the signal source takes the form of very approximate square waves whose fundamental frequency is determined by the width of the original bars. Groups of these patterns with bars of appropriate widths make it possible to explore the video bandwidth. Owing to their resemblance to optical gratings, these are often known as 'frequency gratings' or 'resolution gratings'.

An alternative preferred by some, particularly in the USA where it forms part of the well known RETMA chart (IRE, 1961), turns the discrete steps of spatial frequency provided by the series of gratings into a continuous variation by forming the bars into a fan or wedge shape, the width of the bars being adjusted along their length so as to maintain at all points the equality of the black and white intervals. When it was first introduced, this type of chart was primarily intended for the estimation of limiting resolution by the method of exploring the wedge for the point at which the bars can no longer be resolved. However, this technique is rapidly falling into disuse, first because the point at which the spatial-frequency response has fallen to, perhaps, one-tenth or so of its lower frequency value provides very little genuine information about the picture quality, and secondly because the estimation of the point of limiting resolution is markedly uncertain unless the conditions, i.e. the brightness, contrast etc. are carefully standardised. Apart from the fact that a continuous variation of spatial frequency is available, there are no advantages in this type of grating; indeed, it has additional disadvantages compared with discrete steps.

The bars mentioned so far have had a 'square wave' type of luminance distribution. They were originally adopted, no doubt, because gratings of this kind had been employed long before the advent of television for optical measurements of resolution, but they had also one significant point in their favour, namely that an abrupt transition from black to white and vice versa is not modified by the γ corrector of the signal source. However, it is only recently that the many shortcomings of these, and, for that matter, all types of grating, have come to be recognised.

The Fourier series for a square wave contains only odd harmonics, as is evident from its symmetry. The series is, in fact,

$$E(t) = \frac{1}{2} + \frac{2}{\pi}(\sin \omega t + \tfrac{1}{3} \sin 3\omega t + \tfrac{1}{5} \sin 5\omega t + \dots)$$

Now the gratings normally used do not correspond to a spatial frequency of less than 1·0 MHz, and frequently 1·5 MHz is the lowest provided. Hence the corresponding waveform can never contain more than two harmonics in addition to the fundamental; and, for most of the frequency range, only the fundamental frequency remains. These vestigial square waves have peak-to-peak amplitudes which differ appreciably from the assumed nominal amplitude of the square wave, i.e. the value it would have if all the harmonics were retained; as a result, a measurement using them is no longer a reliable guide to the spatial-frequency response. As can be seen from the Fourier series, the amplitude of a square wave, paradoxically enough, increases when all the harmonics have been removed by a factor of $4/\pi$; for which reason it is often known as the '$4/\pi$ effect'. The maximum error takes the form of a step in the spatial frequency characteristic of about 2 dB; and, if the original bars are fully modulated, i.e. if they extend from black to white level, such an increase in amplitude is sufficient to drive a transmitter into a condition of unduly high modulation depth. The result will be less harmful with a colour transmitter than with monochrome, since the former is designed to handle high video amplitudes during the radiation of saturated colours.

In practice, the error tends to be rather less than the maximum as a result of the measurement of the square-wave amplitude from the peak of the large 'ring' generated by the removal of all but the lowest harmonics (Fig. 10.2); on the other hand, this compensation is less effective when the spatial-frequency response rolls off in a quasi-Gaussian manner, as is not infrequently the case. This is shown in Fig. 10.3, where, for purposes of illustration, the third harmonic amplitude has been reduced by 1·2 dB and the fifth by 3·6 dB.

Corresponding practical waveforms obtained from an optical test card are shown in Fig. 10.4, in which the situation is further complicated by distortions of an electro-optical nature. The lowest-frequency grating here corresponds to a spatial frequency of 1·5 MHz and the highest to 5·25 MHz; so the waveform of Fig. 10.4*a* contains only the third harmonic in addition to the fundamental. The transition between this and the waveform of Fig. 10.4*b*, in which even the third harmonic has been removed and only the fundamental frequency remains, is very clear. The spatial frequency characteristic has a roll off in this instance (Figs. 10.4*c* and *d*).

The uncertainty in the amplitude of a grating resulting from the removal of the square-wave harmonics can be reduced by replacing the black and white strips by a pattern having a sinusoidal distribution of reflectance or opacity, or, to be more exact, having a distribution

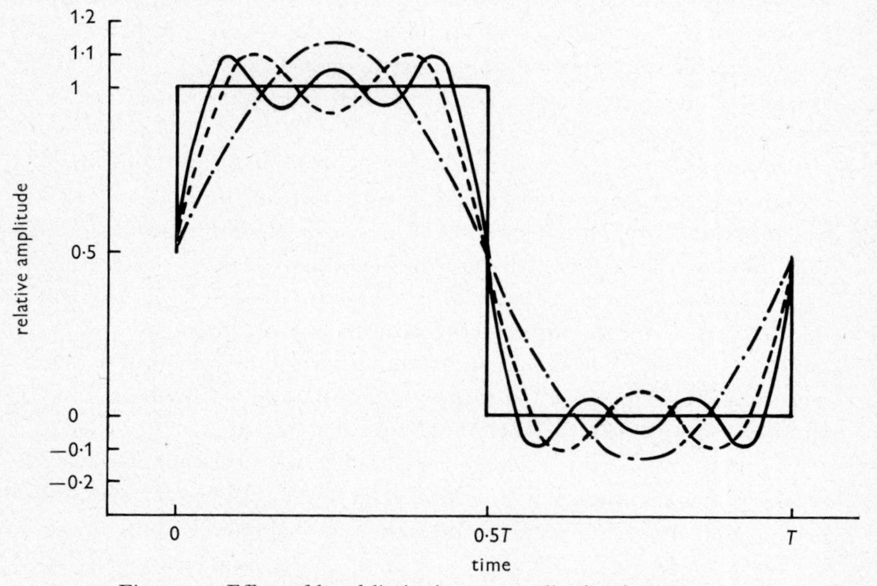

Fig. 10.2 Effect of band limitation on amplitude of square wave
—-— fundamental only
- - - fundamental + 3rd harmonic
—— fundamental + 3rd and 5th harmonics

of these in such a way that the resulting waveform is a sinusoid after the γ correction of the signal source. This is considerably more difficult than providing square-wave gratings from two points of view: a negative has to be produced which has the correct law of variation of opacity, taking into account the characteristics of the photographic emulsion, and this has then to be reproduced accurately, with regard not only to the law of variation of opacity but also to the precise setting of the maxima and minima of opacity or reflectance in the subsequent test card so that the specified modulation characteristic of the grating is maintained.

In spite of the numerous difficulties, this method was successfully used in test cards D as well as 51 and 52, the technique being to feed predistorted sinusoidal waveforms into a telecine machine to obtain a master negative. The standardisation of copies is described by Hersee (1967).

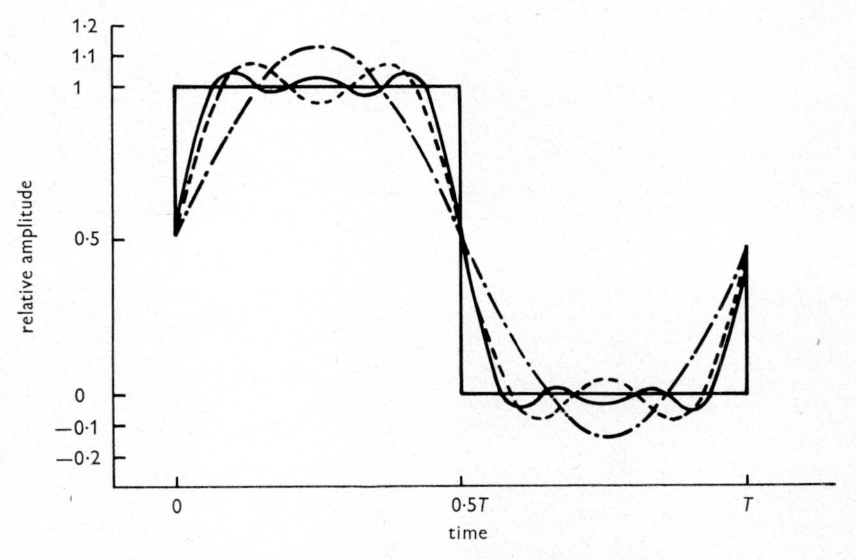

Fig. 10.3 Effect of band limitation on amplitude of square wave with quasi-Gaussian rolloff

—·— fundamental only
– – – fundamental + 3rd harmonic
——— fundamental + 3rd and 5th harmonics

It has since been found that the law of variation of opacity or reflectance need not be quite as accurately sinusoidal as was at first believed; a reasonable approximation will give rise to errors which are not important when the other inevitable errors of the type of measurement under discussion are considered. An ingenious technique for producing such approximate sinusoids, believed to have been devised by A. B. Palmer of the BBC, was actually employed in the design of a test card for the setting of camera exposure which has been in operational use for a number of years (Section 12.3.2).

The basic principle is very simple. A suitable master pattern is needed which consists of a series of parallel, rectangular, opaque strips

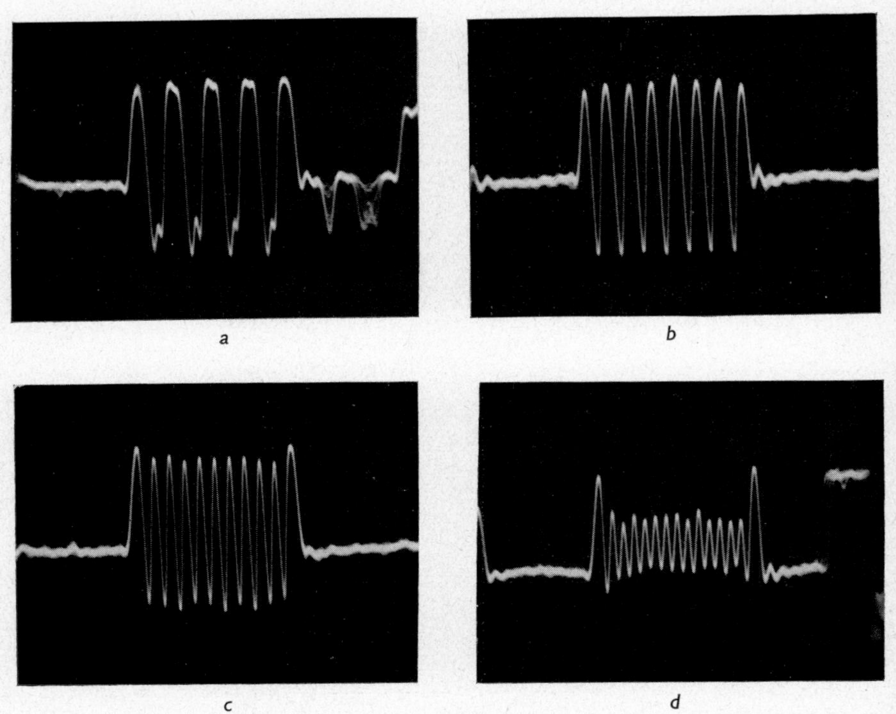

Fig. 10.4 Waveforms corresponding to optical resolution pattern

a Spatial frequency 1·5 MHz
b Spatial frequency 2·5 MHz
c Spatial frequency 4·0 MHz
d Spatial frequency 5·25 MHz

separated by transparent, rectangular strips half the width of the former, thus making a kind of grating. For purposes of description, it will be assumed to be in the form of a photographic negative such as can be produced with a process camera from an oversized original on card. This grating is laid flat on top of a sheet of perspex or glass, which in turn is laid on a sheet of unexposed photographic film, the total spacing between the grating and the emulsion of the film being a predetermined factor of the width of the grating elements. The assembly is then suitably clamped together.

The grating is illuminated by means of a high-quality, plane diffusing surface, such as a piece of ground or opal glass, whose area must be substantially greater than the area to be illuminated and which must be placed parallel with the assembly. Ideally, the light which reaches the transparent areas of the grating has all possible angles of incidence, resulting in a distribution of intensity of illumination on

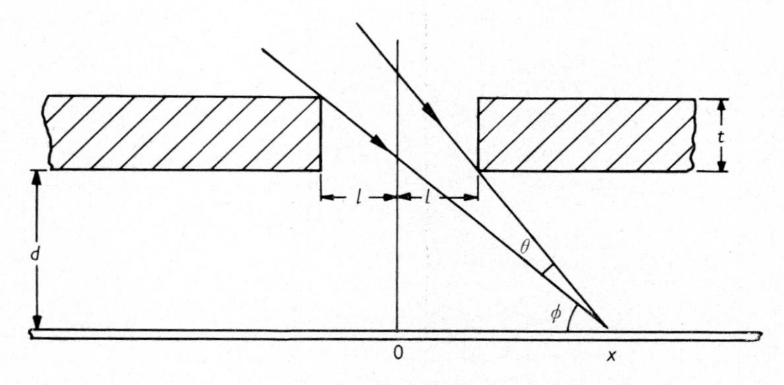

Fig. 10.5 Synthesis of sinusoidal grating using a series of rectangular apertures

the emulsion of the film in a direction normal to the length of the strips which has some quasisinusoidal form. The film, whose characteristic curve should have a long, straight portion, is given a suitable exposure and development, yielding a master negative suitable for use as a test-pattern master. In general, a complete series of resolution gratings of different equivalent spatial frequencies can be generated from a single master negative by photographic reduction.

An approximate analysis of the shape of the resultant distribution of illumination is easily carried out; this will neglect the effects of scattered light, imperfect diffusion, diffraction etc. Fig. 10.5 shows a cross-section of the assembly of grating, spacer and film in the area of one of the transparent strips. The total width of the slit is $2l$, the thickness of the grating is t, and its lower surface is a distance d from the emulsion.

If we consider a point on the emulsion located at a distance x from the centre of the slit, the exposure it receives is proportional to the solid angle which it subtends at the boundaries of the transparent slit; and it is then simple to demonstrate that the exposure I is given by

$$I = \tan^{-1}\frac{2dl+tl-tx}{x^2-l^2+d(d+t)}$$

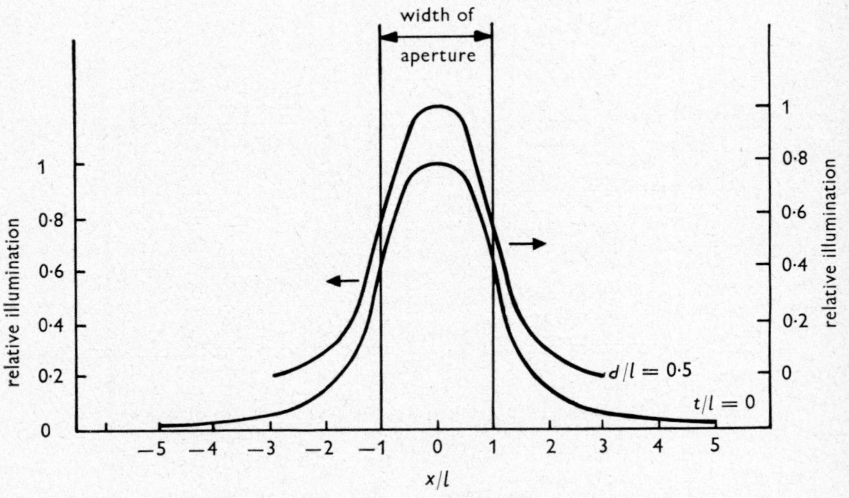

Fig. 10.6 Variation of intensity of illumination across plane surface via rectangular aperture

Width of aperture = 4 × distance of surface from aperture
$d/l = 0{\cdot}5$
$t/l = 0$

where, for convenience, the constant of proportionality has been taken to be unity.

Fig. 10.6 gives curves of the distribution of exposure evaluated from this expression for a spacing d equal to one-quarter of the total width of the grating slit, in the one instance for a grating of negligible thickness and in the other for a grating of thickness equal to the spacing d. For convenience, the independent variable x has been normalised to x/l. As might be expected, when the grating has negligible thickness, the distribution of exposure of the emulsion has a very long 'tail', which is considerably shortened when the grating is sufficiently thick. Otherwise, in each instance, the upper part of the curve is quite a close approximation to a sine wave. Nevertheless, it should not be assumed that the former cannot be used, since, if

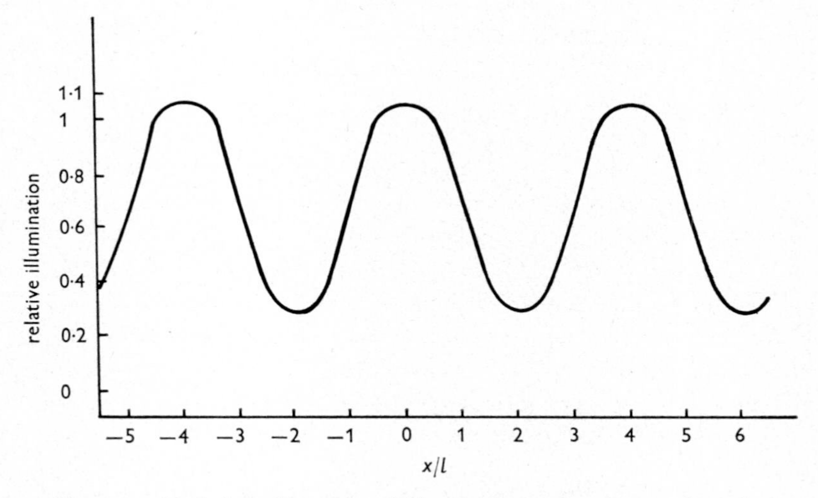

Fig. 10.7 Variation of intensity of illumination across plane surface via a series of rectangular apertures spaced by twice the aperture width

a series of such slits are present (Fig. 10.7), as is the case in the practical grating, the addition of all the curves yields a fair approximation to a sine-wave burst, close enough at all events for the error to be considerably lessened compared with a square-wave pattern. The principal deviation takes the form of a reduction of the effective modulation depth of the sine-wave burst, although, as will be shown below, this is not necessarily a disadvantage. It is very fortunate that this particular pattern should not be unusable, since a very thin master grating is much easier to produce.

Although the ratio $d/l = 0{\cdot}5$ as used in Figs. 10.6 and 10.7 has been found usable, it is not necessarily the optimum. Also, the effect of multiple reflections within the 'sandwich' formed by the grating, spacer and film is to prolong the tail of the distribution somewhat.

There is yet one more possibility of error in the measurement of the spatial-frequency response from the resolution gratings, particularly important with test cards such as C and D, which are most often employed after amplitude modulation. This is harmonic distortion generated by the quadrature distortion resulting from the asymmetry of the two sidebands in the radiated signal which affects the shape of the modulation envelope. It may be considered as being due to the appearance of a quadrature component of the modulating waveform which adds vectorially to the inphase component; hence the name 'quadrature distortion'. Since it is an envelope phenomenon, it is present in the signal recovered by envelope demodulation, but it can be eliminated by synchronous demodulation, which is able to discriminate between the inphase and quadrature components.

A brief discussion of the way in which a sine wave is affected is needed for estimating the extent of the possible error in the measured spatial frequency characteristic. A good general treatment of the effect of asymmetric sideband distortion is given by Cherry (1949).

For simplicity, assume that one of the sidebands has been completely removed, which in fact is so with all but the very lowest-frequency gratings in the test card. Assume also that each grating consists of a continuous sinusoid instead of a relatively short burst.

Let the single-sideband signal be

$$\sin pt + m \sin (p+q)t$$

where

$$p = \text{carrier pulsatance}$$

$$q = \text{modulating pulsatance.}$$

Note that modulation coefficient m is not identical with that used in the double-sideband case, but a different definition has been adopted, more appropriate to the present case, in which the carrier amplitude at maximum modulation depth just becomes zero as m reaches unity.

This expression can be rewritten as

$$\sin pt\,(1 + m \cos qt) + m \cos pt \sin qt$$

where the first term has the form of an undistorted, double-sideband modulation of the carrier and the second term is a quadrature component since it contains cos pt. The amplitude E_m of the modulation envelope is then given by

$$E_m^2 = (1 + m\cos qt)^2 + \sin^2 qt$$

or

$$E_m = (1 + m^2 + 2m \cos qt)^{\frac{1}{2}}$$

This expression has been evaluated for $n = 0{\cdot}5$ and $m = 1$, and the

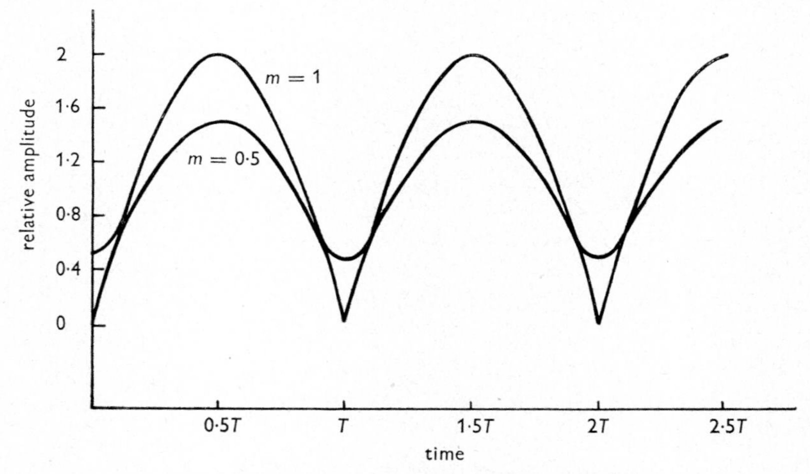

Fig. 10.8 Half envelope of single-sideband modulated sine wave for 50 and 100% modulation depths
$T = 1/f$

resulting waveforms are shown in Fig. 10·8. The harmonic distortion when $m = 1$ is very considerable, but is much reduced when $m = 0{\cdot}5$.

The harmonics in these distorted waveforms will be removed by the receiver cutoff, so that, to estimate the error, we need to find the amplitude of the fundamental frequency in each instance. This can be done by expanding the expression for E_m by the binomial theorem in powers of cos qt and replacing these powers by their multiple-angle equivalents. The terms in cos qt can then be collected, a sufficient number of terms being taken to ensure the required accuracy. This is not unduly difficult since only even powers of cos qt need be considered. In this way, the peak-to-peak amplitude of the fundamental term when $m = 0{\cdot}5$ is found to be 0·968 when the peak-to-peak amplitude of the distorted envelope is unity. The change in amplitude when the harmonics are removed is consequently only 0·3 dB.

The series for the envelope waveform when $m = 1$ can be found rather simply by recognising that it is a half-frequency sine wave with alternate half periods inverted, as in the output from a fullwave rectifier. The Fourier series is therefore the half-frequency sine wave multiplied by a square wave with the same period; i.e.

$$E_m = \sin \tfrac{1}{2}qt \times \frac{8}{\pi}(\sin \tfrac{1}{2}qt + \tfrac{1}{3}\sin \tfrac{3}{2}qt + \ldots)$$

which can be readily transformed into

$$E_m = \frac{4}{\pi} - \frac{8}{\pi}(a_1 \cos qt + a_2 \cos 2qt + \ldots + a_n \cos nqt + \ldots)$$

where $$a_n = (4n^2 - 1)^{-1}$$

Thus, when the harmonics in this instance are removed leaving only the fundamental, the amplitude drops from 2 to $8/3\pi$, i.e. a change of 4·6 dB for $m = 1$ compared with 0·3 dB for $m = 0{\cdot}5$, which underlines the importance with test cards which are to be radiated of keeping the modulation depth as low as is consistent with reasonable contrast on the final picture.

The above analysis assumes that the grating waveform is sinusoidal after γ correction. While this can be achieved, some obvious difficulties present themselves, and it would be profitable to investigate briefly the magnitude of the error. For simplicity, the calculation will assume negligible quadrature distortion; i.e. it can be taken to apply to a single-sideband modulated sinusoid, provided that the depth of modulation is fairly low.

The γ of the picture tube has at various times been assigned values ranging from 2·2 to 3, at the moment a value close to the top end of this range being preferred. Since fractional exponents introduce considerable complication, but integral exponents present little difficulty, the distortions will be calculated for gammas of 2 and 3 as approximately representing the normal range.

A sinusoid varying between the limits of zero and unity is represented by

$$E = \tfrac{1}{2}(1+\cos qt)$$

After γ correction using an exponent of 2, this becomes

$$E^2 = \tfrac{1}{4}(1+2\cos qt+\cos^2 qr)$$

hence the amplitude of the fundamental frequency component remains unchanged.

If an exponent of 3 is used,

$$\begin{aligned} E^3 &= \tfrac{1}{8}(1+3\cos qt+3\cos^2 qt+\cos^3 qt) \\ &= \tfrac{1}{8}(\tfrac{5}{2}+\tfrac{15}{4}\cos qt+\tfrac{3}{2}\cos 2qt+\tfrac{1}{4}\cos 3qt) \end{aligned}$$

The amplitude of the fundamental component drops from $\frac{1}{2}$ to $\frac{15}{32}$, i.e. a change of 0·5 dB.

Unfortunately, we have not yet quite reached the end of the list of potential sources of error with frequency gratings. One further source which affects all types of grating to some extent derives from the fact that the waveform corresponding to the grating is not, as is commonly assumed, a continuous phenomenon but a relatively small number of periods of the waveform concerned, whether sine wave or square wave or whatever it may be. This is equivalent to the gating of the continuous waveform by a narrow rectangular pulse, and hence mathematically the corresponding Fourier series is

$$E(t) = \sin qt\,\{a_0+a_1\sin(\omega_L t+\theta_1)+\ldots+a_n\sin(n\omega_L t+\theta_n)+\ldots\}$$

where the expression in brackets represents the Fourier series for a rectangular pulse. In this f_L is the line frequency; the field component is assumed to be absent for simplicity; and likewise the basic waveform of the grating is taken to be a sine wave. These simplifications do not affect the argument.

The rth term can be expanded to give

$$\tfrac{1}{2}ar\,\{\cos(qt-r\omega_L t-\theta_r)-\cos(qt+r\omega_L+\theta_r)\}$$

which implies that the spectrum of the grating consists of a central frequency, identical with the basic frequency from which the grating is formed, surrounded on either side by a symmetrical set of sidebands. If the grating is formed from a nonsinusoidal waveform, its own Fourier components will give rise to still further sidebands. The shorter the gating pulse which selects the block of frequencies, the less is the decrement of the corresponding sidebands, so that the types of grating in common use in test cards may have sidebands extending over a considerable frequency range. With the higher-frequency gratings, the kind of band limitation of the video signal already discussed will reduce the amplitude of the upper sideband, leading to a decrease in the amplitude of the grating waveform accompanied, almost certainly, by a certain amount of transient

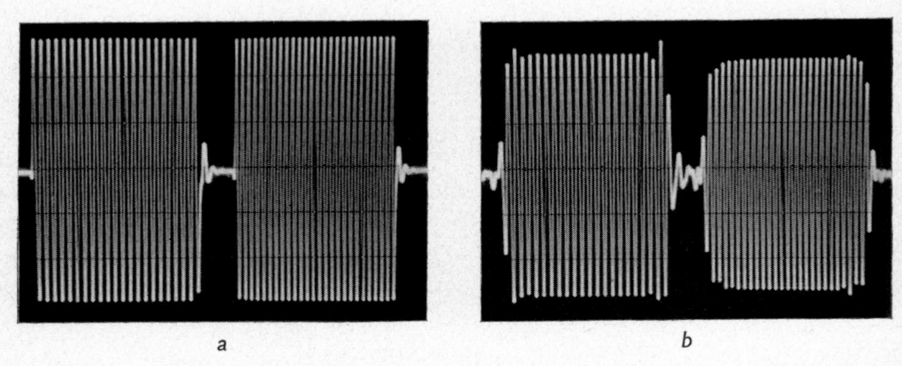

Fig. 10.9 Effect on bursts of 3·8 and 4·8 MHz of lowpass filter with 5·5 MHz cutoff

a Undistorted bursts
b Bursts of (*a*) after passage through lowpass filter

distortion. This point is illustrated in Fig. 10.9; the two isolated flat-topped frequency bursts of 3·8 and 4·8 MHz, respectively, are shown before and after passage through a phase-corrected lowpass filter with a nominal cutoff frequency of 5·5 MHz. The insertion loss at the two frequencies was made the same. Although the degree of band limitation was by no means severe, the drop in level of the higher-frequency burst and the slight distortion are quite evident. The two bursts used were rather wider than the grating waveforms often encountered, otherwise the effect would have been even greater.

This error is aggravated by the use of frequency gratings in the form of a continuous wedge, since the sideband spectrum is still further complicated by the effective pulse-width modulation introduced by the grating shape, added to which the grating width in this form tends to be somewhat less than with purely rectangular gratings, so that the sidebands have a larger amplitude.

The use of frequency gratings is accordingly fraught with a surprising number of potential errors, none of which is perhaps very great in itself, but which may, and often do, add up to an extent which makes nonsense of the reverence with which measurements of this type are sometimes treated. This is not, of course, to deny the usefulness of spatial measurements made with resolution gratings, but to show that they should only be used with a knowledge of, and respect for, their limitations.

10.3.1 Octobar resolution pattern

Chapter 12.5 contains a description of the application of a proposal by Lewis and Hauser (1962) for the measurement of bandwidth by using the response of the system under test, in that instance a television camera, to a rectangular pulse of variable duration. Briefly, the duration of the input pulse is determined for which the output pulse from the system has 87·3% of the amplitude of a corresponding pulse of long duration. If this critical pulse duration is T_p, the pulse bandwidth is defined as $1/T_p$.

In the above application, the pulse of variable duration was derived from a vertical tapered slit which, when scanned, produces a train of pulses whose durations are in arithmetical progression. However, this differs from the original proposal, primarily intended for the determination of photographic resolution, which was framed in terms of a special test pattern. In principle, this should also be applicable to the measurement of the resolution of television picture-origination equipment.

It consists (Fig. 10.10) of two symmetrical sets of eight bars whose durations are in geometrical progression, each set being the negative image of the other; from the 8-bar sequence is derived the name 'octobar'. When scanned, it gives rise to a long train of positive-going pulses followed by a similar train of negative-going pulses, or vice versa. The two sets are required since, with optical conversion devices, the resolution is not necessarily identical in the two cases. The range of pulse durations is 5:1, and, in the practical pattern, the nominal durations are rounded off by the use of the R10 series of preferred numbers.

When a diapositive of this pattern was used as the test object in a caption scanner, the waveform corresponding to the black bars was as shown in Fig. 10.11; note that, in this instance, the shortest bar is on the left and the longest is on the right. The output of this caption

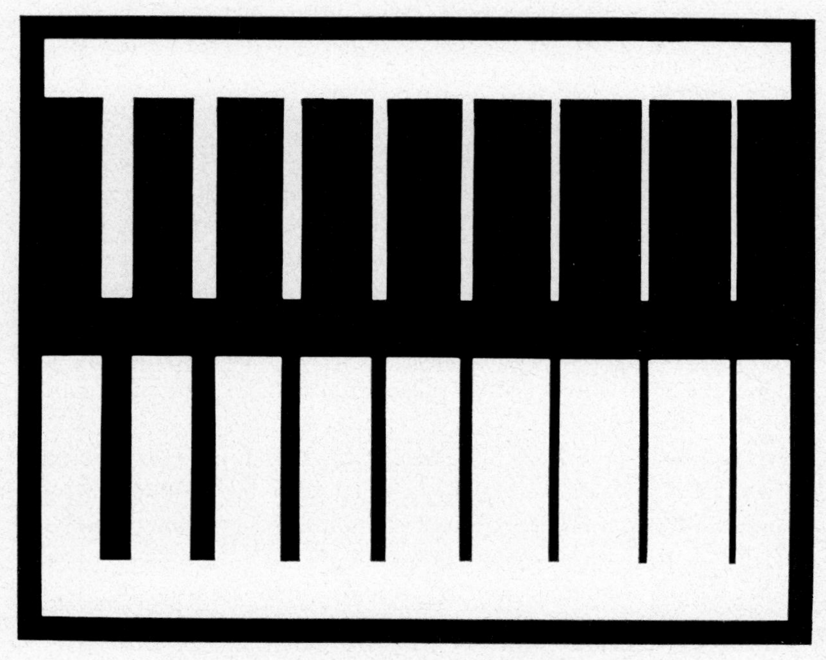

Fig. 10.10 Octobar resolution pattern

Outer dimensions of rectangle are 125 × 100 units
Bar widths are 5, 4, 3·15, 2·5, 2, 1·6, 1·25 and 1 units, respectively, spaced by 15 units

scanner was band-limited by a lowpass filter for two reasons: first, to make the range of variation of pulse amplitude more evident, and secondly to demonstrate the effect of irregularities in the pulse response. The filter gives rise to a highly damped 'ring', which is clearly evident on the base of each pulse at the top of Fig. 10.11. Nevertheless, one is in no doubt where the level of the base of the pulse lies. However, the smaller duration of the top of the widest pulses prevents little more being seen than the initial transient, and the true amplitude of the pulses is not apparent. Admittedly, this effect has been exaggerated for purposes of illustration, and the magnitude of the transient is greater than one would expect to find; but, apart from

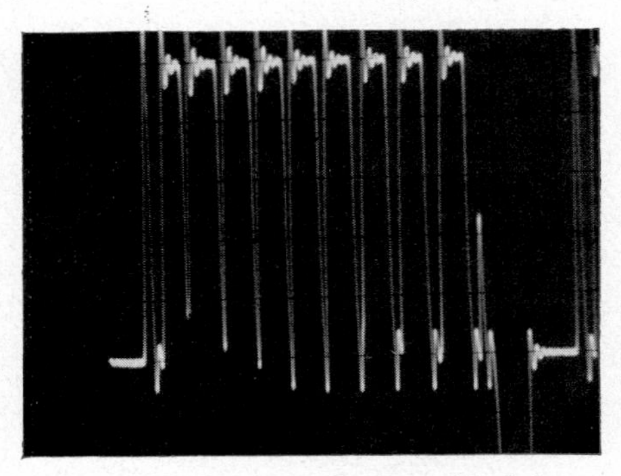

Fig. 10.11 Octobar waveform: black bars

any transient distortion of this or allied kinds, optoelectrical convertors are rarely free from anomalous nonlinear phenomena which can equally well make the pulse amplitudes ambiguous. The situation could not be improved by changing the scale of the pattern, since the filter cutoff frequency was specifically chosen so as to set the amplitude of the narrowest pulse to the critical value of 87·3%. It seems that a reference bar of much greater duration is required in addition.

Another difficulty arises from the quantisation of the pulse durations, which makes it necessary in general to interpolate between the observed amplitudes to find the critical input pulse duration. However, the combined effect of the transient and the reduction in pulse duration actually causes the pulse amplitude to rise slightly before it finally starts to decrease regularly (Fig. 10.11). Hence, unless the test pattern can be tailored to suit the bandwidth to be measured, it may occur that very few values of pulse amplitude are in any case available for determining the critical value. From this, it would seem that, for television purposes, it would be advisable to provide a greater range of pulse durations. On the other hand, for photographic purposes, for which it was originally designed, the octobar pattern is doubtless very suitable.

For these reasons, the use of a tapered slit was preferred for the measurement of camera resolution in Section 12.5. Nevertheless, this type of resolution pattern has certain merits, and further investigation into the optimum form is warranted on the above lines. Another likely improvement lies in the reduction of nonlinearity effects and high-contrast phenomena in the equipment under test by the use of a pattern with a suitably decreased contrast range, as is recommended in Section 12.5.1. It would then be advisable to add a step wedge to the pattern for standardisation purposes.

10.4 Test cards 51 and 52

These are identical except that card 51 is intended for 405-line and card 52 for 625-line standards. They are highly specialised patterns intended purely for camera testing, and as such were designed to the highest possible standards. Only a very brief description can be given here, but the details of the preparation of the masters as given by Hersee (1967) are well worth reading.

The principal requirements which these cards had to fulfil can be summarised as follows:

(*a*) measurement of the frequency response using sinewave bars, to avoid the anomalous effects described in Section 10.3
(*b*) measurement of the contrast law over a minimum range of 60:1
(*c*) measurement of the definition in the corners as well as in the centre
(*d*) the provision of a picture with a contrast range of about 30:1 and suitable for a general visual assessment of picture quality, since certain spurious effects in cameras make it necessary to avoid a total reliance on resolution and similar measurements.

These requirements are so severe that it was found impossible to use a standard photographic master. The master positive was formed as a montage of individual portions, each of which was individually standardised; e.g. the step wedge is a modified commercial type intended for photometric and photographic measurements. Note that, in this instance, the generation of subjectively equal visual steps of brightness was not adopted, since the purpose of this test card is measurement and not the adjustment of receivers or monitors as with test card D. The resolution gratings were accurately prepared sine-wave bars furnished by a telerecording channel.

Fig. 10.12 Test card 52

Various other details of the card are visible in Fig. 10.12. In particular, it should be noticed how suitable is the choice of the background picture. It has good areas of detail in the shadows and the near-white areas as well as in the middle tones; the checked skirt is useful for focusing as well as for assessing the detail resolution over a wide area including the edge of the picture as well as the centre. In addition, the general composition and appearance are pleasing. This card has been very successful and has been widely used.

10.5 Synthetic test patterns

The process used for producing test card D, i.e. by starting with an optical master and from that making a quite large number of transparencies to a high degree of accuracy is difficult, and a very great deal of care and skill have to be used to assure a sufficiently high standard in the finished product. And this is not the end, since the required video signal has then to be obtained from the transparency by a flying-spot slide scanner, a complex instrument which necessarily needs care in adjustment as well as occasional maintenance.

In spite of the many difficulties, a very good and consistent test pattern can be produced, and yet one is forced to recognise that certain elements of the pattern could have been directly generated by electronic means more easily and with considerably higher reliability; the most outstanding example is the pattern of resolution gratings, of course.

This approach has been adopted by some administrations; e.g. the useful test pattern designed by the Bayerischer Rundfunk is extensively used during European programme exchanges; but until recently, the design of such test patterns has been severely restricted by the difficulty of generating certain types of pattern element such as the circle and the diagnonal grids in test card D, which has resulted in test patterns being built up principally from resolution gratings, tone wedges and grid patterns. Even then, the designer could not afford to be too ambitious or the switching-logic circuitry became unduly complicated.

The previous limitations have now been considerably eased by the availability of compact magnetic stores which can supply the information for the synthesis of, say, the circle pattern, or rather a quadrant of a circle, since, for reasons of economy, one would only store the minimum amount of information, and a quadrant supplies all that is necessary for the purpose.

In the future, an increasing use will probably be made of test pattern generators which synthetise the signal directly. The exceptions are complicated line drawings, like the well known 'Indian head' in the RCA test pattern, and photographic inserts containing halftones, which will still have to be produced by photoelectronic means and added to the synthetic pattern. Block-lettering, required for station identification and similar purposes, can be supplied by a character memory store, or alternatively by waveform-processing methods. In the latter case, the characters tend to be more natural and pleasing.

10.6 Colour test patterns

10.6.1 Test card F

This is an extension of test card D, for checking and aligning receivers, and the basic structure is unchanged (Fig. 10.13, frontispiece). The grid pattern now serves for the adjustment of convergence and for picture linearity. The castellations are retained as a check of picture synchronisation and size, but are modified to contain blocks of colour. The top castellations are cyan, since this has a high luminance component and shows up decoder errors quite effectively. The castellations on the left-hand side of the picture are red and blue, to check the accuracy of burst gating, since these cause the greatest disturbance of the regenerated phase if the burst gate is passing picture information. Depending on the decoding method, either bands of saturation changes or 'Hanover bars' will be visible on coloured areas in the picture.

The right-hand castellations are yellow and white, to check synchronising-pulse separator operation with and without a high subcarrier level at the end of lines. To avoid confusion between effects due to the coloured castellations on the left-hand and the right-hand sides of the picture, the relevant castellation blocks are staggered vertically with respect to one another. The lower castellations are coloured green to provide a means of judging the performance of the subcarrier regenerator at the end of the field.

The upper castellations are virtually replaced at present by 20 lines of the standard UK colour-bar signal to allow simple measurements to be carried out on receivers, if required. In this position, they can be viewed on a waveform monitor with fairly simple synchronising facilities by using the field pulse as a trigger. Such colour bars are quite impossible to obtain accurately by optical means, and are accordingly added from an electronic generator (Section 10.8).

An ingenious feature of this test card is the picture within the circle. It was required to retain the circle for checking picture geometry while including a pattern at the centre of the picture suitable for checking static convergence, which would preferably contain a white cross on a black ground. This need was met by the central picture of a child drawing a noughts-and-crosses pattern in the centre of a blackboard. The use of a child instead of an adult for a model ensures that the picture will not become outdated so quickly as a result of changes in fashion.

The resolution gratings correspond to frequencies of 1·5, 2·5, 3·5, 4, 4·5 and 5·25 MHz. The use of 2 and 3 MHz bars has been carefully avoided, since their harmonics can cause 'buzz on sound' with receivers using intercarrier sound. The bars are now made rectangular in cross-section to avoid the production difficulties associated with sinusoidal bars, but their amplitude has been limited to 78% of white amplitude to avoid overmodulation due to the $4/\pi$ effect mentioned above.

The tone wedges have been increased in number to six, principally to match the number of blocks of resolution gratings. The luminance steps are still chosen to give equal subjective steps of brightness. These also play a dual role, since they are required for adjusting the grey-scale tracking of the receiver.

10.6.2 WDR test card

An all-electronic test card of considerable interest, which perhaps sets a precedent, has been utilised by the Westdeutscher Rundfunk in the Federal Republic of Germany (Schedel, 1967). The basis is their test card B, which consists of a standard linearity-test grid, in the centre of which is a rectangle containing horizontally disposed areas of black and white, a continuous tone wedge and five blocks of resolution gratings.

On each side of the central rectangle is a vertically disposed block of four squares, each of which carries a signal designed so that the major faults or misalignments of the colour received can be corrected by reference to the screen of the receiver only. These eight signals constitute the simpler of two sets of signals designed by Mayer and Holoch (1967) for this purpose.

On the left-hand side, in order, are:

(*a*) cyan with the luminance value for saturated green
(*b*) 12% luminance
(*c*) subcarrier corresponding to $G - Y = 0$
(*d*) $+(R - Y)$.

On the right-hand side are:

(*e*) saturated green
(*f*) $\pm(R - Y)$
(*g*) $+(B - Y)$
(*h*) $\pm(B - Y)$.

The order of these signals is not arbitrary, but is chosen so that, in certain instances, an adjacent reference signal is provided for a signal which is being adjusted.

A full description of the manner in which these test signals can be used for the accurate adjustment of each of the principal parameters of the colour monitor or receiver would require far more space than can be allowed, but some very brief examples will be given so as to convey the principles.

The balance of the colour demodulators is checked by means of the 100% saturated green. If the alignment of the green matrix is correct, which is checked with another signal, a saturated green should give rise to zero signal to both the red and the blue guns of the tube. The green gun is then switched off and the relative gains of the two demodulators are adjusted until the green block appears black. It is located immediately next to a black area, which serves as a comparison reference. At the same time, the saturation control is operated; the setting is correct when the true reference block is reached.

If there is a delay error in the $B - Y$ output, a horizontal striping (Hanover bars) appears in the $\pm(R - Y)$ block, which can be taken out by adjustment of the delay, provided that, at the same time, the phase of the regenerated carrier is adjusted so that no colour appears in the $+(R - Y)$ block.

Certain difficulties can arise if the signal suffers from differential phase. This can be recognised by superimposing subcarrier with an angle along one axis on the tone wedge. When demodulated along the other axis, this should appear colourless, and the presence of any colouring along the wedge is an indication of differential phase, provided the receiver is otherwise correctly aligned.

The use of these test patterns is neither particularly quick nor easy, but the process can be reduced to a routine, so that a competent serviceman can align both the decoder and the display with the minimum of equipment.

10.7 Pluge

Pluge (pronounced *plooj*) is an acronym for 'picture lineup generating equipment'. It is the universally used name for a special test-pattern generator for the accurate luminance adjustment of picture monitors. Originally designed by the BBC for its own use, it has proved so useful that it is now very widely used, sometimes with minor variations.

Picture monitors in production and operational areas are used, not only for checking the technical quality of pictures, but also for judging the visual effect of the pictures considered from an artistic point of view. It is accordingly very important that they should be set up, not just for the approximately correct brightness and contrast range, but in such a way that all monitors are as alike as possible. This requirement is particularly stringent when a number of monitors are grouped within a single field of view, as, for instance, on a control desk.

It is not very difficult to adjust the peak white luminance of a monitor to a standard value. This can be done with a fullfield signal containing peak-white bars, or a 'window' signal, with a photometer having a reasonably narrow acceptance angle. The contrast control is then varied until luminance is 70 cd/m^2, or whatever the standard figure may be. However, such a setting does not uniquely determine the contrast setting, because there is a certain interaction between the contrast and brilliance controls.

The chief difficulty lies with the adjustment of the contrast range, as it effectively is, by the adjustment of the brightness control so as to give the correct value in the near-black region. In principle, this could be achieved with a special signal containing a component with a suitably low signal level, but such a measurement requires a spot photometer accurate to very low light levels and with correspondingly low internal scatter; and, in any case, it would be operationally inconvenient. Furthermore, it does not take into account the effect of the ambient lighting on picture quality.

The Pluge signal avoids these difficulties by using the picture monitor, in a sense, as its own extinction-type photometer during the adjustment of the near-black region. The combined line signal of the

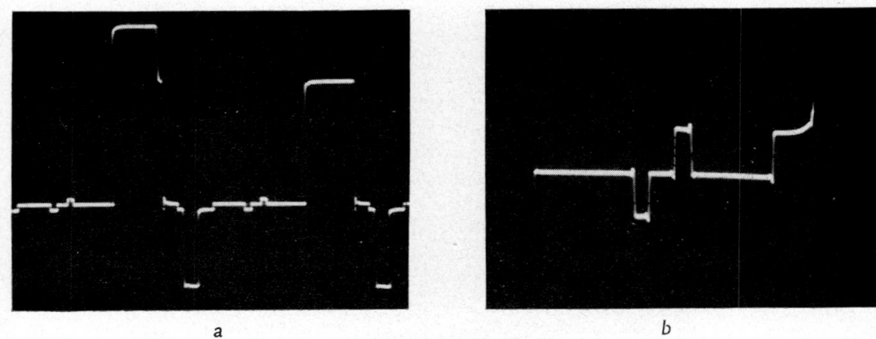

a b

Fig. 10.14 Pluge waveform

a Line signal
b Detail of 'ribbons'

two principal areas of the waveform is shown in Fig. 10.14*a*, and the picture monitor display for different states of monitor adjustment are given in Fig. 10.15. The right-hand side of the screen contains two large uniform areas of tone, one at white level and the other at mid-grey. On the left-hand side are two narrow ribbons corresponding to two short rectangular pulses in the waveform, one erect and one inverted, situated on a small pedestal; each is 2% of white level. The pedestal amplitude is fixed for a given set of circumstances, but can be modified to cater for a change in the ambient illumination.

To use this signal, the luminance of the white area is first set to the reference value with a spot photometer. The brightness control is then

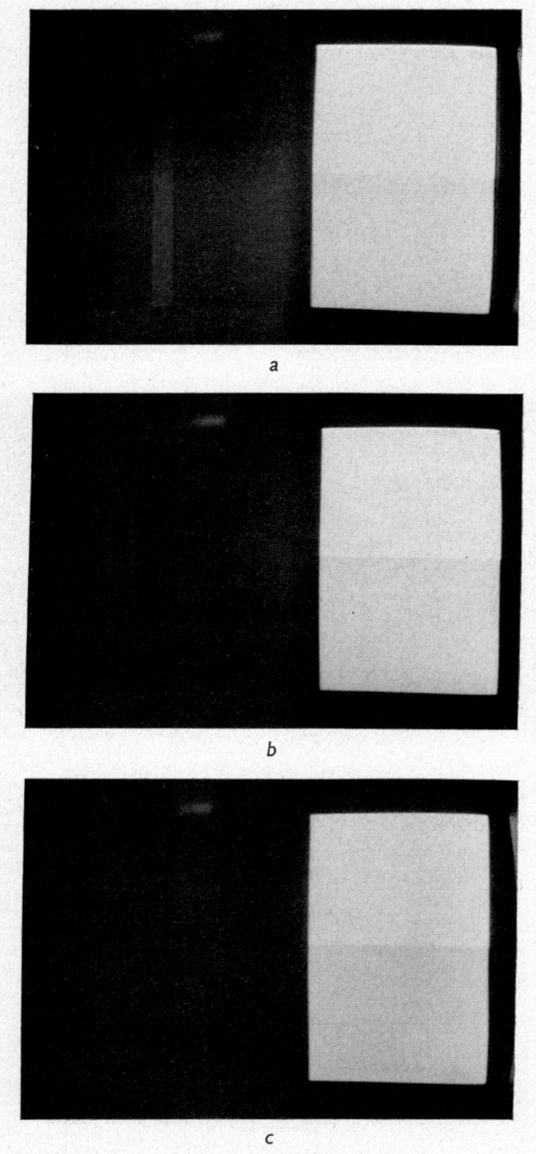

Fig. 10.15 Pluge waveform observed on picture monitor

a Brightness too high
b Brightness too low
c Brightness correct

varied while the narrow ribbons on the left-hand side of the picture are carefully watched. If the brightness is too high, both strips are visible, as in Fig. 10.15*a*; if it is too low, neither is visible, as in Fig. 10.15*b*. At the correct setting, the left-hand strip just disappears, with the right-hand strip still just visible. An attempt has been made to demonstrate this in Fig. 10.15*c*, but such borderline phenomena do not reproduce well. The exact point of disappearance is easily found, and can be accurately repeated.

The brightness of the white area is usually slightly altered by the adjustment of the brightness control, and needs resetting by the contrast control, after which the black level should be readjusted. The process is highly convergent, and the correct adjustments are reached expeditiously and with certainty. The technique is simple and can very quickly be used reliably even by comparatively unskilled staff.

The correct setting for the near-black region is not completely independent of the ambient light, and it is necessary to select a value of the waveform pedestal for given ambient lighting conditions. This relationship between the pedestal and the ambient illumination has been investigated by Quinn and Siocos (1967), who publish a curve for low-light conditions. They further show that over the measured range of ambient illuminations there is a constant range of ambient illuminations for each value of pedestal over which the appearance of the pictures is acceptable. Hence, for any given lighting conditions, an optimum value of pedestal is selected which must be adhered to.

Two precautions must be taken. The first, and more important, is to ensure that the signal levels supplied to all picture monitors are correct and stable in amplitude. A change of only a small fraction of a decibel in the signal amplitude supplied to one picture monitor is sufficient to cause a noticeable difference between the pictures. The second is to ensure that the displayed peak white has not been falsified by any possible clipping of the white level, and the midgrey area is therefore included as a check. A single bar could be drastically limited from some cause without its being very apparent on a visual observation of the waveform. However, such effects should admittedly rarely occur, and, for simplicity, some users prefer to omit the grey bar.

10.8 Colour-bar signals

Colour-bar signals form a group of electronically generated colour-test signals, some varieties of which are in very widespread use. They consist of a number of uniform blocks of high-saturation chrominance signals together with a reference luminance level and possibly also black level, their arrangement and specification depending on the use to which the particular signal is to be put. The great advantage of such synthetic test signals is the possibility of precise standardisation, together with the ability to incorporate in them chrominance components corresponding to colours of a degree of saturation which it would be difficult, or even impossible, to achieve by any other means.

It is worth briefly noting in passing that colour transparencies, which have been extensively employed as standard test objects, have a number of inherent limitations, and should only be used with these strictly in mind. They derive from the deficiencies of the available colour emulsions, where not only is the range of chromaticity values less than can be achieved with a colour camera, but the maximum available colour saturation is also very significantly less. Such test objects are not capable of exploiting the full potentialities of modern colour systems.

Colour-bar signals may be most broadly classified into horizontal colour-bar signals and vertical colour-bar signals, according to the orientation of the resulting areas of colour on a picture monitor. Horizontal signals, however, may be dismissed fairly summarily, since they are normally encountered only in research and development work, where they are employed to a very limited extent for the investigation of phenomena concerned with vertical transitions. No generally standardised waveform exists, and generators are usually designed to fulfil some specific purpose.

On the other hand, vertical signals are encountered so frequently that the qualifying adjective is seldom or never employed in practice. Several versions with precise specifications exist. In all cases, the line signal consists of a luminance bar followed by six uniform subcarrier blocks, each of which corresponds to a uniform area of highly saturated colour; in most, the sequence is closed by a black bar. The six hues are, in all these signals, the three primary colours and their complements, i.e. the hues corresponding to the six possible combinations of the colour-separation signals when these are either zero or unity, thus:

Hue	*R*	*G*	*B*
Yellow	1	1	0
Cyan	0	1	1
Green	0	1	0
Magenta	1	0	1
Red	1	0	0
Blue	0	0	1

The order of the bars is not arbitrary, but follows a logical sequence. That given immediately above is known as the 'amplitude step' sequence, since their corresponding luminance components form a descending series of steps similar to a staircase waveform, but 'phase step' bars are also known in which the bars are arranged in the order of the angles of the subcarrier vectors. The argument for this is the smaller resulting transitions when the signal is analysed with a vector monitor, but in practice it is found that the advantage is more theoretical than real, and the amplitude-step sequence is now practically universal.

The very great advantage of signals of the type of the colour-bar signal is that they can be derived from three colour-separation signals, which are simple rectangular waveforms whose amplitudes may be measured very precisely by methods such as the calibrated square wave (Section 2.4). When these three signals are applied to the three inputs of a coder, the output signal, if the coder is correctly aligned, should agree precisely with the specified coded signal. Coding errors give rise to characteristic changes in the waveform which can be identified.

A considerable improvement in the analysis of the coded waveform can be obtained with a vector monitor (Section 11.5), which effectively plots on a polar diagram the points corresponding to the tips of the subcarrier vectors, including the burst vector. These points may then be compared with their correct positions as indicated on a graticule,

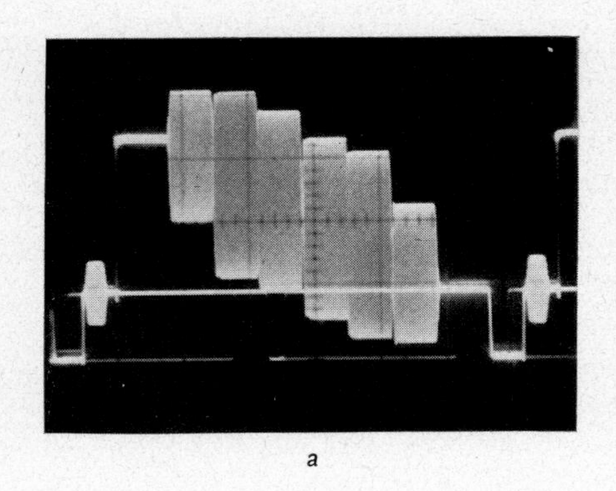

a

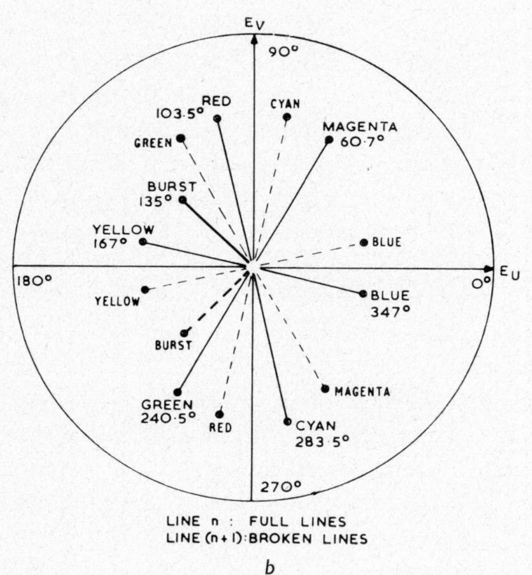

b

Fig. 10.16 Colour-bar signal

a Colour-bar waveform (Pal)
b Corresponding vector-monitor display

thus making it possible almost at a glance to determine whether or not the coded colour-bar signal is within specification. A photograph of a colour-bar signal and a diagram of the corresponding polar display are given in Figs. 10.16*a* and *b*; the signal is according to the Pal standards.

The converse of the check of the coding process may also be carried out. If a colour-bar signal known to be correct is available, it can be applied to the input of a decoder, when the waveforms produced by the three colouring channels should have amplitudes corresponding to the specification values.

Colour-bar signals are also of considerable importance for the operational checking of videotape recorders and transmitters which involve modulation processes, since it is found that such signals, simulating as they do camera signals of scenes containing high-amplitude, highly saturated colours, are most useful dynamic tests of performance, particularly since, in such instances, the results cannot always be deduced from signals of low amplitude and saturation.

10.8.1 Colour-bar nomenclature

The system of nomenclature formerly employed for the colour-bar waveform was never very satisfactory, a matter of some consequence in the UK where three colour-bar signals must be distinguished, although of lesser importance in some other areas where only a single standard signal is employed. A decision was jointly taken, therefore, by the UK broadcasting authorities, programme companies and television manufacturers to adopt a system which, while logically unexceptionable, is unfortunately rather clumsy. Nevertheless, it is now the subject of a CCIR recommendation (1970).

According to this, each colour-bar signal is specified by a sequence of four numbers which define the percentage amplitudes of its corresponding colour-separation signals, the reference being the amplitude of any one of the γ-corrected colour-separation signals forming a white bar. These are, in order:

(*a*) the maximum percentage amplitude of E'_R, E'_G, E'_B in a luminance bar
(*b*) the minimum value of the same quantity
(*c*) the maximum percentage amplitude of E'_R, E'_G, E'_B in a coloured bar
(*d*) the minimum value of the same quantity

Using this convention, the four most common colour-bar signals become:

Former name	*a*	*b*	*c*	*d*
100% bars	100	0	100	0
95% bars	100	0	100	25
EBU bars	100	0	75	0
75% bars	77	7·5	77	7·5

These would be written, for example, as 100–0–100–0 bars, and so on. Although the intention behind this later convention is very commendable, one suspects that the old names, in spite of being quite inaccurate in one instance, will not easily be superseded.

10.8.2 100–0–100–0 colour-bar signal

The bars in this variant are chrominance signals which correspond to 100% amplitude, 100% saturated colours, from which is derived the former name of '100% bars'. They form the most severe test signal one can devise for the Pal and NTSC systems; the corresponding signal is not employed with Secam. The field of application of this signal cannot be unrestricted, since the peak excursion of the yellow and cyan bars in the coded video signal reaches 1·234 V, which corresponds very nearly to zero carrier in the radiated signal, and some distortion would be expected in that region not only from the transmitter but also from the envelope detector in the receiver. By the same token, it is not reasonable to expect it to be handled impeccably by a videotape recorder, all the more since camera subjects corresponding to 100% amplitude, 100% saturated colours are not attainable under practical television conditions.

The peak excursions of the yellow and cyan bars would, in fact, be considerably greater if standard weighting factors were not introduced into the $R-Y$ and $B-Y$ components during the coding process. These factors also result in equality of the maximum excursions of the yellow and cyan bars together with equality of the minimum excursions of the red and blue bars (Fig. 10.17). A table of the amplitudes is given in Appendix 16.1.1. These equalities constitute a useful quick guide to the accuracy of the coded video signal, but such indications are more qualitative than quantitative, and should not be expected to replace an analysis using a vector monitor.

The most appropriate application of the 100–0–100–0 colour-bar signal is as an operational check of the alignment of coders and decoders, where its larger amplitudes give it an advantage over other

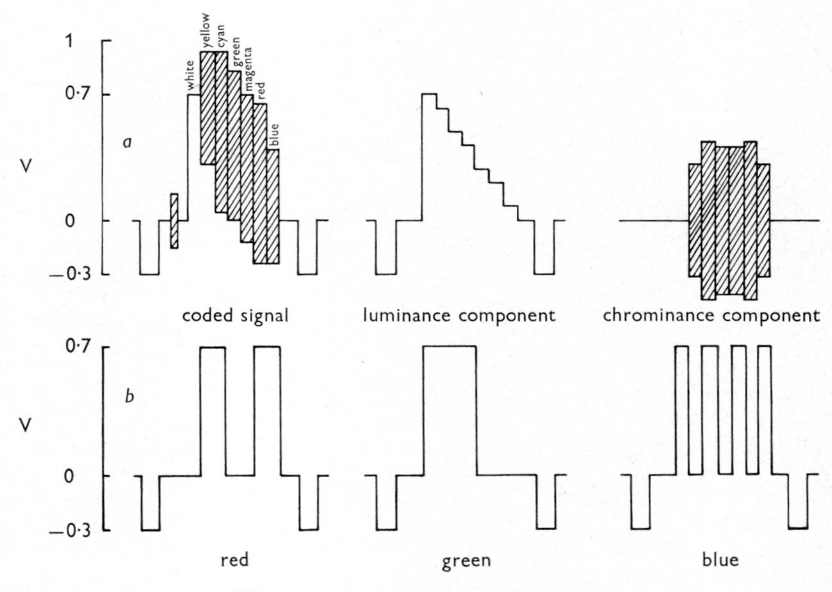

Fig. 10.17 100–0–100–0 colour bars
a Detail of coded signal
b Corresponding colour-analysis signals

colour-bar signals. Even though it is in very widespread use for the precise alignment of coders and decoders, it is not necessarily the most suitable signal; but, of course, its general availability in operational areas gives it an advantage.

An application of some interest is the use of this signal as an insertion test signal for the continuous monitoring of the performance of a studio technical area (Section 7.4.5).

10.8.3 100–0–100–25 (95%) colour-bar signal

This signal (Fig. 10.18) is of practical importance, since it has been found to correspond fairly closely in its characteristics to the signals obtainable from modern colour cameras viewing very bright, highly saturated colour scenes. One would therefore expect that any equipment handling the colour signal should be able to deal with this colour-

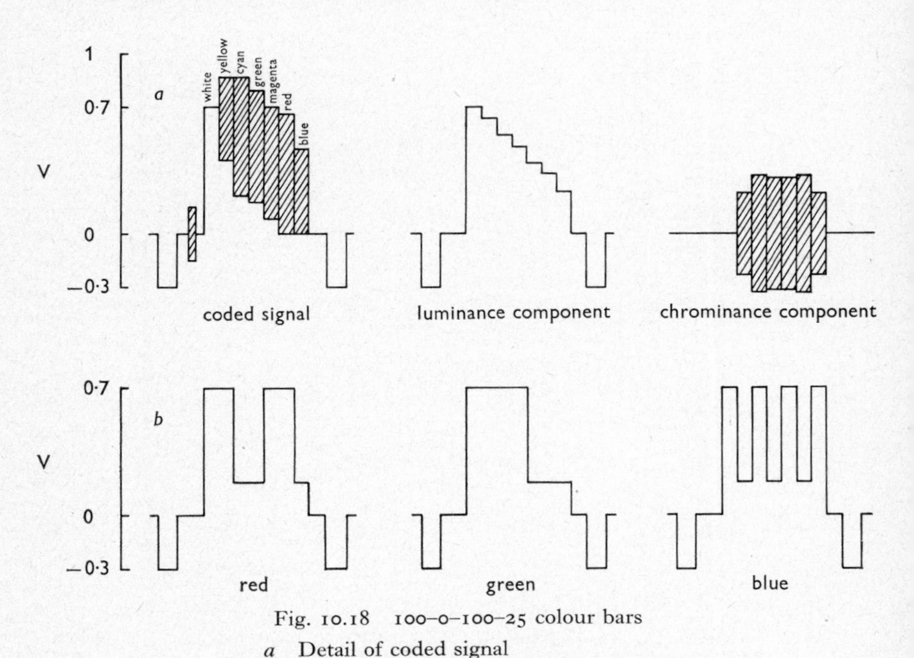

Fig. 10.18 100–0–100–25 colour bars

a Detail of coded signal

b Corresponding colour-analysis signals

bar signal without significant distortion. The maximum yellow and cyan excursion is now 1·175 V, i.e. some 6% lower than with the 100–0–100–0 bars; not a large amount, perhaps, but of importance when one is concerned with peak-modulation conditions.

Although the name '95% bars' is commonly encountered for this signal, it is to be deprecated, since it is no longer valid. The specification of the coded colour-bar signal is framed in terms of the

γ-corrected values of the colour-separation signals, whereas the percentage saturation is calculated from the values before correction. Consequently, any change in the accepted value of γ for the picture tube results in a change in the nominal percentage saturation, even though the characteristics of the coded video signal remain unchanged. Originally, it was assumed that $\gamma = 2{\cdot}2$, but measurements on more modern display tubes have made it necessary to revise this to a mean value of 2·8. The saturation of the 100–0–100–25 colour-bar signal is now more correctly 98%.

A convenient feature of this form of the signal is the equality of the minimum red and blue excursions with black level, which are consequently provided with a built-in reference for checking purposes. A table of the amplitudes of the encoded video-signal components is given in Appendix 16.1.1.

The similarity of this variant of the colour-bar signal to a practicable camera signal suggests that it might be profitable to combine it with a test card; in fact, this is made use of in test card F (Section 10.6.1), in which 20 lines start the top of the picture. With videotape machines, it is useful to combine it in the same field as a large area of saturated red, serving for the measurement of velocity errors (Section 13.7.2).

10.8.4 100–0–75–0 (EBU) colour-bar signal

This signal was originally devised by the European Broadcasting Union as a standard signal for the subjective testing of colour television systems. It is still in extensive use in Europe for a variety of purposes, and is, in fact, regarded as the general-purpose colour-bar signal.

It is apparent from the designations of the two signals that the amplitudes of the colour-separation signals in this instance are identical with those in the 100–0–100–25, signal; but, in the latter, they are effectively provided with a 25% pedestal. Consequently, the amplitudes of the chrominance components are also equal, but the maximum excursions of the yellow and cyan bars in this signal are now 1·00 V, i.e. the same as white level, instead of 1·175 V. Likewise, the minimum excursions of the red and blue bars are now brought well down below black level. These dissimilarities can be verified by a comparison of Figs. 10.18 and 10.19.

The decrease in the peak-to-peak video signal amplitude by $17\frac{1}{2}\%$ makes the 100–0–75–0 signal significantly less stringent as far as transmitters and other equipment employing modulation are con-

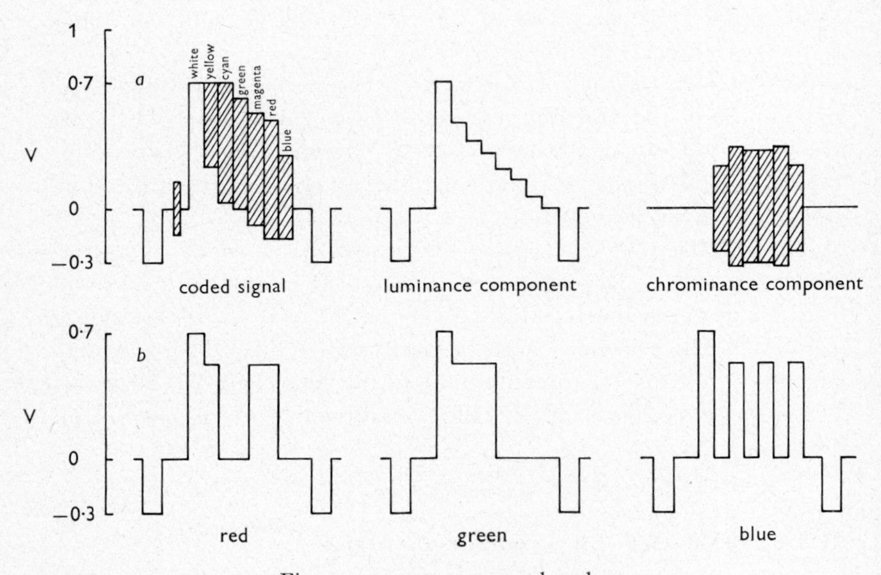

Fig. 10.19 100–0–75–0 colour bars

a Detail of coded signal
b Corresponding colour-analysis signals

cerned. However, this is not necessarily an advantage, and many engineers feel that it fails as a general-purpose signal by not being sufficiently representative of the signals which may be obtained from a colour camera. The amplitude range between white level and, say, 1·175 V is of great importance with transmitters employing negative modulation and with the corresponding receivers, particularly with respect to phenomena such as 'buzz on sound'; and this area is left untested by the 100–0–75–0 (EBU) signal. Instances are known where a change from this type of bar to the 100–0–100–25 bar signal has revealed distortions which previously had passed unnoticed, and which would undoubtedly have degraded camera pictures.

10.8.5 525-line '75%' colour-bar signal

In the USA, Canada and some other countries employing 525-line standards, this is in very widespread use as a general-purpose colour-test signal, but is seldom encountered elsewhere. It is essentially the 100–0–75–0 (EBU) signal with the amplitude of the luminance bar reduced to 75% of white level, and consequently would be designated 77–7·5–77–7·5 in the UK notation.

The amplitudes of the signal components are given in Fig. 10.20, but in this instance not in volts, but in the IEEE units preferred in the countries where the '75%' signal has been standardised; it will be recalled that, according to this convention, the peak-to-peak video-signal amplitude is expressed as 140 IEEE units, the interval between blanking level and white level being 100 units. The signal is always superimposed upon a 7·5 unit pedestal (signal setup).

In its most usual version, it is a fullfield composite signal in which the colour bars proper occupy the upper 75% of the field, while the lower 25% contains, in order, a $-I$ bar, a white bar, a Q bar and finally a black bar. In a line-repetitive display on a waveform monitor, these latter bars appear superimposed on the vertical colour-bar waveform, so that the white bar is available as a reference for the tops of the yellow and cyan bars, which should also be located at 100 IEEE units, as well as serving as a clear indication of white level. The 75% luminance bar (77 IEEE units) may be utilised as a reference for the top of the magenta bar, which should also be at a level of 77 units. The $-I$ and Q bars are useful signals for checking both encoders and decoders.

The fullfield version is well designed and practical, converting the straightforward colour-bar signal into what is almost a test card. It has, however, the disadvantages associated with the 100–0–75–0 (EBU) signal, in that it lacks the large chrominance amplitudes at high luminance levels required for the proper testing of the limiting modulation capabilities of transmitters, microwave links etc.

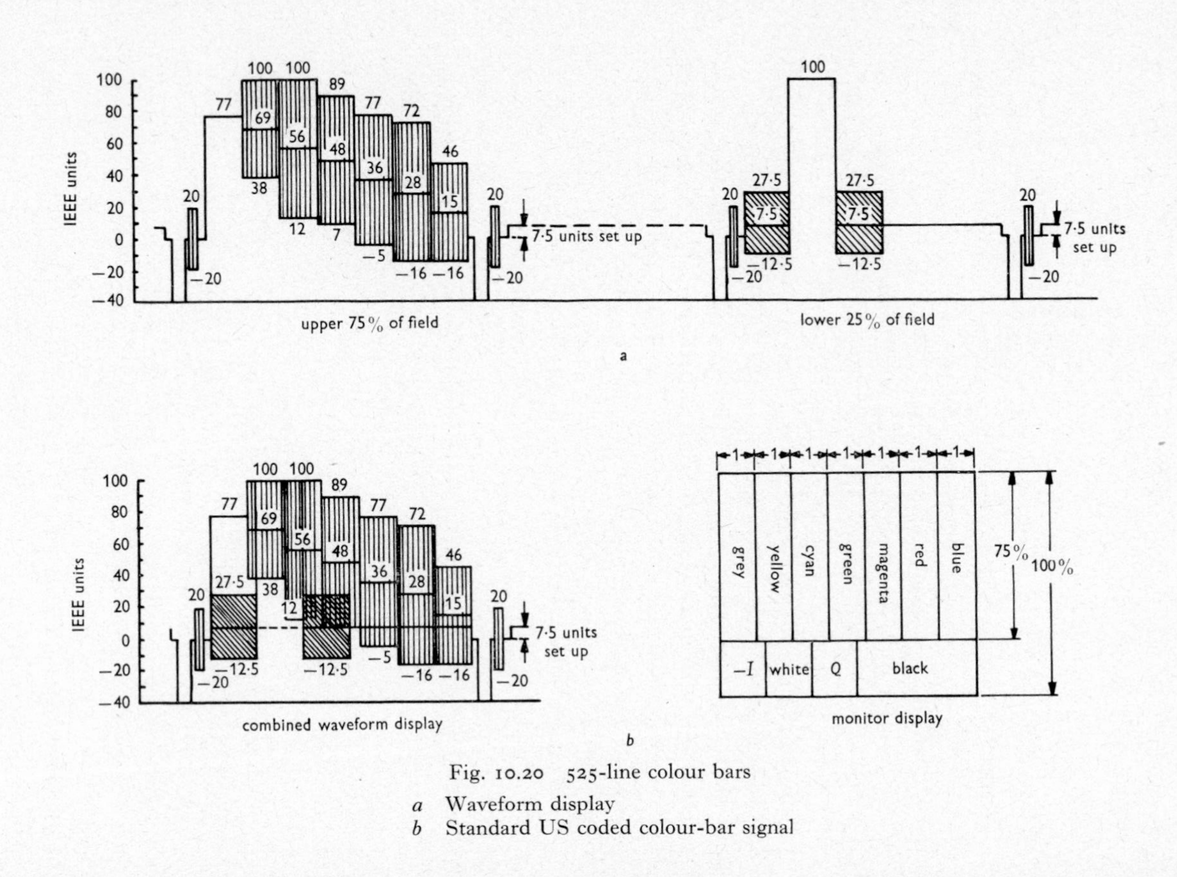

Fig. 10.20 525-line colour bars

a Waveform display
b Standard US coded colour-bar signal

10.9 Test chart 57

This is a grey-scale reflectance chart intended primarily for the very important task of setting the tracking of a colour camera, i.e. the adjustment of the optoelectrical-transfer characteristics of the colour-separation channels to equality, within close limits, at all points within a contrast range of 40:1, or at least of 30:1. If such a per-

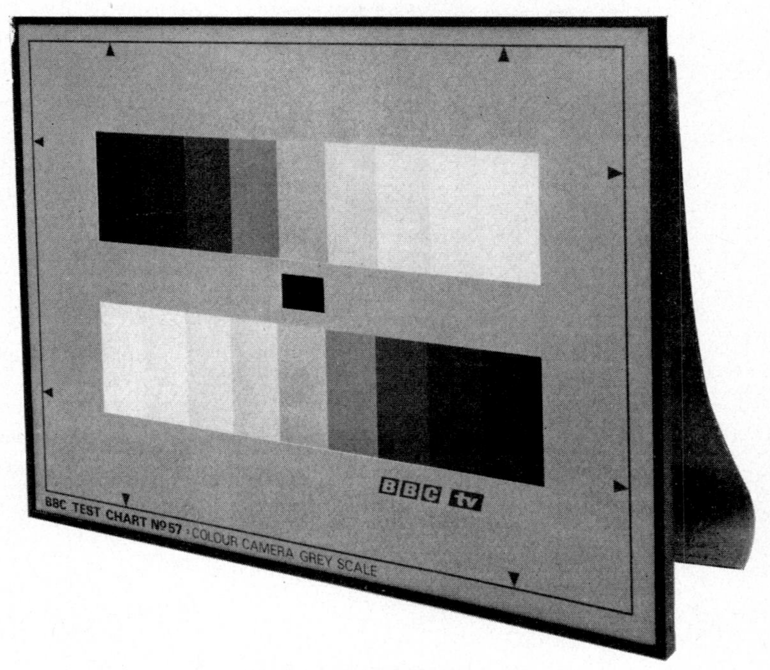

Fig. 10.21 Test card 57

formance is not achieved, the hue of the corresponding picture will depend noticeably on the intensity of the illumination.

Although there are numerous ways of checking the tracking of a colour camera, the most generally suitable has been found to be the use of a test card having an overall mean reflectance close to that of an average scene, with the range of luminance values provided by centrally disposed grey scales having a total area of the order of 30%

of that of the area viewed by the camera. This presents the whole of the transfer characteristic in quantised form and minimises difficulties due to secondary effects within the camera. It further becomes possible to adapt the test card for other important adjustments without prejudice to its primary function.

Although originally designed by the BBC for its own internal use, this card has received such wide acceptance by other broadcasting authorities that a brief description would seem justified. More detailed information is given by Lent (1970). The practical details of its use are given in Chapter 12.3.2.

The standard version (Fig. 10.21) is based on a stiff, self-supporting card of stable material specially printed so as to have the required background reflectance with great uniformity and with a total absence of specular reflection, the usable area being 610 × 455 mm. When not in use, it is protected by a transparent plastic cover, but a raised rim is also provided along the outer edges to avoid damage if it should carelessly be placed face downwards without the cover. Such precautions are highly necessary in view of the nuisance value of scratches and abrasion marks. Apertures are made in the card to receive the grey scales, which are separately prepared, and an additional small central aperture is covered at the rear by a 4 in-square box lined with light-absorbent material, thus providing a close approximation to a true black in the classic 'black body' manner.

Except for the two steps of lowest reflectance, the steps of the grey scale are prepared by a variable-area printing process, somewhat similar to the familiar halftone printing process, which permits the reflectance to be set in production with great precision; in prototype models, a specially chosen photographic paper was used. The last two steps are made from selected materials having low, standardised reflectance. All of the reflecting surfaces are controlled, so that at no wavelength between 400 nm and 700 nm does the reflectance change by more than ±4%, thus ensuring an adequate grey-scale neutrality.

In principle, the reflectances of the successive steps would most conveniently follow the reciprocal of the camera γ corrector, thus furnishing an output waveform consisting of a number of equal increments in voltage; but, owing to differences between cameras in the contrast range over which the γ correction is maintained, some modification is necessary. From measurements on a range of cameras, a satisfactory compromise has been established by making the γ of the steps 2·5 in the 'white' region, decreasing to 2·2 in the dark-grey region. The reflectance of the white step is made precisely 60%, since it has been found by experience that, in studio work, a limitation of the highest reflectance to that value furnishes the most acceptable rendering of facial tones. With a 40:1 contrast range in the grey scale,

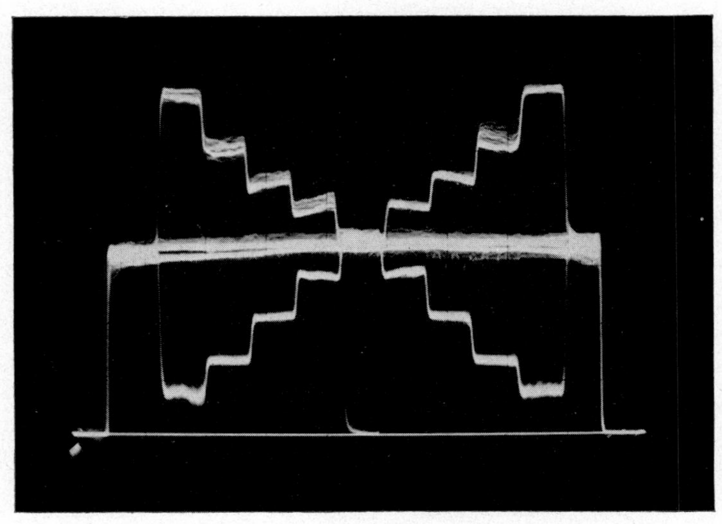

Fig. 10.22 Typical waveform display from test card 57

the reflectance of the lowest step becomes $1\frac{1}{2}$%. A typical waveform display is shown in Fig. 10.20.

The central 'superblack' area provides a means for the adjustment of flare correction; and moreover, since typical γ correctors do not conform to a true power law in the near-black region but tend to assume a constant slope, the superblack may also be utilised to represent true black level when the contrast range is 100:1, usually accepted as the norm for modern cameras, with insignificant error. In fact, even with a contrast range of 30:1, the error is not serious.

The superblack waveform can be seen at the bottom centre of the waveform in Fig. 10.22.

A very interesting paper by Royle (1970) gives a clear and detailed account of the way in which the original BBC design was used as a basis for the commercial manufacture of test card 57. The amount of professional expertise and knowledge involved, not to speak of the specialised equipment, is impressive, and should effectively deter any one who believes that this type of test card is easy to make.

10.10 References

CCIR (1970): Documents of the XIIth Plenary Assembly, New Delhi, Recommendation 471

CHERRY, C. (1949): 'Pulse and transients in communication circuits' (Chapman & Hall)

HERSEE, G. (1967): 'A survey of the development of test cards used in the BBC', BBC Engineering Division Monograph 69

IRE (1961): 'IRE standards on video techniques: measurement of resolution of camera systems', *Proc. Inst. Radio Engrs.* **49**, (2), pp. 599–601

LENT, S. J. (1970): 'BBC test chart 57: new grey scale reflectance chart for colour cameras', *IERE Conf. Proc.* 18, pp. 95–102

LEWIS, N. W., and HAUSER, T. V. (1962): 'Microcontrast and blur in imaging systems', *J. Photogr. Sci.*, **10**, pp. 290–293

MAYER, N., and HOLOCH, G. (1967): 'Test Signale zur Einstellung des Pal-Farbfernsehempfängers nach dem Fernsehbild', *Radio Mentor*, **33**, Pt. I, (10), pp. 781–785, and Pt II, (11), pp. 857–859

QUINN, S. F., and SIOCOS, C. A. (1967): 'The Pluge method of adjusting picture monitors in television studios', *J. Soc. Motion Picture Televis. Engrs.*, **76**, (9), pp. 925

ROYLE, J. T. (1970): 'The production of the BBC test chart 57: colour camera grey scale', *IERE Conf. Proc.* 18, pp. 103–111

SCHEDEL, W. (1967): 'Test-Bild Vorschläge für das Farbfernsehen', *Funktech.*, **13**, pp. 474–476

11 WAVEFORM MONITORS

11.1 General

The waveform monitor is probably the most ubiquitous and frequently used item of television test equipment. Indeed, it is difficult to conceive of a television service under present conditions without a large number of waveform monitors.

It is convenient to classify them as follows:

(*a*) observation monitors
(*b*) television waveform monitors
(*c*) general-purpose waveform monitors
(*d*) vector monitors.

Any such classification must be somewhat arbitrary, since it inevitably depends on the operational practices of an organisation, and individual uses occur which do not strictly fall under any of the above headings; but it probably covers most practical cases reasonably well.

11.2 Observation monitors

This group includes those monitors whose primary purpose is to indicate the broad general features of the video waveform. Sometimes almost their only function is to indicate the presence or absence of the video waveform at a given point.

The most important type is the channel monitor, which checks that the principal characteristics of the video signal from a given picture source are correct. It will usually be closely associated with a picture monitor, and may even be designed to match it in dimensions so that such pairs can be conveniently located in studio desks.

An essential feature of these instruments is the reduction of the number of facilities to the absolute minimum, to make it possible to select any required display immediately by means of a small number switches, the triggering being completely automatic. Another feature is the attempt to provide a display as large as possible, for the salient features of the video waveform to be recognisable quickly, even when a number of channel monitors and their associated picture monitors are mounted side by side.

The function of a channel monitor is to check that the chief characteristics of the video signal are correct, e.g. that signal white occurs at the correct level with the signal blacks at a position corresponding to the correct dynamic range of the video signal. If the signal is composite, the synchronising information must follow the specification. A watch is kept on 'speculars', i.e. narrow pulses produced by specular reflection from metal objects, buttons, eyeglasses etc., which may have an amplitude greater than signal white. Undesirable effects such as line or field distortion, hum and noise are also monitored. In colour transmissions, a check is also kept on the amplitudes of the signals from the colouring channels, i.e. R, G and B, together with the luminance or Y channel in 4-tube cameras.

A brief review of the specification of a model designed by the author's laboratory in close association with planning and operational personnel can be a guide to the functions to be provided in a channel monitor, although individual requirements and preferences will, of course, vary, but this may be taken as a guide.

This monitor is completely solid-state. The cathode-ray tube is mounted with the normal x scan vertical to obtain a display height of 8 cm. This is less than is really desirable, since this type of monitor may have to be situated some feet from the observer, but it was the best achievable with the available cathode-ray tubes suitable for transistor drive at the time of design. A black-level clamp is incorporated to ensure stability of the display against the edge-lit graticule provided; its time constant can be changed on a switch from short to long to show up field-time distortion or interference. All facilities are immediately available from push-button switches on 405-line, 525-line or 625-line standards.

These facilities are:

(*a*) 2-line display
(*b*) 2-field display

(*c*) *RGB* or *YRGB* displays from a signal provided by an external sequential-gating unit

(*d*) white-level cursor; i.e. a horizontal line indicating the precisely correct location of white level with respect to the signal black level, generated sequentially to the display, although, of course, it appears superimposed. It is available on a switch. This facility is invaluable, since it completely avoids parallax errors.

It is essential, with this type of monitor, that the gain should be stable over long periods at the value required to fit the graticule precisely when a video signal at standard level in 75 Ω is applied to the input. The only control provided here was a preset with a very restricted range to standardise the gain.

The triggering was completely automatic under all conditions, and was operated from either normal station pulses or the incoming signal. Change of standard was effected by a pushbutton.

The above monitor has proved excellent in practice, but this version is not necessarily the only one possible. Extra facilities may be incorporated; e.g. an indication of the amplitude of the colour burst by means of an additional marker multiplexed into the display. Alternatively, the essential information can be extracted from the signal and displayed in some easily legible or 'condensed' form.

Most of the information in the display of the video signal is redundant for the purpose of displaying the chief amplitude parameters of the signal, i.e. synchronising-pulse bottom, blanking level, peak signal level, minimum signal level, and also black level if the monochrome video signal includes a pedestal ('setup'). It is not difficult to extract these amplitudes from the signal and display them in a simplified form, possibly as horizontal lines capable of being used against a graticule. However, a critical choice of time constant must be made when the maximum and minimum signal amplitudes are measured. It is possible, and indeed quite common, for very narrow pulses having extreme amplitudes to occur, which are not representative of the range of excursions of the video signal and should accordingly not be counted, so that the time constant is not too short. On the other hand, if the time constant is too long, small areas of white or black which rightfully should be considered will be excluded. A discussion of a weighting network of a suitable type is given by Grosskopf and Schaumberger (1965). Another, somewhat less accurate, solution used is to provide waveform monitors for television with a restricted bandwidth of Gaussian shape.

When the required information has been reduced to such a basic form, it is possible to utilise it in various other ways; e.g. it may be made available in a suitable area of a picture monitor. It may also be measured and displayed on indicating devices such as a meter or a digital voltmeter, with the further possibility of recording or remote-monitoring facilities.

11.3 Television waveform monitors

These are much more comprehensive instruments than the channel monitors described in Section 11.2. They are located, for example, in apparatus rooms, mobile control rooms, switching centres and similar positions. Their task is the precise measurement of the video signal, signal levels, insertion test signals, line-repetitive test signals, hum, interference, and often random noise as well.

This demands a very high standard of performance and stability, together with a degree of organisation of the functions and the controls which clearly distinguishes a television waveform monitor from a versatile high-quality waveform monitor. It carries out a number of specialised functions under operational conditions, and it is consequently imperative that no time should be lost by the operator in having to work out how to use the instrument for a particular purpose, or in precise adjustments of the controls to obtain the required display. Anything which simplifies the operation is likely to reduce errors, particularly since it may have to be used on occasions under conditions of stress or local difficulty.

Nevertheless, for reasons of economy or availability, it is sometimes expedient to put in the position of the television waveform monitor an instrument not designed specifically for this use but which consists of a high-class general-purpose waveform monitor, perhaps with certain modifications. In this case, some of the remarks made in Section 11.3.1 will also be applicable.

The complete use of solid-state devices for the active elements in these waveform monitors is of great benefit, not only on account of the improved stability and reliability, but also because it then becomes possible to design the instrument to cope with a wide range of supply voltages.

However, the relatively small power dissipation within the instrument means that the waveform monitor may be in difficulties when the ambient temperature is low, owing to the lack of internal heating and the effect of the cold on the performance of components, particularly transistors and electrolytic capacitors. Such conditions can occur during outside broadcast operations. Even a mobile control room or other apparatus-equipped vehicle, which is normally air-conditioned, may have to be brought into use almost immediately after standing overnight in subzero temperatures.

As a guide, the instrument should be expected to operate within its specification at least over the range 5–50° C. In case of difficulty in low temperatures, a small internal heater can be provided which is switched off automatically by a bimetallic-strip as the temperature rises.

11.3.1 Display

The size of the display is preferably as large as possible, but one is usually limited to a maximum horizontal deflection of 10 cm, in some instances to 8 cm by the bezel required for the mounting of a camera. Rather more important is the size of the vertical deflection, which principally governs the apparent display size. This should be at least 6 cm for a single-beam display; furthermore, the 'window', i.e. the vertical distance between the two areas of beam cutoff, should be at all times adequate to provide a reasonable margin with respect to the normal vertical deflection.

With improvements in amplifier design and sweep circuitry, the limitation on the display linearity should arise from the geometry of the cathode-ray tube rather than from any shortcomings in the electronics. It is remarkable how tube designers have contrived to combine high deflectional sensitivity with large angles of deflection and still maintain good deflectional linearity, but nevertheless the cathode-ray tube should be the factor which determines the overall linearity.

A simple method of checking the linearity where an external X input can be accepted is to apply a separately generated sine wave to both the X and the Y inputs, thus producing a Lissajous figure in which the relationship between the two component waveforms is arbitrary. If the ratio of the two frequencies is fairly large, the envelope of the display draws a figure which should be truly rectangular (Fig. 11.1*a*). The two input amplitudes and the shifts are adjusted so that the figure exactly coincides with the working area, then the

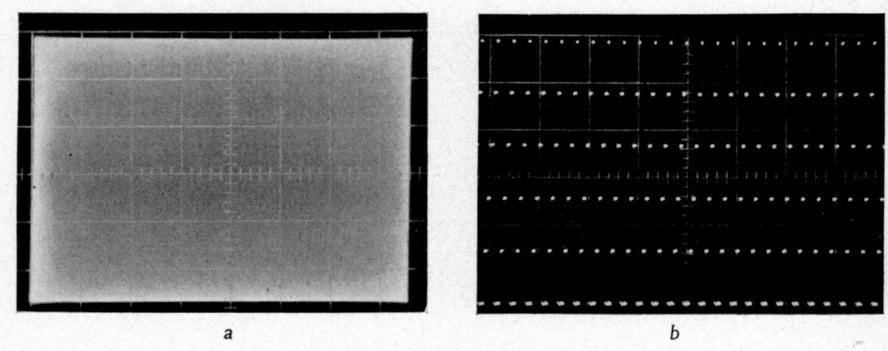

Fig. 11.1 Checking display linearity
a Using Lissajous figure
b Using staircase waveform

perimeter of the display is examined against a graticule for deviations from linearity. This test has the advantage that it is independent of the sweep linearity, so that, assuming the X amplifier is linear, which should be sufficiently true for all practical purposes unless a fault condition is present, the resulting nonlinearity of the display is that attributable to the cathode-ray tube alone.

It may be possible to remove, or at least to reduce, any curvature present in the display adjusting the relative gun potentials as directed by the manufacturer, either with an internal preset potentiometer or by selecting a resistor in the potentiometer chain. Where a truly rectangular display cannot be achieved, parallelism of the horizontal, rather than of the vertical, lines should be preferred. In any case, the opposite sides of the figure should be substantially symmetrical. Where there is any doubt about the residual distortion, reference may be made to the maker's specification for the cathode-ray tube.

A further check, which in this instance includes the linearity of the sweep at a fairly low rate, consists in applying to the vertical input a line staircase waveform of the type used for the measurement of line-time nonlinearity (Chapter 5). With a suitably low sweep rate, it is possible to obtain the stationary pattern of Fig. 11.1*b*, in which the horizontal steps of the staircase appear as a rectangular pattern of equidistant bright dots. It then becomes possible to set the dots against the intersections of a rectangular grid pattern placed in contact with the face of the tube or, if desired, an optical image produced by a projector, and the lack of coincidence can be measured as a function of the grid spacing (Chapter 12).

It is not practicable to lay down firm limits for the linearity required in a waveform monitor, since much depends on its function. A useful guide is that the linearity should be virtually impeccable over a centrally positioned rectangular area whose vertical dimension is the maximum specified vertical deflection and whose horizontal dimension is equal at least to that same length. Furthermore, any marked errors should occur only at the extreme edges.

The sweep linearity can be checked with a train of narrow rectangular pulses, the frequency in each instance being chosen so as to display an edge of each pulse against a vertical line of an equidistantly divided graticule. If the graticule lines are at 1 cm intervals and, at the same time, the pulse-repetition frequency is monitored with a digital counter, the sweep rates may also be calibrated. If appreciable nonlinearity of the sweep exists, the central area of the screen should be chosen for measurement or adjustment of the sweep rates.

With instruments not specifically designed as television waveform monitors, particular attention should be paid to the initial portion of the trace, since not only is this the region where nonlinearity is very likely, especially at high sweep rates, but it is also the area where the start of any waveform internally synchronised will appear. This region should also be closely examined while the brightness is varied over a wide range. Any variation of the apparent start of the trace indicates a brightup (unblanking) pulse with a poor leading edge, which, if present to any appreciable extent, is an annoying fault. An oscillation at the beginning of the top of the brightup pulse can give the corresponding part of the trace a beaded appearance. Any irregularities of the start of the trace which persist when the signal is removed and the sweep is externally synchronised are probably due to breakthrough from the trigger circuitry into the vertical amplifier.

The effect of any such faults is minimised if a delaying sweep is provided, or, with a true television waveform monitor, when special triggers derived from the video waveform are utilised for all displays, since, in each instance, the portion of the waveform to be examined can be brought to the centre of the screen, where the sweep waveform is unlikely to be other than impeccable.

An important factor with television waveform monitors displaying insertion test signals is the brightness at low screen occupancy and a high sweep rate, as when the sine-squared pulse in that signal is expanded for K rating against a graticule. The sweep rate is likely to be, say, 100 ns/cm, whereas the waiting period between sweeps is 20 ms or even 40 ms. Under such conditions, it is required that the expanded pulse be clearly visible without the use of long or face-fitting hoods in a reasonable working ambient illumination. It is also essential that the trace definition be high under these conditions. This is a formidable requirement, but nevertheless a number of waveform monitors will meet it very adequately.

The achievement of a high trace brightness is not in itself sufficient, since one is fundamentally concerned here with the subjective brightness, which is a function of the adaptation of the eye. The working point of the eye is determined by the mean brightness over the field of view, so that irrelevant light sources should, as far as possible, be excluded; this refers not just to lamps and bright picture monitors but also to reflections of light sources in the screen of the tube and from the surrounding metal work. Operator fatigue is reduced if the waveform monitor and the surrounding equipment are finished in matt black. This is particularly easily arranged when the waveform monitor is mounted in a desk.

The contrast between the trace and the screen is also very important, and several devices are available to improve this, the most common being a sheet of coloured transparent material mounted in front of the screen. Ideally, this should transmit the principal spectral components of the screen emission with negligible attenuation but introduce a high loss elsewhere. Naturally, this is unattainable, but in practice even an approximate match between the colours of the filter and the trace can be of considerable help, although care in selecting the optimum filter material is very worth while if one is aiming at the maximum contrast and brightness.

However, this still does not eliminate the problem of reflections from the graticule or the front surface of the contrast screen. Two further aids are available, supplied as accessories by certain manufacturers. The first is a sheet of fine mesh, perhaps embedded in a transparent synthetic-resin plate, mounted outside the graticule between the user's eye and the screen. Preferably, it is constructed from thin, flat ribbon, so that each component rectangular cell formed by the mesh acts as a miniature hood through which part of the screen is viewed. The device is coated in matt black to prevent reflections from the mesh itself. A cheaper form utilises wire of circular cross-section, but this is much less efficient because of the loss of light from the greater area of the wire and also because the display tends to be marred by annoying diffraction effects. These are not entirely absent with flat-section mesh, but can be reduced by good design to insignificant proportions and, of course, the loss of light is appreciably reduced.

The second type of contrast filter has the advantage that it is not difficult to contrive if a commercial model is not available for a particular instrument. It consists of a thin sheet of commercially available circular polariser, mounted within a short hood attached to the bezel of the waveform monitor. The polariser is retained within the hood, so that it takes up a curve concave to the outside, the chord joining the ends of the curve being at 45° or so to the vertical.

The polarisation of circularly polarised light is reversed by reflection from a dielectric surface, such as the graticule and the face of the tube; in this condition, the light cannot return through the polariser, thus eliminating reflections from these sources. Reflections from the front surface of the polariser are rendered invisible by the curve into which it is formed. In theory, it should be parallel to the face of the tube and the curvature of its surface should be correctly proportioned; but it is found in practice that these conditions are by no means stringent. In particular, the polariser, which is in the form of a thin flexible sheet, can be cut to a length greater than the diagonal of the hood and sprung into place between two clips, the exact length being determined by experiment.

Ideally, the polariser would reduce the brightness of the trace by a factor of 50%, and in practice the figure is over 60%, but the removal of the reflections makes the subjective brightness at least as great as it appears without the contrast filter, and at the same time the trace is more clearly viewed against the darker background.

The circular polariser consists, of course, of a laminate of a plane polariser and a quarter-wave plate, i.e. a sheet of anisotropic material of such a thickness that the phase shift introduced by it into the two mutually perpendicular vector components of the incident plane polarised light is 90°. Since it is necessary for the operation of the contrast filter that the stray light incident on the screen of the waveform monitor be circularly polarised, the quarter-wave plate should be mounted towards the screen, otherwise the contrast is not increased.

The orientation of the circular polariser can very quickly be determined by holding it up to a source of light and inspecting it through a small piece of plane polarising film; Polaroid sunglasses are also suitable. When the quarter-wave plate faces the light source and the circular polariser is rotated, maxima and minima of the transmitted light appear, which are absent when the quarter-wave plate faces the observer.

Loss of contrast can also be caused in other ways. One of these is the edge lighting of the graticule (Section 11.3.2). Leakage of light around the edges of the graticule is unlikely in a standard commercial monitor, but possible if it has been modified, say, with a replacement graticule which is not a good fit. More common is a haze from the graticule surfaces due to scratching, particularly likely when graticules are changed for different measurements. They should always be treated with great respect, and kept clean but not with abrasive substances, and should never be left lying about when not in use. The author has seen instances of waveform monitors blamed for poor brightness when the trace could barely be discerned through the general glow arising from multiple scratching of the graticule surfaces.

The cathode-ray tube itself can also cause reduced contrast, and, as nothing can be done about this, it should be taken into account in the choice of a waveform monitor. The principal causes are haloing and electron scatter. The former takes the form of a bright area surrounding the actual trace, and the latter of background illumination from the screen, sometimes general and sometimes in the form of bright patches. Electron scatter is aggravated by high beam currents, which makes its effect most evident under conditions such as insertion-test-signal inspection where the brightness control is fully advanced but the actual trace brightness is relatively low. Occasionally, it only appears when the sweep magnifier is brought into operation or when the shift controls are away from their normal positions for a centred display.

Even experienced personnel can overlook the loss in brightness from the buildup, on the screen or on transparent surfaces near it, of an almost invisible film of very fine dust particles due to electrostatic charges. The effect varies widely from one type of waveform monitor to another because of differences in tube construction and in the sealing of the bezel. Measurements have shown that a reduction in brightness by a factor of four or even more is possible. Such dust films should be removed with a detergent or a proprietary cleaner and the surfaces polished with a soft cloth. An antistatic liquid has been found useful.

The aim so far has been the attainment of a sufficient brightness, but this can be an embarrassment when an operator is examining waveforms over a range of sweep rates. Switching directly from an insertion-test-signal display to a field or even a line display can result in a trace brightness positively painful to the eye unless care is first taken to reduce the brightness appropriately. A device, fitted to at least one commercial waveform monitor, links the beam current with the display selector in such a way that the brightness remains roughly constant. It is really only suitable for professional television waveform monitors equipped with a limited number of standard sweep rates, but it is a worthwhile and useful refinement.

11.3.2 Graticules

The general principle should be to design test methods not requiring graticules, e.g. the method of level measurement against a square wave described in Chapter 2; of course, this is not always practicable. It is then essential that the design and construction of the graticules used should be such that the inherent accuracy of the waveform monitor is degraded to the least possible extent, and that operationally they are as convenient as possible. External graticules are the more versatile, and are considered here in detail.

The most common form is a sheet of transparent material bearing a pattern which is held against the face of the cathode-ray tube. This has some obvious objections; on the other hand these graticules can be designed for easy removal, so that not only can an adequate number be provided for the tasks to be carried out, but they can be added to or brought up to date as required.

The best of this type are engraved on a stable transparent synthetic resin such as perspex, thick enough to resist handling and prevent deformation. In place of engraving, silk-screen printing is sometimes used. This is somewhat cheaper and can give excellent results, but the printed pattern is considerably more vulnerable to mechanical damage. Incidentally, the term 'silk screen' is really a misnomer, since nowadays the traditional material is replaced by a very fine stainless-steel mesh to ensure high mechanical stability and accuracy of the pattern. Such graticules are normally edge-lit; i.e. light is directed from small bulbs along the engraved face of the graticule and practically parallel with it. The direct light from the bulbs is shielded from the user, so that no light is directed forwards except where the engraved lines intercept and diffuse it, giving the appearance of a self-luminous pattern. The number and the disposition of the bulbs are such as to provide a sufficient and uniform illumination. It is not uncommon to find that coloured bulbs or coloured filters are used to improve the contrast by making the graticule illumination approximately complementary to that of the trace. However, the improvement is not as great as one might expect, and whether the device is employed or not is principally a question of individual preference.

Graticules for special purposes can be made to order by specialist firms if a suitable drawing is supplied; but small numbers are somewhat expensive in view of the amount of preparatory work needed before engraving. The line width is important, since too fine a line gives good resolution but a reduced brightness, whereas a thick line gives good brightness but a coarse pattern with a tendency to 'blobbing' at intersections. A suitable width has been found to be $0{\cdot}006 \pm 0{\cdot}001$ in for graticules for cathode-ray tubes with a screen of 5 in diagonal.

The graticules are engraved from a template; and, if a high accuracy is required, it is essential that not only an initial specimen but also subsequent random samples be very carefully checked. In the author's experience, this is best done by having an exact fullsize photographic transparency made by a firm specialising in high-class process work, this transparency being taken from an accurate overscale drawing of the graticule pattern and printed as a mirror image. If the transparency and an engraved graticule are placed in contact face to face and aligned at reference points, any errors can be seen very quickly, particularly if the pair are projected optically on to a screen. Any distortion from the projector is, of course, immaterial.

The method of retaining the graticule should be designed in such a manner that graticules can be changed rapidly, preferably with captive screws or nuts to avoid their inevitable loss to the most inaccessible area of the floor at critical moments.

A device much used operationally for changing rapidly from one graticule to another consists of a plate engraved with a different graticule pattern on each side. Two sets of lamps are used, positioned so that one set illuminates the outer graticule only and the other the inner graticule only. In a particular realisation, the normal variable resistor for adjusting the lighting intensity was replaced by a centre-zero type, so that a transition could be made smoothly from one pattern to the other, but a switch would have been adequate. The interference between the two graticules is negligible in practice. Several commercial waveform monitors have been modified for this facility.

The great disadvantage of removable graticules is, of course, parallax error, and many attempts have been made to overcome it. The simplest measures such as double-sided graticules and the use of hoods which constrain the movement of the head are not very convenient for use, and cannot be guaranteed to give the same answer. At first sight, it seems a good idea to utilise a half-silvered mirror on the 'Pepper's ghost' principle to produce an optical overlay of the graticule and the display. Such a device, designed in the author's laboratory some years ago, gave excellent results, but proved unpopular with all except a minority of operational staff. Similar devices can be obtained comercially.

Another approach, adopted by some manufacturers of waveform monitors, consists in constructing the cathode-ray tube with the graticule fused into the faceplate just in front of the phosphor, so close that the parallax error is negligible. The graticule can be illuminated by a normal type of edge lighting, or by an electron 'flood gun', which causes the whole area of the phosphor to emit evenly. The edge lighting, if properly designed, does not degrade the contrast of the display noticeably, although unfortunately the reverse is the case with the flood gun.

It is very desirable to provide adjustable edge lighting, so that the intensity of the graticule can be varied to suit the conditions of use. If photographs are required, it is essential. It is then worth while to feed the bulbs from the emitter of a cheap transistor, so that the variable resistor has only to cope with the base current of the transistor instead of the full current of the bulbs.

11.3.3 Vertical deflection

A master waveform monitor for use with colour signals must have vertical-deflection circuits which are exceptionally stable both for gain and for amplitude/frequency response, since it is desirable to have not only an impeccable response both to 1 T and 2 T sine-squared-pulse and bar signals, but also to have a luminance–chrominance gain error not exceeding $\pm 0{\cdot}1$ dB. In principle, the relative delay error should also be specified, but it is highly unlikely with a high-class instrument that any delay error would even be perceptible.

When assessing the performance of a waveform monitor for precision television measurements, the effect of the vertical-shift control on the waveform response should always be checked. While a great deal of shift and a great deal of vertical overscan are not required, at least any change in the waveform response should be extremely small. This is best checked with an augmented pulse-and-bar signal or chrominance–luminance pulse-and-bar signal and also with the 1 T pulse-and-bar superimposed so that the peak of the pulse just coincides with the centre of the bar. One example was once found by the author where the 1 T pulse/bar ratio could be varied over a considerable range merely by using the vertical-shift control.

An invaluable feature for certain purposes is the provision of differencing facilities, often called 'differential' facilities, although this name is to be deprecated, since it describes the function of the amplifier in a misleading manner. This makes it possible to compare very accurately the levels of a number of signals, e.g. by displaying the difference between each in turn and a standard.

The preferred gain control circuit is a correctly matched 75 Ω attenuator which is in circuit under all conditions. Even then it is difficult, considering all the possible sources of error, to make the gain stability of the waveform monitor a true $\pm 2\%$ over reasonably long periods, excluding parallax errors. The use of a high-impedance input tends to worsen this figure, but unfortunately it is often demanded by the equipment with which the master waveform monitor is used.

The measurement of signal level ought to be possible with an error no greater than $\pm 1\%$, so additional means are needed. One solution is to provide an accurate calibrating waveform, usually a relatively low-frequency square wave, by which the gain can be checked. This is used by setting the display size against the graticule; even if the square-wave amplitude is adequately stable and accurate, serious parallax errors may occur when an external graticule is provided. The method of Section 2.4 is very greatly to be preferred. Provided the random-noise level is low, an accuracy within $\pm 0{\cdot}1$ dB or better is easily obtained.

Master waveform monitors for colour work are sometimes provided with a lowpass and a highpass filter which may be switched in series with the vertical amplifier when required. The former removes the chrominance information and displays the luminance, and the latter removes the luminance information and displays the chrominance, thus making it possible to measure differential gain directly. Their principal function, however, is in measuring colour-bar waveforms when checking colour coders, and it is important that the gain of the vertical-deflection circuit should be the same whether the filters are inserted or not.

11.3.4 Black-level clamp

The provision of a good black-level clamp is essential in a television waveform monitor. Apart from the fact that signals of varying picture content may have to be examined and it is inconvenient continually to be readjusting the vertical shift, the test signals received from a long circuit are not infrequently accompanied by a 'd.c. wander', which makes accurate measurements impossible if no clamp is available.

The stability of the potential to which black-level is clamped should be good, especially with respect to thermal effects, to avoid drift after switching on. The change in the position of the clamping point with average picture level is readily investigated with the CCIR non-linearity-test waveform (Chapter 5). The maximum change should not be more than a very few millivolts between the '3 blacks' and '3 whites' conditions. At the same time, no transient effects should be noticeable either in the waveform or in the synchronisation as the test waveform is changed between the two conditions.

It is essential, however, that the clamp should still behave perfectly with signals having the worst signal/noise ratio likely to be used, say, 26 dB unweighted. A signal having this amount of random noise could not be tolerated for entertainment purposes, but would be acceptable for programmes having a high degree of topical interest.

11.3.5 Triggering

The triggering circuits require the greatest degree of organisation to make the television waveform monitor easy to use under operational conditions. The following facilities are desirable, and should be available on a switch or switches, with a continuously variable control for precisely locating the required feature of the video waveform:

(*a*) 1- or 2-line display
(*b*) 1- or 2-field display with positive selection of the field
(*c*) line selector with picture-marker output
(*d*) automatic selection of insertion test signal for a choice of numbered lines in either field
(*e*) automatic choice of component of insertion test signal, e.g. sine-squared pulse, bar, or pulse and bar double-triggered
(*f*) automatic selection of sine-squared pulse, bar and double-triggered pulse and bar with sine-squared-pulse and bar signal.

As with the black-level clamp, the triggering should be unaffected by the random noise in any signal likely to be used for programme purposes.

11.3.6 Sweep

The minimum number of sweep rates needed for the various functions of the television waveform monitor should be provided, to make it easier to select the correct rate for a given application. The simplest way is to have multiples of the field and line rates together with special settings for sine-squared-pulse and bar work, so that the displayed waveforms fit the graticules provided without further adjustment.

A sweep magnifier, i.e. extra amplification inserted in the *X* amplifier path, would seem to be an attractive means for changing the sweep rate, since the magnification factor applies equally well to both field and line displays, thus simplifying the switching. The usual objection to a high magnification, i.e. the degradation of the trace linearity at the extreme start and finish of the sweep, is less important here, since the part of the display of interest is normally centred on the screen. However, care must be taken that the horizontal overscan does not cause a poor background with loss of display contrast.

An accuracy within ± 3% is typical for the sweep rate, and is adequate. A higher accuracy is hard to maintain over long periods.

Owing to the closeness of the line periods of 525-line and 625-line systems and the fairly close ratio of the field periods, it is not unduly difficult to make a given monitor operate on both systems, and automatic triggering is quite practicable. This is a further advantage of a calibration in terms of field and line periods instead of microseconds per centimetre. 3-standard operation, i.e. 405-, 525- and 625-line operation, is often required in the UK, and this can also be arranged at the expense of introducing further complication.

11.4 General-purpose waveform monitors

These are far less specialised than the television waveform monitor since they may assume a wide variety of roles. They therefore require a wide range of sweep rates, extensive synchronising facilities, a greater bandwidth to deal with the fast waveforms encountered in the maintenance and alignment of equipment, and at least one high-impedance input suitable for a voltage probe. The provision of an optional double-beam display is very desirable for checking divider chains and for ensuring the correct synchronisation of the various waveforms to be found in video apparatus. Most usually they consist of, or are based on, standard commercial waveform monitors, which can in some instances provide most, if not all, of the required facilities. Many are fitted with interchangeable plug-in subunits to extend the range of facilities without unduly complicating the basic instrument; this has the advantage that the facilities provided may readily be modified or added to, but on the other hand, the convenient storage of more than a very small number of such units can present problems.

One absolutely basic requirement, not found on many general-purpose instruments, is a field-synchronising separator, preferably designed so as to furnish separated field pulses with random-noise levels equivalent to a signal/noise ratio up to, say, 26 dB unweighted, which, as a rule of thumb, may be taken as the worst signal/noise ratio at which video signals are likely to be examined. Without separated field synchronising pulses, it is not, in principle, possible to display waveforms at field rate, although some oscilloscopes nevertheless contrive to do this tolerably well as a result of the shaping in the trigger circuits. However, unless the circuits have been deliberately designed for displaying field waveforms, the locking is unlikely to be as solid as is required for operational purposes, and displays of the individual fields are not possible. The availability of field synchronising pulses also makes it possible to utilise a delaying sweep as a form of line selector, a facility indispensable for general video work.

An efficient black-level clamp or d.c. restorer is also highly advisable where measurements have to be made on video signals or insertion test signals during programme hours. Although it is possible to do without this, the frequent vertical movement of the display due to the inevitable changes in the average picture level is most annoying and frustrating during measurements; and, of course, the accuracy is bound to be affected.

Where a commercial oscilloscope is suitable apart from the lack of some or all of the specialised facilities required for video measurements, an acceptable solution is an auxiliary panel specifically designed to provide these in the exact form required. A brief description of such a unit is given in Section 11.4.3.

11.4.1 Vertical deflection

It is very often assumed that if a waveform monitor is specified to have, say, a bandwidth of 15 MHz, it must naturally be good enough for colour work as far as bandwidth is concerned. This is by no means necessarily the case.

The bandwidth of oscilloscopes is now universally quoted as the frequency at which the response falls by 3 dB compared with the low-frequency value. While this is an arbitrary convention, it has the advantage that, with many simple lowpass networks including amplifiers, the 3 dB frequency is approximately equal to the equivalent rectangular bandwidth, i.e. the noise bandwidth, and the risetime of an ideal step function passed through such a network is approximately given (Wolf, 1963) by time of rise = $350/F_{3dB}$ nanoseconds, in which F_{3dB} = 3 dB frequency in megahertz.

However, to obtain well shaped transitions, the vertical amplifier of a general-purpose oscilloscope is not designed to have a band which is flat as far as it can be held, but one which resembles a Gaussian (normal-probability) curve, since such a shape in a minimum-phase-shift system ensures a linear phase/frequency relationship. Usually the lower part of the band is somewhat flatter than a true Gaussian curve. As a result, the gain is found to be low at 4·43 MHz compared with the lower video frequencies; and, what is more, this error is not always particularly stable with time. This is not to be regarded as a shortcoming of these instruments with regard to the purposes for which they are originally designed.

Another possibility of error arises from the signal-delay circuits, particularly if they are constructed in the form of lumped-constant delay lines. It is difficult to adjust these to give no internal reflections, and aging or maladjustment can produce a considerable train of echoes. Now it is known from the theory of paired echoes (Wheeler, 1939) that each echo is equivalent to a cosinusoidal ripple in the gain curve and a sinusoidal ripple in the phase-shift curve, and it is therefore possible to have a considerable irregularity in the gain/frequency curve in the region of subcarrier frequency even when this is only a small fraction of the total 3 dB passband.

Distributed–constant delay lines of several kinds are also used, and are a considerable improvement in general as far as internal reflections are concerned, but terminal reflections are still a possibility, particularly when, for reasons of economy, only one end of the delay line is terminated.

The author has for a long time used a step waveform generated by a mercury-wetted relay for examining the vertical-deflection systems of waveform monitors for such faults. This is a reed relay which relies on capillarity to maintain a thin film of mercury between the contacts, which prevents contact bounce. As it is contained in a small-diameter glass ampoule and driven by an external magnetic field, it may be mounted coaxially inside a metal tube and used to make or break a direct voltage to form a close approximation to an ideal step waveform. The principal drawback is the mechanical inertia of the reed, which limits the repetition frequency to a maximum of about 400 Hz. The risetime has never been measured, but is certainly very much less than 1 ns. Solid-state pulse generators are now available which have a performance not much inferior to the mercury-wetted relay, with the advantage of a much higher pulse-repetition rate.

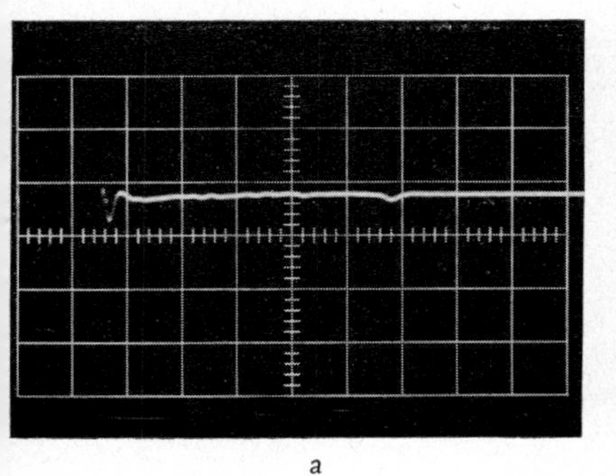

a

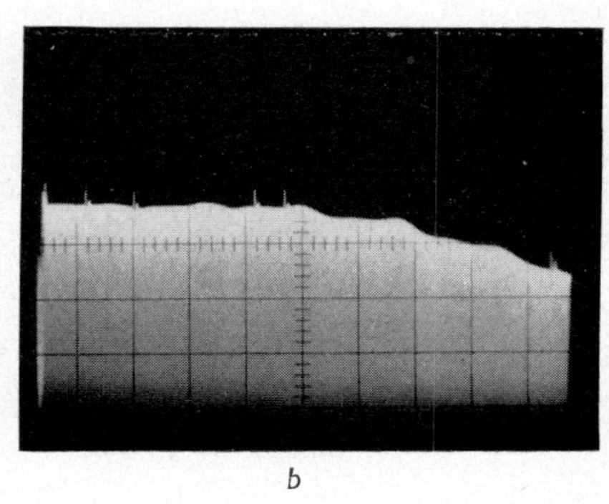

b

Fig. 11.2 Vertical-amplifier response of waveform monitor

a Step response
Time scale = 20 ns/division
b Sweep corresponding to (*a*)
Central markers are 4·43 and 5 MHz

Figs. 11.2 *a* and *b* show the transient response of a waveform monitor and the corresponding frequency sweep in which the irregularities resulting from the multiple reflections can clearly be seen. Careful adjustment of the trimming capacitors of a lumped-constant delay line may enable an appreciable improvement to be effected, but anyone attempting this for the first time is advised to proceed very cautiously and systematically; otherwise, the final state may well be worse than the first. Detailed instructions are usually given in the handbook of the instrument and should be followed closely.

Even when the maximum possible improvement has been effected, one may still be left with a luminance–chrominance gain inequality, from this or some other cause, which is unacceptable. While any error can be taken into account in a measurement, this is not really feasible operationally, and should be avoided if at all possible. An extremely acceptable alternative found is a small equaliser left connected to the *Y*-amplifier input socket, or, if space is available, installed inside the monitor. It is intended to be connected to a high-impedance input, and then provides a very good termination, better indeed than a good 75 Ω resistor across the same point, since a partial

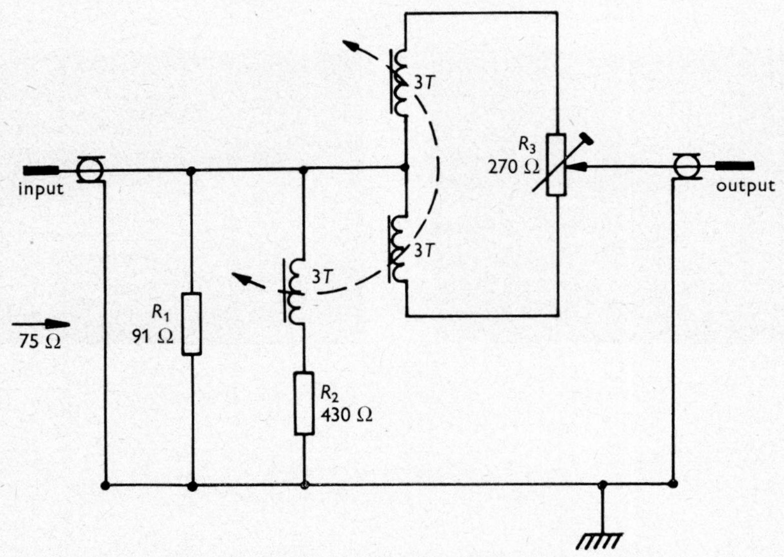

Fig. 11.3 Circuit of luminance–chrominance equaliser

correction is achieved of the input capacitance. The circuit is given in Fig. 11.3, which should be self-explanatory. Variation of the small preset potentiometer provides a chrominance–luminance gain correction of about $\pm 0{\cdot}5$ dB with negligible relative delay error.

The adjustment of this equaliser or any other device for improving the gain inequality should always be carried out by some means accurate and reliable enough to ensure consistency and reproducibility; otherwise, the purpose is frustrated. The most easily available method, although not the most convenient operationally, is a direct comparison of the waveform-monitor deflectional sensitivity at two frequencies, a low video frequency and subcarrier frequency. The exact frequency of the former is not important, about 10–100 kHz is usually suitable, and the latter frequency need not be precise. Both signals, however, should have a very low harmonic content, particularly with regard to even harmonics. A simple and reliable method for ensuring such a low harmonic content is to pass them through suitable lowpass or bandpass filters. The requirements are not severe; a loss at the second harmonic frequency of 20 dB is adequate and can easily be furnished by a single prototype lowpass-filter section.

The obvious, pedestrian method of comparing the deflectional sensitivities by successive application of the two frequencies is far too insensitive, and, with many waveform monitors, the additional parallax error from the graticule is unacceptable. A better technique uses the simultaneous application of the two signals. It requires two oscillators with 75 Ω output impedances and, of course, fitted with suitable filters. The outputs of the filters are connected to the two inputs of a 75 Ω splitting pad of the usual form of a delta of 75 Ω resistors. The output of the pad is connected to the monitor input which, as has already been stated, presents a good 75 Ω termination.

The two signals, which are mixed in the pad, must now be individually adjusted by some means guaranteed to be free from amplitude/frequency error over the range concerned. An instrument found effective contains a thermocouple intended for high-frequency applications, which has its heater impedance built out to 75 Ω at an input of 1 V pk–pk, with which the applied alternating voltage can be compared with a known 1 V d.c. (Chapter 2.5).

This value is chosen to be the same as the standard video level, since it is important for the adjustment of the equaliser that the sensitivity controls of the waveform monitor should be at, or very close to, the settings at which the equalisation is required to be exact. This applies particularly to any variable gain controls fitted, since these not infrequently modify the gain/frequency characteristic as a function of their position. The attenuator settings ought to be distortionless if they are in correct adjustment (Section 11.4.2); but, if the vertical amplifier is overdriven, the gain/frequency response may be affected.

When the 2-frequency signal is applied to the input of the waveform monitor to be equalised and the sweep is suitably double-triggered, the display will have the appearance of Fig. 11.4*a* if the luminance–chrominance gain inequality is zero, as is indicated by the exact

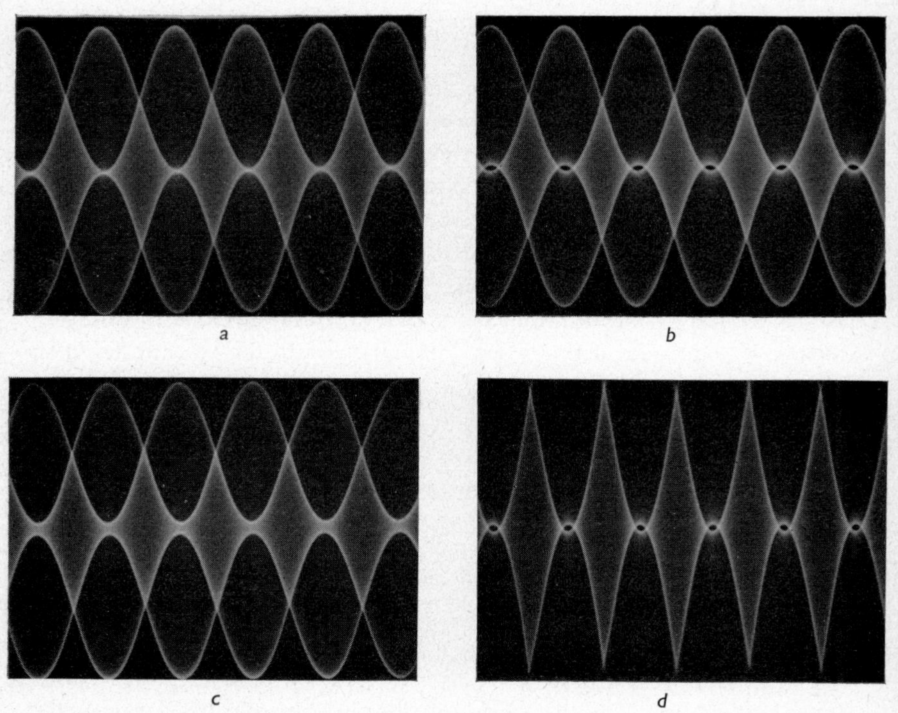

Fig. 11.4 Check of luminance–chrominance gain equality using two sine waves
a No gain inequality (trace thickened centrally by halation)
b Chrominance 0·5 dB low
c Chrominance 0·5 dB high
d Chrominance 0·2 dB low (gain of monitor increased by factor of 2·5)

coalescence of the loops between the lozenge-shaped figures in the centre. The thickness of the trace in this region is exaggerated in the illustration owing to the slight over-exposure required for reproduction purposes; visually they appeared just to touch. Fig. 11.4*b* shows the same waveform with the relative subcarrier amplitude reduced by 0·5 dB; and, in Fig. 11.4*c*, it has been increased by 0·5 dB. The setting can usually be made within 0·05 dB or even less, depending on the definition of the trace. If increasing the gain with the attenuator does not affect the response, increased sensitivity may be obtained. A negative luminance–chrominance gain inequality of 0·2 dB with the gain of the waveform monitor increased by a factor of 2·5 is illustrated in Fig. 11.4*d*; even allowing for the increased apparent thickness of the trace due to the photographic process, it is clear that a discrimination of better than 0·05 dB is possible.

While the above method is useful to the extent that it is not unduly difficult to improvise, it is not in the most suitable form for operational use. Accordingly, the principle has been adapted somewhat to produce a compact, stable and operationally extremely useful instrument. The component waveforms are now as shown in Figs. 11.5*a* and *b*. They consist of chrominance subcarrier with an amplitude of 0·5 V pk–pk (Fig. 11.5*a*), and a square wave also with an amplitude of 0·5 V pk–pk (Fig. 11.5*b*). These amplitudes are set by temperature-compensated Zener diodes, and can be arranged to remain within 0·5% or even less over a period of some months, depending on the amount of money which can be spent upon stable components. As usual, the stability and accuracy of the instrument are very much a function of its cost (Section 2.5).

The result of adding these two waveforms together is the composite waveform of Fig. 11.5*c*, in which the correct adjustment is indicated by the precise alignment of the blocks of subcarrier thus formed at the centre of the waveform. Any deviation of the luminance–chrominance gain ratio from unity is indicated by either a gap or an overlap at the centre (Figs. 11.6*b* and *c*). The adjustment of the oscilloscope equaliser then merely amounts to a rotation of the potentiometer until the condition of exact alignment at the waveform centre is obtained, which is possible with a degree of discrimination greater than might be suggested by the waveform photographs, particularly when the sweep rate is kept fairly low.

The use of the square wave instead of the previous sine wave is less to improve the sensitivity of the setting than to increase the versatility of the instrument, since it is then simple to provide separate outputs of the square wave and subcarrier at precise amplitudes of 1 V pk–pk as standards for other purposes. In particular, the square wave is available for the measurement or adjustment of signal level (Chapter 2). It might be thought that relatively low-frequency distortion of

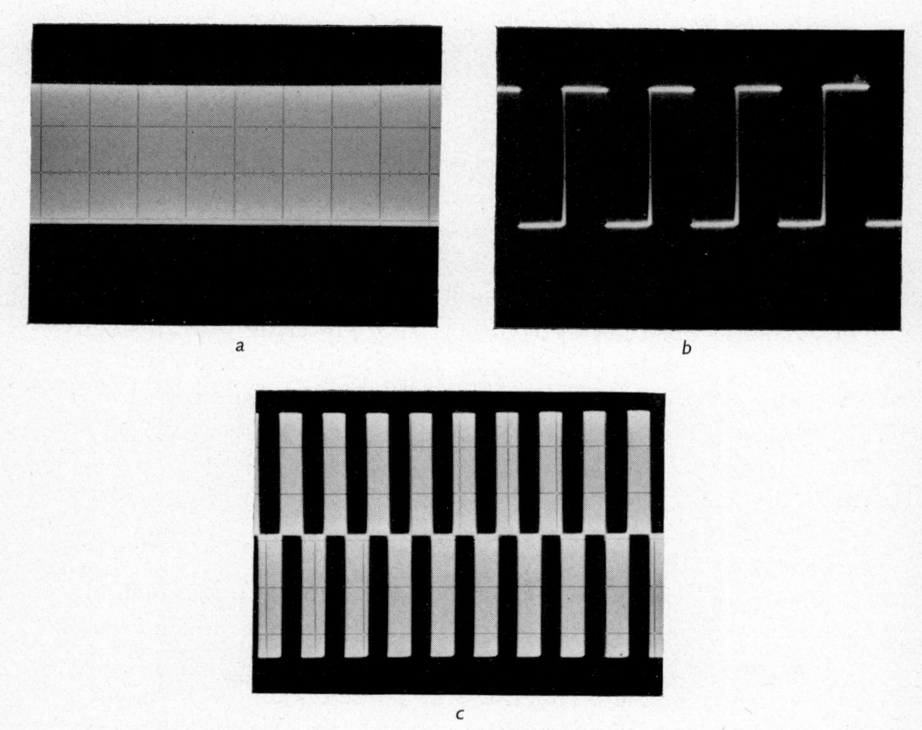

Fig. 11.5 Preferred waveform for luminance–chrominance gain-equality check

a Subcarrier component with amplitude of 0·5 V pk–pk
b Square-wave component with amplitude of 0·5 V pk–pk
c Standard waveform produced by addition of (*a*) and (*b*)

the square wave of a type such as to give rise to tilt or overshoots on the top or bottom of the wave would degrade the accuracy of setting the calibrator waveform; but, in fact, this need not be so. How little the measurement is affected by such distortion is demonstrated very clearly by Figs. 11.7*a*, *b* and *c*, in which the square wave has been deliberately distorted. These are analogues of Figs. 11.6 *a*, *b* and *c*, except that the waveform monitor gain has been increased and, at the same time, the amount of gain inequality introduced has been decreased. Provided that a fairly low sweep rate is employed (Figs. 11.7 *a*,

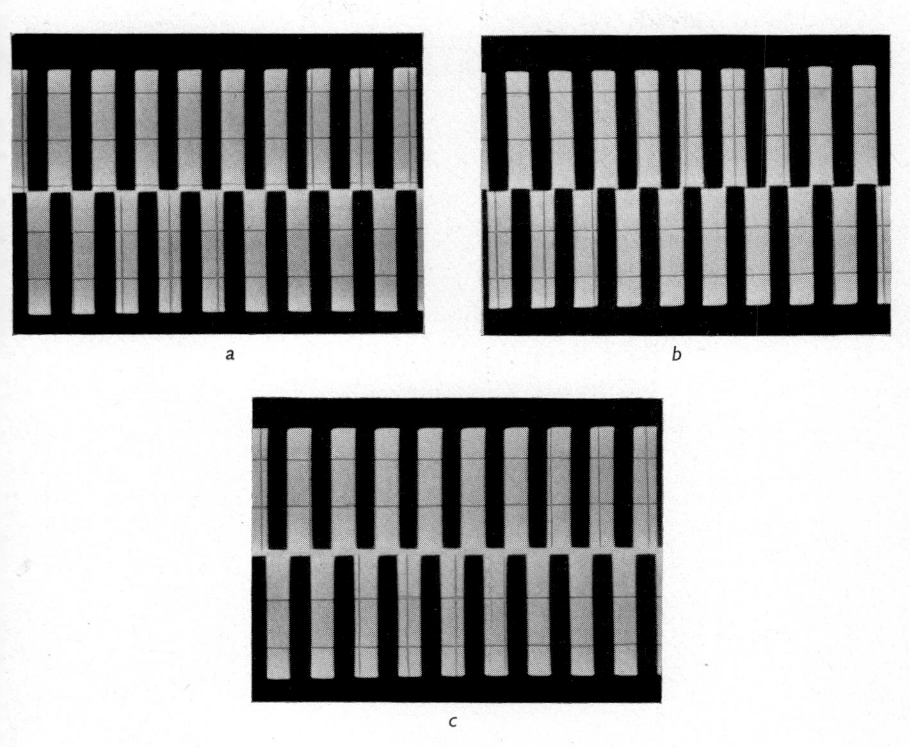

Fig. 11.6 Effect of luminance–chrominance gain inequalities on test waveform

a Correct adjustment
b Chrominance 0·2 dB low
c Chrominance 0·3 dB high

b and *c*), the change in the central region for a gain inequality of only 0·1 dB is quite marked; and, with a further increase in gain, the discrimination could be correspondingly improved. In particular, at the setting for the precise equality of the luminance and chrominance gains, a fine, continuous line appears at the centre of the waveform which is unmistakable, even when the square-wave distortion is severe.

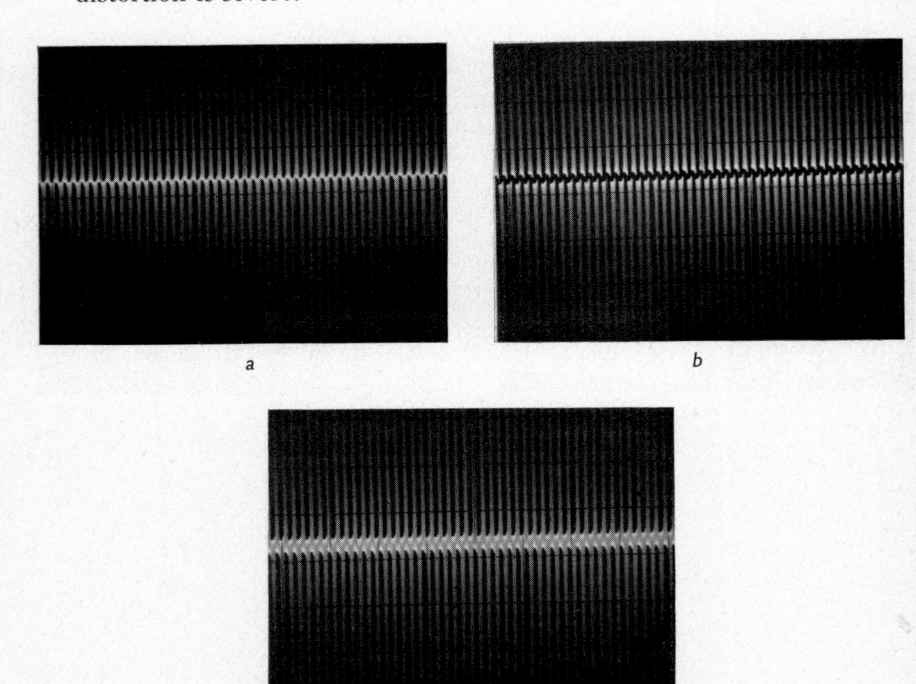

Fig. 11.7 Preferred test waveform with square-wave distortion

a Correct adjustment: slow sweep rate and increased gain
b As (*a*), but with chrominance low by 0·1 dB
c As (*a*), but with chrominance high by 0·1 dB

11.4.2 Attenuators

The high-impedance attenuators fitted to some oscilloscopes can give rise to an annoying form of distortion. This is because the attenuator steps are not constant-impedance pads but simple resistive potential dividers, the total resistance of which is kept constant, frequently at 1 MΩ. The lower branch of the potential divider is inevitably shunted by the capacitance of the input stage of the vertical amplifier which, together with the stray circuit capacitance, can easily amount to 30 pF or even more. Since the resistance values are quite high, the stray capacitance produces a division ratio which is no longer constant with frequency, together with a step in the phase-angle/frequency curve.

The distortion can be completely corrected by simply shunting the upper branch of the potential divider with a capacitor to equalise the time constants of the two arms; i.e. $R_1C_1 = R_2C_2$ (Fig. 11.8).

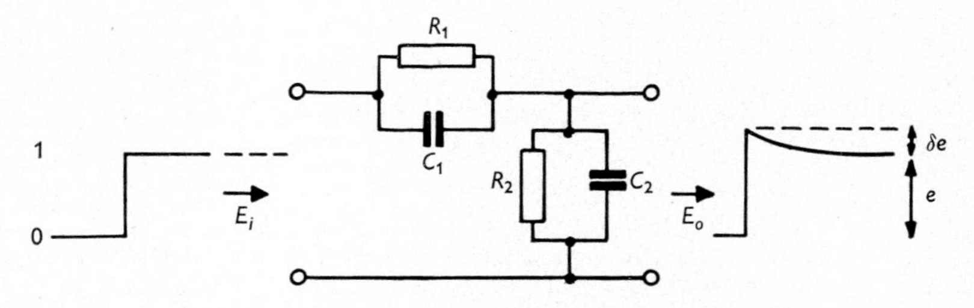

Fig. 11.8 Effect of incorrect attenuator compensation

$$e = \frac{R_2}{R_1+R_2} E_i$$

$$\delta = \frac{C_1R_1}{C_2R_2} - 1$$

Unfortunately, the component values are such that, unless special forms of construction are adopted, it is not easy to maintain the equality of the two time constants over long periods. The result of an inequality is a hook or rounding at the leading edge of a long-duration bar or a tilt over one of shorter duration, these effects being all the more unwelcome since one may be unable to decide whether the fault lies with the waveform under examination or with the oscilloscope.

If the instrument is always to be used to terminate the input circuit in 75 Ω, this distortion can be avoided by replacing the high-impedance attenuator with a 75 Ω constant-impedance attenuator; alternatively, the high-impedance attenuator may be left in its most sensitive position and the input connection made through a terminated 75 Ω attenuator. Where alternative A and B inputs are provided, one of them may conveniently be permanently terminated in 75 Ω for this purpose.

To demonstrate the distortion produced by an error in the attenuator compensation, the effect of a given difference between the time-constants of the two arms of Fig. 11.8 may be calculated.

It is easily shown that the output transient resulting from the application to the input terminals of a step waveform of unit amplitude is given by

$$E_0/r = 1-(1-c)\exp(-t/T_c)$$

where r = division ratio $R_2/(R_1+R_2)$

$$c = R_1R_2(C_1+C_2)/(R_1+R_2)$$

$$T_c = R_1C_1/c$$

If, now, $C_2R_2 = T$ and $C_1R_1 = T(1+\delta)$,

$$\frac{E_0}{r} = 1 - \frac{\delta(rT-1)}{1+8rT}\exp\left(\frac{-t}{1+\delta rT}\right)$$

which, since T is always very small compared with unity, can be approximated with negligible error to

$$E_0/r = 1+\delta\exp(-t/T)$$

Hence, the error waveform is an exponential whose maximum amplitude is the same fraction of the final amplitude of the output step waveform as the fractional difference in the two time constants; i.e. a 1 % time-constant error will give rise to a 1 % exponential, either positive or negative according to the sign of the error. The effect of compensation errors of this type on relatively long bar waveforms are shown in Figs. 11.9*b* and *c*. With shorter duration bars, such as the bar in the sine-squared-pulse and bar signal, the effect is more in the nature of a tilt, depending on the magnitude of the time constant T.

11.4.3 Auxiliary units

As already mentioned, the triggering of many standard commercial oscilloscopes, i.e. instruments not specifically designed for television, lacks a number of facilities very important for the examination of video and test waveforms. A field synchronising-pulse separator is occasionally provided, which does facilitate many of the operations required, provided that a delaying sweep is also available. This should preferably be of the gated type, so that the main sweep can be initiated by a transition in the waveform close to the area under observation,

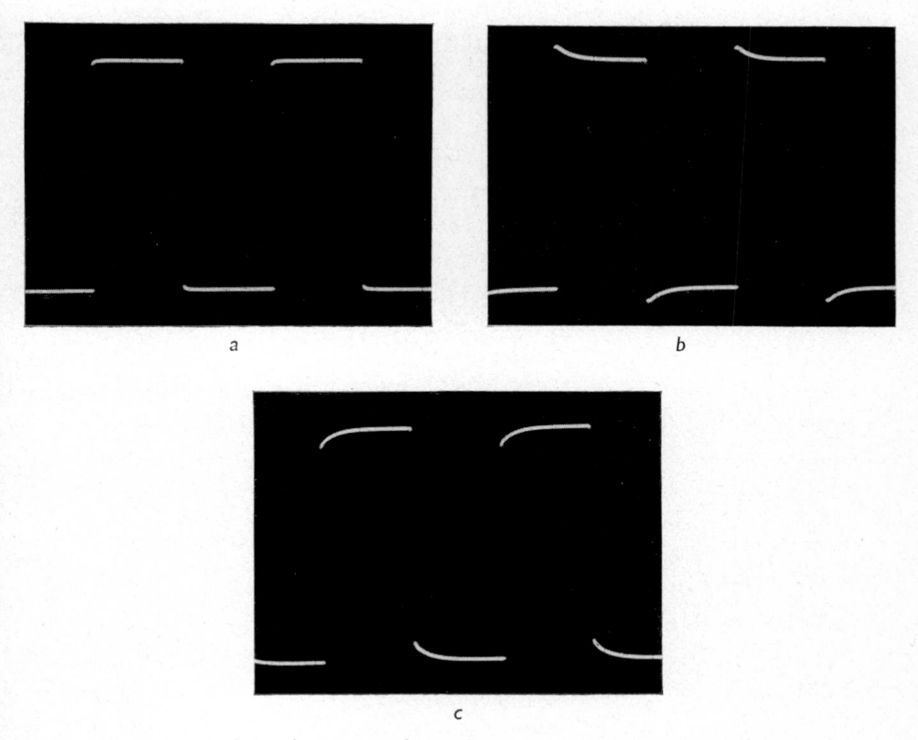

Fig. 11.9 Waveform error due to incorrect attenuator compensation
a Correct adjustment
b Attenuator overcorrected
c Attenuator undercorrected

thus avoiding the jitter which otherwise inevitably occurs with long delays. If field triggers can be supplied as an external trigger source, a synchronising-pulse separator may be dispensed with; but, in general, it is not wise to assume that the oscilloscope will always be operated in areas where these are available.

A practicable solution, but which may require a certain amount of ingenuity, is the design of a small transistorised synchronising-pulse separator which, if at all possible, should be incorporated with the oscilloscope in some manner. The two chief difficulties lie in providing the video pickoff and the power supply; both these are minimised if the unit can in some way be added internally.

However, such a solution may not be at all adequate where the oscilloscope is required to fulfil the functions of a master waveform monitor in addition to general and maintenance work. It has then been found well worth while to design a series of self-powered external units which are used in conjunction with the oscilloscope, and with which all the functions the monitor has to perform can be carried out in the best and most convenient manner. Naturally, this entails a considerable amount of design and manufacturing effort, and can only be justified if a sufficient number of oscilloscopes are to be equipped. Nevertheless, this has been put into practice very successfully with a good-quality, but inexpensive, commercial oscilloscope as the basis, the low cost of this instrument compensating to some extent for the cost of the auxiliary units. Section 11.3.5 will suggest what facilities might be provided.

11.5 Vector monitor

This type of monitor differs completely from those described above, since its primary function is to provide a visual display of the amplitudes and relative phase angles of the chrominance components of NTSC and Pal encoded colour-bar signals, thus making it possible to determine, with a standardised signal, whether the encoding process has been correctly carried out. It is also frequently employed for a number of other purposes, such as the correct phasing of a number of colour video signals at a studio mixer, and the measurement of differential phase.

The chrominance components of a colour video signal are commonly called 'colour vectors'. The name 'phasor' would, strictly speaking, be more accurate, but 'vector' has become so established that it would be pedantic to attempt to change the terminology;

hence the term 'vector monitor'. 'Vectorscope' is also encountered, but this is in fact the proprietary name of the product of a single manufacturer.

An NTSC signal corresponding to a uniform field of constant hue gives rise on the polar display of a vector monitor to two vectors, one for the burst and one for the colour field, the angle between the vectors giving the hue. The chrominance components of this signal are flat-topped, so that the display has the appearance of two bright dots, with the transitions indicated by much fainter lines joining these dots to the centre of the display.

On the other hand, the equivalent Pal signal, displayed in the same manner, appears as four dots: two for the $\pm 45^\circ$ positions of the colour burst relative to the $B-Y$ axis and two for the chrominance vectors in successive lines. These also appear to be imaged in the $B-Y$ axis as a consequence of the phase reversal of the $R-Y$ component from line to line. Such a display with each colour presented twice would be confusing for most of the purposes for which the vector monitor is used, and measures are taken in the Pal version of the instrument to deal with this.

Apart from the polar display, most vector monitors are provided with a line display, i.e. a Cartesian presentation against time of the components of the colour vectors in a chosen direction. This display is useful for the precise measurement of the vector angles, and can also be adapted for the measurement of differential phase. Both these displays are described in greater detail below.

11.5.1 Circuit features

The circuitry of a vector monitor is highly complex, and it would not be feasible to discuss it in detail, but two principal points have a bearing on the operation of the instrument.

The first concerns the provision of a phase reference. Since a vector monitor measures phase angle, it must be furnished with a source of coherent subcarrier. From this and the colour vector, a voltage or current is derived proportional to the angle between the two for use in constructing the display. In some instances when local measurements are being made, a common source of colour subcarrier may supply both the test-signal generator and the vector monitor. Otherwise, the reference must be regenerated from the colour burst contained in the test signal by standard and well known techniques.

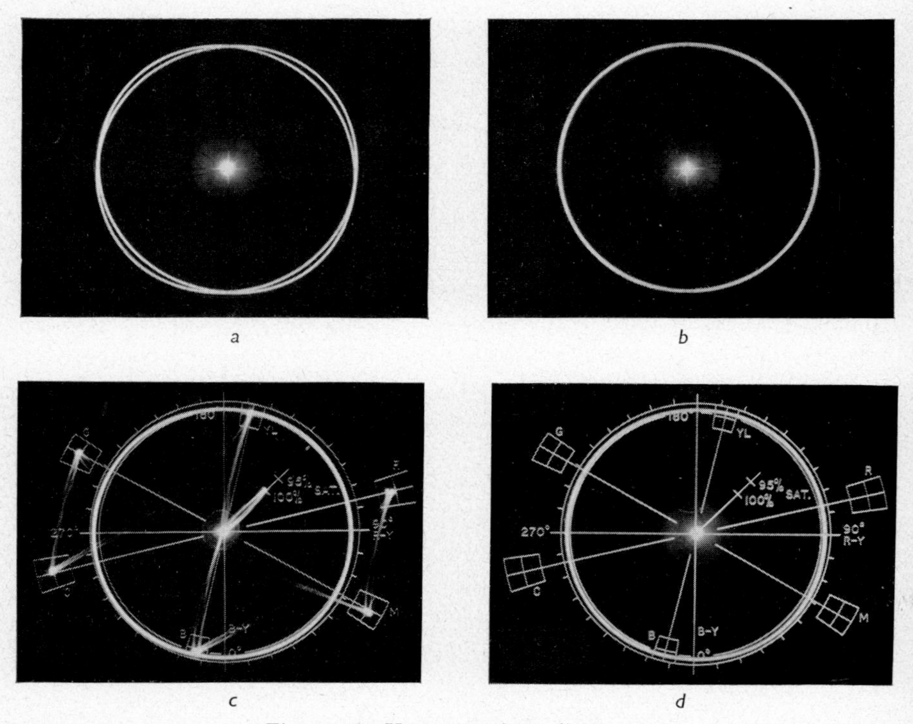

Fig. 11.10 Vector-monitor adjustment

a Quadrature error indicated by noncoincidence of test circles
b Correct adjustment: circles superimposed
c As (*b*), but with circles aligned against graticule
d Test circles time-shared with vector display
Pal system; tolerance boxes are $\pm 5^\circ \times \pm 10\%$

The reference subcarrier is passed within the instrument through a phase shifter variable over a wide range of angles, such as a goniometer or variable delay line, with which the display can be adjusted to a standard position. Adjustment of the phase shifter rotates the polar display and changes the line display in the vertical direction. A calibrated vernier phase shifter and/or calibrated steps of phase angle enable a more accurate measurement of the vector angles to be made than is possible from the display.

The second point concerns the so-called 'test circles'. The polar display is naturally derived from a simultaneous display of the horizontal and vertical components of the colour vectors, which in turn are derived from phase detectors with reference frequencies in quadrature. These are usually in the form of bridge modulators whose carrier supplies are derived either from a regenerated subcarrier frequency or from an external source, the two feeds being given the appropriate 90° phase shift by means of some adjustable network. The polar display will only possess truly circular co-ordinates when (*a*) the angle between the two carrier supplies to the modulators is precisely 90° and (*b*) the overall deflectional sensitivities of the two channels displaying the components of the colour vectors are precisely equal. The test circle provides a very elegant method of ensuring that both conditions are simultaneously met.

As with many other ingenious devices, it is basically very simple. An auxiliary oscillator is provided whose frequency differs from subcarrier frequency by a few kilohertz. This is fed into the two modulators along the same path as the chrominance signal derived from the normal video signal, with the result that difference frequencies are generated, the phase shift between which is equal to the phase shift between the carrier supplies to the modulators, and whose amplitudes are proportional to the gains of the channels. It then follows that, if these two low-frequency difference signals can be made to obey the above two conditions, these latter will also be obeyed by the desired output signals from the two modulators.

When the chrominance signals are removed from the modulators and the auxiliary oscillator is substituted, the display on the screen of the tube becomes an ellipse in general. Presumably, the ellipse would have quite a small eccentricity, since the requisite adjustments should be roughly correct; but nevertheless, without some guide, it would not be possible by eye to find the required point where the ellipse becomes a circle.

This is achieved by the elegant device of periodically reversing one of the modulator inputs, which yields a pair of superimposed ellipses with opposite eccentricities (Fig. 11.10*a*). The two ellipses coincide to form a single ellipse when, and only when, the phase shift between the channels is precisely 90°. Equalising the gains of the channels then turns the pair of coincident ellipses into the pair of coincident circles of Fig. 11.10*b*. A pair of concentric circles provided on the graticule (Fig. 11.10*c*) help considerably in the adjustment to circularity. Usually provision is made for the test circle to be time-shared with the polar display of the video signal, so that the quadrature adjustments can be constantly monitored during use. Fig. 11.10*d* shows the appearance of the time-shared display.

11.5.2 Polar display

One of the chief operational uses of the vector monitor is the panoramic display in polar form of the amplitudes and phases of the chrominance components of the colour-bar signal (Chapter 10), which provides an immediate indication of the state of adjustment of the apparatus from which the signal has been derived. For this purpose, a suitable graticule is needed, such as that in Fig. 11.10*c*.

Its principal features are:

(*a*) a pair of concentric circles with a scale divided into degrees or convenient multiples of degrees

(*b*) a line at the angle of the colour burst on lines *n*, calibrated for the lengths of the burst vector with the colour-bar signals in common use

(*c*) the locations of the $B-Y$ and $R-Y$ axes

(*d*) tolerance boxes for the lengths and angles of the colour-bar vectors. It should be noted that the positions of the $B-Y$ and $R-Y$ axes are rotated through 90° compared with the more usual convention. The directions of these axes are purely arbitrary; and, in the design of the vector monitor with which these photographs were taken, this made much more effective use of the screen area of the rectangular tube employed.

A Pal vector monitor will incorporate the line-sequential waveform for the $R-Y$ axis which, when the phase of the local subcarrier is made equal to the mean burst phase by adjusting the carrier phase-shifter provided, should result in the complete superposition of a pair of successive line waveforms (Fig. 11.11 *a*). This is achieved by turning the phase shifter to the point at which the two short lines are superimposed. If the coded signal and the vector monitor are free from error, this should also result in the superimposition of the tips of the colour-bar vectors; and, if the horizontal and vertical shifts are properly set, the corresponding line should lie along the graticule

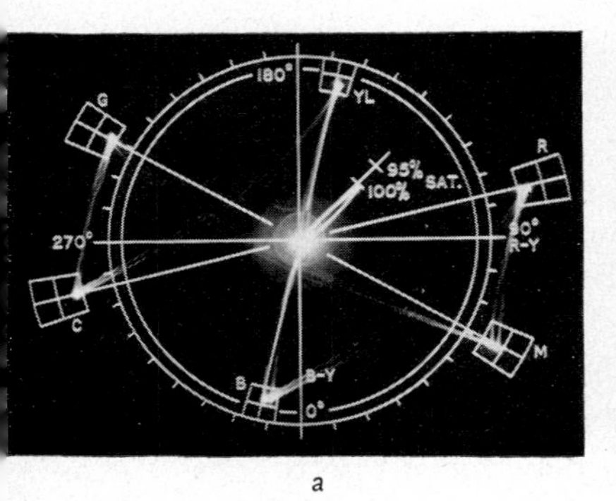

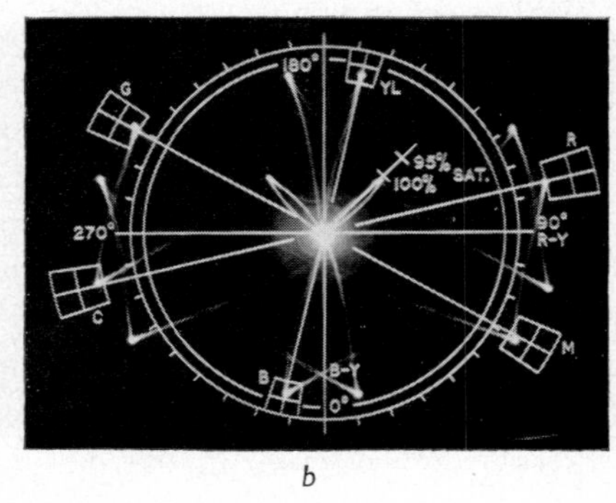

Fig. 11.11 Effect on Pal display of $R-Y$ axis switching

a Colour-bar signal with $R-Y$ axis switching
b As (*a*) with switching removed

line. The amplitude of the signal is then adjusted so that the burst length is correct for the colour-bar signal in use; and one then ought to be able to read off the errors in the colour-bar vectors from the tolerance boxes. However, Fig. 11.11 *a* demonstrates that the precaution should also be taken of checking that the burst amplitude is correct with respect to the video-signal amplitude. In the example, the burst amplitude was slightly high, and in consequence the colour-bar vectors all appear to be somewhat short.

The setting up of the display with an NTSC signal is simple, since no line-sequential switching is involved, and a single pattern is obtained which rotates as a whole as the carrier phase is varied. It is then only necessary to rotate the display until the burst vector lies along the graticule line, and then adjust the gain as with the Pal display.

It is also possible to remove the switching waveform with a Pal signal, giving the display of Fig. 11.11 *b*, where the colour-bar vectors appear to be mirrored in the $B-Y$ axis as a result of the switching of the $R-Y$ axis. One application of this will be given below.

The rather impressive polar display of a vector monitor seems to engender in many users a degree of confidence in the accuracy of its readings which is not always justified in practice. The number of possible sources of error contributing to the production of a polar display is rather large; and, even with a high-grade instrument, it is difficult to guarantee a true overall accuracy within $\pm 2°$ at any part of the scale over long periods. The polar presentation is therefore recommended only for operational checking purposes. For precision measurement, the line display is much to be preferred. By the same token, the vector monitor is not the best instrument for the alignment of coders, although often employed for that purpose. More accurate and reliable methods are available (Section 12.9).

The line display is a presentation against time of the components of the colour vectors in the direction determined by the phase of a reference-subcarrier voltage. Since, in a given display, the measured colour vector amplitudes remain constant while the phase angle varies, the vertical deflection for each colour vector is proportional to the phase angle between the vector and the reference subcarrier. Hence, for simplicity, it will be assumed that the vertical deflections are proportional to the phase angles without further reference to the vector amplitudes. In fact, this is further justified by the normal use of the line display, not as a direct means of measuring phase angles, but as a null indicator, where of course, the amplitudes play no part. An additional use of the line display is for the measurement of differential phase, but here the subcarrier amplitude may be somewhat modified by any differential gain present unless internal limiting is provided.

The line display cannot be considered as a complete alternative to the polar display, since it provides no means for determining the amplitudes of the chrominance components of the input signal. In principle, this is also possible, but the bandwidth requirements of the two types of display are so different that it is not very convenient. Some commercially available monitors, however, provide a display of the luminance components.

The operation of the line display will be explained with respect to one example of vector monitor. Since this was designed to fit in with a particular operational philosophy, it differs in some relatively unimportant respects from other instruments, but the basic principles are much the same.

The time-linear horizontal deflection is conventional, but the vertical-deflection amplifier is supplied with a voltage proportional to the component of each colour vector in the direction of the phase angle of a reference subcarrier which, exactly as with the polar display, passes through a wide-range phase shifter for the purpose of setting up the display. This phase-proportional voltage is very conveniently supplied by one of the modulators which generates the two components of the colour-bar vectors, in this instance the $B - Y$ axis modulator, since it is a property of a linear modulator that it is capable of acting as an approximately linear phase detector, the degree of approximation being quite small for angles between zero and, say, 10°. This comes about because such a modulator fed with two inputs in quadrature gives zero output voltage, and if the 90° relationship is altered by an angle of y degrees, the output voltage is proportional to sin y degrees, which, for small values of y, means that the output voltage is proportional to the angle, assuming that the two input amplitudes remain constant.

It is consequently possible to use this modulator circuit in two ways. First, it can be set up with the local subcarrier exactly 90° out of phase with the zero phase condition of the input chrominance signal when it will operate for small relative phase shifts as a linear phase detector; or alternatively, calibrated phase shifters can be inserted in series with one of the inputs in such a fashion as to restore the 90° phase shift, indicated by zero output level from the modulator. The second mode is considerably more satisfactory, since the modulator is only utilised as a null indicator, and stable passive circuits can be employed to produce the known phase shifts.

The Pal vector monitor considered here by way of example goes rather further in this direction than most since it offers not only a calibrated variable phase shifter but also four fixed phase shifters which may be selected by a pushbutton switch. These latter are 90° and the three angles corresponding to the colour-bar vectors; it will be recalled that the six colour-bar vectors can be grouped into three sets having angles equal and opposite in sign, so that only three angles are needed for a null measurement.

Before a measurement is attempted, it is necessary to set the phase of the reference subcarrier to a standard condition, which is derived from the burst of the signal under examination. With a Pal signal, the standard is the mean burst angle, and the adjustment of the reference subcarrier to this angle is indicated in the line display by the appearance of two burst components of the same sign and equal amplitudes (Fig. 11.12*a*), where the two are precisely superimposed. The small transient to the right of the burst amplitudes comes from the luminance bar, and is of no consequence. The mean burst angle ought to coincide with the $B-Y$ axis, but in practice it may not do so. However, the exact superposition in Fig. 11.12*a* of the conjugate waveforms of the colour-bar vectors demonstrates very clearly that the two coincide completely in this instance.

When a null measurement is to be made, it is necessary to offset the reference carrier by 90° so as to fulfil the condition for zero output from the modulator. Accordingly, the 90° phase shifter is inserted before the adjustment of the reference subcarrier to the mean burst phase. The insertion of the phase angle corresponding to one of the pairs of colour-bar vectors should then, if their angles are correct, set the flat tops of their line-display waveforms to the zero axis. Any lack of coincidence can then be put right with the calibrated variable phase shifter, from which the error in the colour-bar angle can then be read. In Fig. 11.12*b*, the insertion of the green–magenta phase shifter brought the peaks of a central pair of waveforms to the axis line, showing that there was negligible error, although, for the sake of clarity in the photograph, they have deliberately been offset by 2° to make them visible.

In the corresponding measurement with the NTSC colour-bar

waveform the reference is naturally still the burst phase angle, but it happens that, in this instance, the adjustment of the subcarrier to the burst phase angle shifted by 90° can be carried out very conveniently without recourse to a 90° phase shifter, simply by setting the sub-

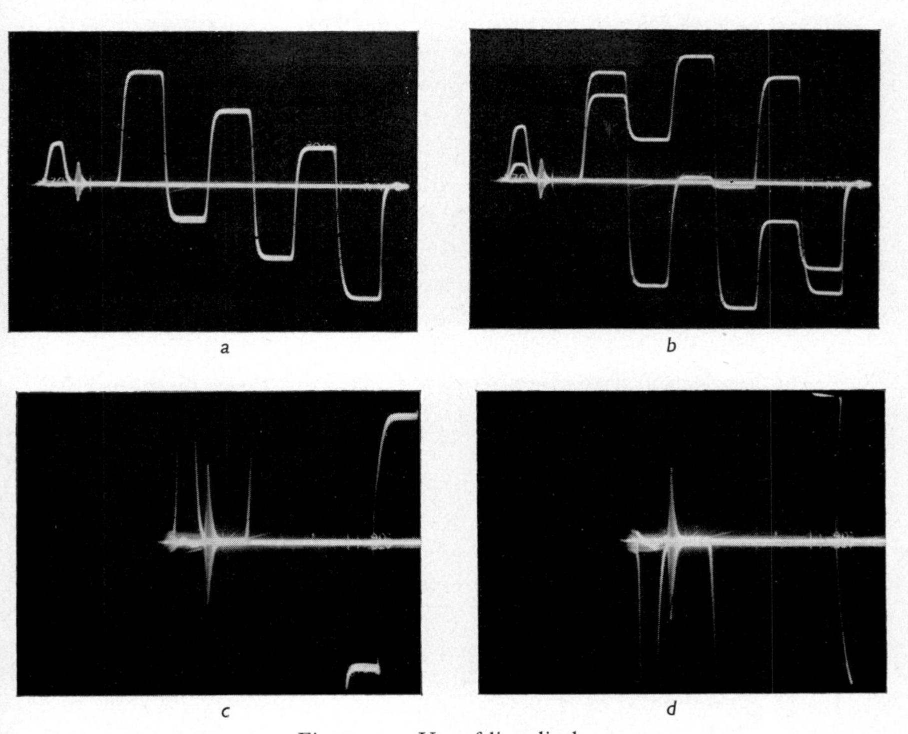

Fig. 11.12 Use of line display

a Reference subcarrier at mean burst angle
b As (*a*) with added phase shift corresponding to green–magenta bars
c One burst vector adjusted to zero amplitude
d As (*c*) after insertion of 90° phase shifter
Pal system: tolerance boxes are $\pm 5° \times \pm 10\%$

carrier angle to the position where the burst waveform disappears. This is evident, since, with NTSC, there should be no burst component at right angles to the $B-Y$ axis.

With a Pal signal, the 90° phase shifter can also be utilised for measurements in its own right, e.g. for the measurement of the angle between the two burst positions in successive lines. This is illustrated in Fig. 11.12*c* and *d*, in which additional vertical gain available by means of a switch has been inserted to improve the sensitivity, with the result that the tops of the burst vectors now fall outside the screen area.

The reference-subcarrier phase is adjusted until one of the conjugate burst vectors is made to coincide precisely with the zero line (Fig. 11.12*c*). The 90° phase shifter is then inserted, so that the situation is completely reversed, except that the conjugate burst vector, which should be zero, is, in fact, slightly below the zero line. This, unfortunately, is not as clear in Fig. 11.12*d* as one would wish, owing to the slight overexposure required to define the leading and lagging edges of the burst waveforms. An exact setting to zero by means of the variable phase shifter showed that the error in the nominal 90° between the two burst positions was 0·8°.

Another virtue of the line display is that it can also be employed for the measurement of differential phase, in which case the reference subcarrier will be set so as to bring to zero the portion of the waveform corresponding to the subcarrier phase angle at black level. To make the measurement of differential phase possible, the vector monitor must also include suitable amplification and filtering for the test waveform. One commercial instrument includes a method for doubling the sensitivity by the use of the switching waveform to reverse the polarity of the display on successive lines. When the test waveform is a staircase, the variable phase shifter can be used to bring into coincidence in succession the flat portions of the corresponding positive and negative displays. From the phase-shifter readings, the differential phase is then derived.

The great advantage of the line display over the polar display lies in its use to indicate a null setting, rather than the phase angle directly, so that a very large number of the inherent errors are removed from the equation, and the resulting accuracy of measurement is very largely dependent on the accuracy and stability of precision phase-shifters constructed as passive networks and reliable to one-tenth or two-tenths of a degree over quite long periods of time.

On the other hand, with the polar display, the accuracy is dependent on the internal quadrature relationship between the two carriers applied to the modulators, the adjustment of the gains in the two channels, the linearity of the channels, the orthogonality error in the display tube, and a number of other factors too numerous to detail.

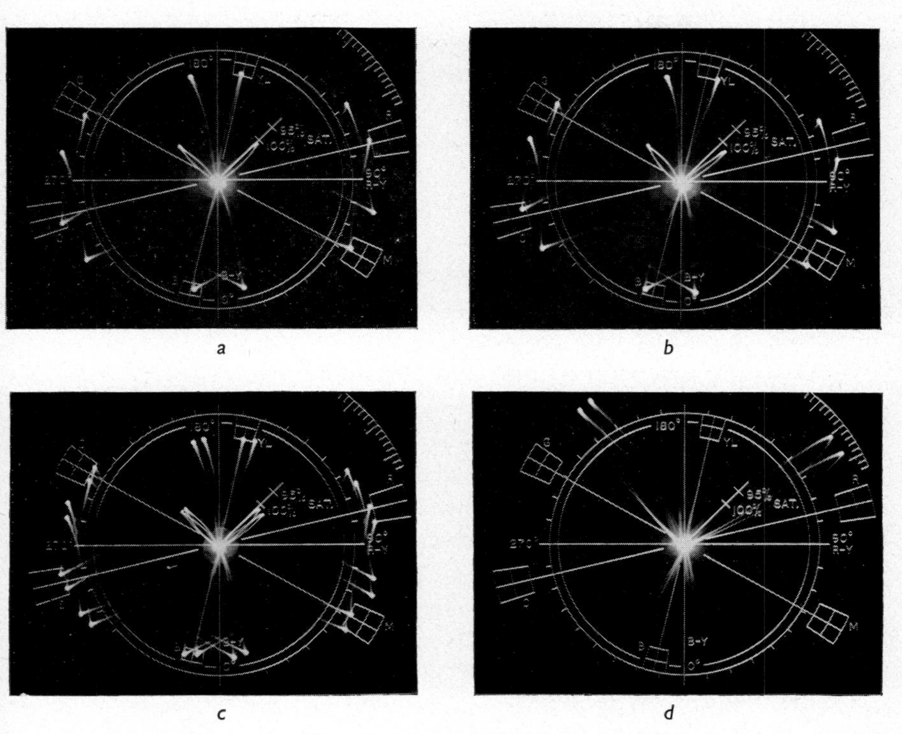

Fig. 11.13 Use of vector monitor for source phasing

a Source A
b Source B
c Sources A and B time-shared
d Use of auxiliary scale for increased reading accuracy

Precision phase-measurements should therefore always be made with the line display in preference to the polar display whenever possible.

Another common use of the vector monitor, particularly important in television studios, is source phasing, i.e. ensuring that, at every point at which picture sources must be mixed or interchanged, the phases of the colour bursts are the same within close limits. The associated problem of ensuring that the burst remains in the correct relationship with the chrominance components of the signal is dealt with by the methods described above.

Source phasing is most usually carried out with the time-sharing facility built into most vector monitors, although it is feasible to manage with a means of switching between the two sources. A disadvantage of the time-shared display is its rather complex appearance, and it is usually desirable to display Pal signals in the unswitched or 'NTSC' mode.

Let us assume that Fig. 11.13*a* corresponds to source A, where the phase of the reference subcarrier in the vector monitor has been set so that the burst vector corresponding to the alternate lines coincides with the graticule line. Then source B is evidently from Fig. 11.13*b* advanced in phase with respect to source A by a few degrees. Time-sharing the two sources gives the display of Fig. 11.13*c*. It is, of course, not essential to use a colour-bar signal for source phasing alone, but the presence of the colour-bar vectors is useful as an indication of the behaviour of the system in other respects.

The method of measuring the angle between the two sources, which seems to be preferred operationally, consists in amplifying the burst vectors, so that the angle between them can be read from a calibrated scale of as large a diameter as can conveniently be engraved on the graticule. If a complete circle cannot be accommodated, an arc of reasonable length will suffice (Fig. 11.13*d*), where evidently the angle between the sources is 6°. Alternatively, the angular difference can be measured against an arbitrary mark on the graticule scale by means of the calibrated phase shifter; this is the more accurate and reliable method, particularly if the angle is more than a few degrees.

11.6 References

GROSSKOPF, H., and SCHAUMBERGER, A. (1965): 'Verbesserung der Bildqualität durch automatische Korrektur der Signalamplituden', *Rundfuhktech, Mitt.*, **9**, (1), pp. 26–32

WHEELER, H. (1939): 'The interpretation of amplitude and phase distortions in terms of paired echoes', *Proc. Inst. Radio Engrs.*, **27**, p. 359

WOLF, H. (1963): 'Über den Zusammenhang zwischen Bandbreite und Anstiegzeit', *Elektron.*, **12**, (10), pp. 303–308

12 CAMERAS

12.1 General

It is difficult with cameras, as with other picture sources, to make a clear distinction between measurements, operational tests and alignment procedures, and much depends on the operational philosophy of the user and the nature of the equipment. This, and the fact that little collected information on the subject has been published, seem to call for some general information on good operational practice, particularly with regard to colour cameras, without, however, making detailed reference to any particular make or type of equipment.

At the time of writing, colour cameras in general use are of two principal types, 3-tube and 4-tube, all employing photoconductive pickup tubes, although, in the latter, one of the tubes may be an image orthicon. Colorimetric questions apart, the alignment of the tubes themselves follows much the same lines whether or not they are used in colour cameras. The principal difference with colour cameras lies in the great pains which have to be taken to ensure that the tubes in the colouring channels behave precisely identically under all conditions, and in the need for the exact registration of the images. A further problem is the matching of a number of cameras used on the same production, since the sensitivity of the eye to certain hues, especially to flesh tones, is so high that even extremely small colorimetric differences are perceptible if pictures of the same scene from different cameras are shown in quick succession.

12.2 Linear waveform distortion

This is best measured in all instances by sine-squared-pulse and bar methods (Chapter 6). In view of the need with monochrome and 3-tube colour cameras to ensure that the behaviour of the channel is adequate up to the limits of the videoband, it has hitherto been customary to supplement or even replace the $2T$ pulse-and-bar tests with identical test carried out with $1T$ pulse-and-bar waveforms. This becomes possible because of the very slow rolloff of all the amplifiers, except the head amplifier where excess bandwidth is harder to come by. The rating factor thereby obtained has been called the K_T rating.

However, there seems to be every good reason for replacing this waveform by one of the more modern chrominance–luminance or augmented pulse-and-bar waveforms whose special property it is that it may be passed through an ideal lowpass filter, with its cutoff frequency at the upper nominal limit of the videoband, without distortion. In this, it resembles the $2T$ sine-squared-pulse and bar signal, but unlike that waveform, these newer signals also provide very useful information about the conditions existing near the top of the band, and so overcome the principal drawback of the $2T$ signal.

The need for the revised testing method has arisen since the introduction of colour cameras, not because of the presence of coded signals which, in any case, do not occur in colour-camera channels, but from the increasing use of lowpass phase-corrected filters in the various signal paths to define the channel bandwidths and to remove out-of-band noise; also, to some extent, because of the increasing use for the adjustment of signal timings of delay networks whose behaviour is impeccable over the videoband but which may show sharp changes outside. The $1T$ pulse depends for its use on a slow, Gaussian-type rolloff in the apparatus under test, and is quite unsuitable for use where the transfer characteristic changes rapidly beyond the limits of the videoband. On the contrary, the newer waveforms are quite suitable for testing under these conditions, since their spectra are confined within the nominal limits of the videoband (Section 6.3.9).

In routine measurements through the channel, it is customary to switch out the aperture corrector, since the rising gain with increasing frequency which this furnishes will apparently lead to distortion of the test signal. Measurements can only be made through the aperture corrector by the use of special means described below.

Routine measurements are not usually made through the head amplifier, but, if the response of this must be included, as during acceptance testing, means must be provided for injecting the signal at the input to the head amplifier such that the impedance conditions existing at that point when the camera tube is connected are accurately maintained. This requires some special test adaptor whose precise form can only be determined by measurement. The principles involved are quite straightforward.

12.2.1 Aperture correction

All camera channels contain facilities for aperture correction, i.e. a means for compensating overwidth of the nominal videoband for the droop in the effective frequency response of the tube which is a consequence of the finite scanning spot size. In simple theory, the current distribution over the scanning spot should be Gaussian and, consequently, the effective frequency response should also be Gaussian; this is a unique property of the Fourier transform of the Gaussian curve (Campbell and Foster, 1931). In practice, this is not quite true, but the deviations are not usually very great, since the magnitude of the droop at the top of the videoband must be fairly restricted if the signal/noise ratio is not to be unduly worsened, so that any discrepancies from the theoretical curve are minimised.

Aperture-correction networks should ideally provide a rising amplitude characteristic of the correct shape, variable over a sufficient range to take up individual variations in tubes, and without appreciably modifying the phase response of the channel. Measurements must accordingly be made at some stage to ascertain whether these conditions are fulfilled, usually during acceptance testing but possibly also after maintenance.

A device which has proved very effective measures the channel with a network connected in tandem so designed as to simulate the aperture distortion of the tube. A series of these networks designed by the author consists of pairs of lowpass filters having, respectively, the minimum likely and maximum tolerable droops for each type of tube at the nominal limit of the videoband. The aperture corrector is then set to equalise each of these in turn, and measurements are made through the tandem connection of the network and the camera channel, preferably using a chrominance–luminance or augmented pulse-and-bar waveforms (Section 12.2.)

Since the aperture distortion characteristic is the initial portion of a Gaussian curve, it may be simulated either with a Gaussian filter (Dishal, 1959) or a sine-squared shaping network, i.e. a Thomson filter (Thomson, 1952). As far as the 6 dB point, there is very little difference between the two networks. Both give negligible phase distortion, provided that the order of the Gaussian filter chosen is not too low.

The practical adjustment of the aperture correction on a camera channel is always carried out by the use of a standard test pattern containing resolution gratings such as that in the central area of the exposure chart of Fig. 12.1 *a* (Section 12.3.2) or test card 51 (Chapter 10.3). The test card is preferably set up at a distance from the lens which ensures good resolution, and the optimum stop is selected. The camera is focused as well as possible, then the modulation depth of, say, the 5 MHz grating is examined on a waveform monitor, and the aperture corrector is adjusted accordingly. The difference between full and a low modulation depth is clearly brought out in Figs. 12.1 *b* and *c*. The adjustment sometimes has to be a compromise between resolution and random noise, since, as the aperture correction is increased, the noise spectrum may rise rapidly towards the top of the band, with a consequent degradation of the signal/noise ratio.

12.2.2 Camera-cable equalisation

This is another measurement usually only carried out during acceptance testing, but which might also be required for maintenance purposes.

The object is to transmit a test signal, preferably one of the combined chrominance and luminance pulse-and-bar signals, through the series combination of the camera cable and the equaliser; if one wishes to be thorough, all lengths of cable with their appropriate equalisers will have to be measured. The experimental details will have to be omitted, since they depend so much on the facilities provided in the camera channel, but this need present no difficulty. The distortion of the output waveform ought to be very low if the equalisers are in correct adjustment.

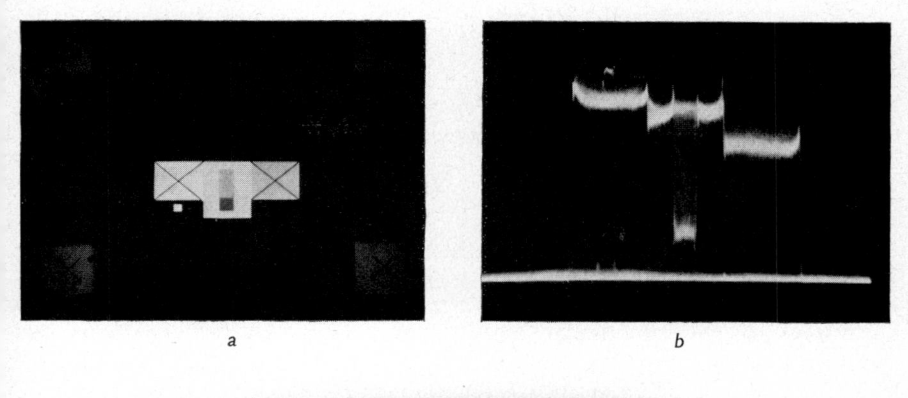

a b

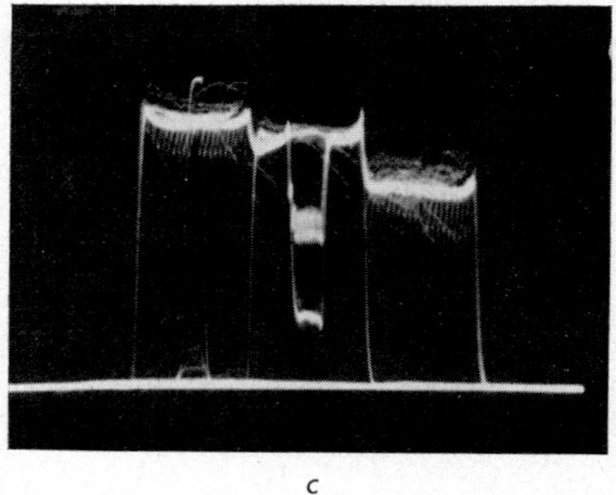

c

Fig. 12.1 Setting exposure of image orthicon

a Test chart
b Signal directly from camera
c As (*a*) after passage through 1 MHz lowpass filter

12.3 Exposure

This is normally of the greatest importance when dealing with image-orthicon tubes which have a pronounced knee in the transfer characteristic, so that the picture reproduction is a critical function of the location of the usable range of scene brightness on the transfer characteristic with respect to the knee. To specify the position of the knee, it is necessary to define the point gamma, $\dot{\gamma}$.

12.3.1 Definition of $\dot{\gamma}$

The transfer characteristic is the relationship between the photocathode illumination L and the signal voltage V_s. If the curve of the relationship between L and V_s is plotted on double-logarithmic scales, $\dot{\gamma}$ is the gradient of the curve at a given point; i.e.

$$\dot{\gamma} = \frac{d(\log V_s)}{d(\log L)}$$

$\dot{\gamma}$ is unity over the linear part of the transfer characteristic, and decreases around the knee where the curvature of the transfer characteristic increases. The knee is, of course, no sudden break in the curve which can easily be identified; so a useful arbitrary choice was made by Brothers (1959), who identified the knee with the point at which $\dot{\gamma} = 0{\cdot}5$ precisely.

The laboratory measurement of $\dot{\gamma}$ can be made by a method described by Brothers (loc. cit.). The camera, previously set up to a standard channel gain, is positioned to view a dark area in the centre of which is a small, square illuminated area, with a side approximately one-thirtieth of the line scan width. A calibrated step wedge can be moved over this window, so that a known range of values of illumination can be provided. The output of the channel is fed into one input of an accurately balanced difference amplifier.

Into the other input of the difference amplifier is fed, via an adjustable attenuator, a flat-topped line pulse synchronised with the camera scans. The output of the difference amplifier is viewed on a waveform monitor using a line selector, so that any given line or sequence of lines can be examined.

The camera is panned, or the pulse is moved along the line, until the output waveform from the camera and the line pulse are in approximate time coincidence; then the attenuator is adjusted until the two waveforms cancel; any spikes on either side of the central portion due to lack of coincidence are ignored. A neutral filter with a small, known value of density is then placed in front of the window, and attenuation is taken out of the variable attenuator until the balance is restored. If the change in the attenuator reading is A

decibels and the density of the filter is D, then $\dot{\gamma} = A/20D$ approximately.

The approximation is more accurate, of course, as the density of the filter is made smaller. In this way, the complete transfer characteristic can be plotted.

12.3.2 Practical method

The method just described is hardly suitable operationally, and an alternative simplified scheme due to A. B. Palmer and described by Anstey and Ward (1963), or some version of this, is usually preferred. By this method, a camera can be set up very quickly to a standard sensitivity and standard gradation. It is based on a special transparency, in the form of a $2\frac{1}{4} \times 3\frac{1}{4}$ in glass slide, attached to a small light box, which in turn is fastened to the front of the camera (Fig. 12.1 *a*). The background of the slide is black, and in the centre is a 3-step grey scale with successive densities of 0·15 between the steps, the end step being nominally clear transmission through the glass. The light source is run from a stabilised supply, and its brightness is adjusted photometrically against a standard. For other purposes, four patches having densities of 0·6 are placed in the corners of the slide, and the central patch includes a 5 MHz grating pattern. To indicate the maximum possible contrast range, two very small patches are made as white and as black as possible ('superwhite' and 'superblack').

The gain of the channel having previously been adjusted by injecting a sawtooth of the correct amplitude into the head-amplifier test point, the output voltage from the test card is set against a special waveform monitor graticule, so that the part of the waveform corresponding to black lines along the 0% horizontal line. The gain of the oscilloscope is set so that white level would fall on the 100% calibration line.

The channel is then adjusted by means of the iris and the dynode gain so that the two lower horizontal levels in the waveform correspond to amplitudes of 66·7 and 80%, respectively. The adjustment is more accurate if the waveform is previously passed through a 1 MHz lowpass filter to reduce the random-noise level. The visual improvement is greater than is suggested by Figs. 12.1 *b* and *c*, owing to the integration of the noise by the film. The brightest patch does not then correspond to 100 % as might be thought, but to 90 % approximately. It is found empirically that this compensates for the edge effect of the image orthicon, and the whites on normal scenes reach 100 %.

What has been done, in effect, is to set a predetermined point on the transfer characteristic to the knee of the tube characteristic, taken to correspond to $\dot{\gamma} = 0{\cdot}5$. This is an arbitrary setting, but one which has been found by long experience to give very good and consistent results. An improvement would be an increase of the number of steps to improve the accuracy of location of the knee with tubes where the knee is fairly long. The use of two identical step wedges running in opposite directions has been tried with the object of reducing the effect of shading, but has now been abandoned owing to ambiguities in signal level, although this device is proving useful for colour cameras where a compromise setting for the three (or four) tubes is needed. A larger transparency is required for use with zoom lenses.

In all measurements or adjustments where test patterns are used, great care should be taken not to leave the pattern fully exposed for longer than is necessary; if the pattern is not actually in use, the iris can always be stopped down until work is recommenced. If this precaution is not taken, 'sticking' of the image may occur, although some modern image-orthicon tubes are relatively free from this defect. An effective preventative is 'orbiting', i.e. a continuous rotary movement of the image with respect to the target.

12.3.3 Colour cameras

The chief difficulty with colour cameras lies not so much in the setting of the exposure itself as with the tracking, i.e. the adjustment of the three colour-separation channels in such a way that the exposure is identically the same for each channel over an adequate contrast range. For this, the three colour-separation channels must have precisely identical lifts, contrast laws and gains, not only at the normal setting of the master gain control, but also over the whole range of adjustment. It follows that it is not possible in this case to make a clear distinction between exposure and the setting of the contrast laws of the channels, although, for the sake of clarity, these must be given separate treatments.

An extremely useful general-purpose test signal for the alignment of colour camera channels takes the form of a line-repetitive waveform containing a sawtooth and a bar in each line, the amplitude of each being maintained precisely at 700 mV. The white bar then serves as a reference for the signal amplitude, and the sawtooth allows contrast correctors to be checked while providing a critical indication of peak clipping from the action of the channel limiters. A 5-way signal gate directs the first three lines of a sequence of five into the colour-separation channels in the order red, green and blue, and the next line into the luminance or Y channel if present; the fifth line is unused, to serve as a control and also to form a break between the sequences of four measurement lines for identification. All these lines can be added at the output of the channel to give a display which, when suitably triggered, has the appearance of Fig. 12.2*a*. The bar transitions are faintly visible on each side of the staircase.

Before proceeding to the adjustment of the exposure, it is desirable to set the contrast-law (γ) correctors by the techniques of Section 12.4. Now, in the alignment of a colour camera, it is customary to consider the green channel as a master and to align the other two channels to it. Accordingly, the gain of the green channel is first set by injecting the waveform described above into the test point of the green head amplifier, and the green gain and lift are made to give a waveform at the output with the correct amplitude and sit. The other colour-separation signals are then brought into precisely the same condition by examining in turn the discrepancies between the red and green, and blue and green, at the output of a high-gain difference amplifier and adjusting for minimum residual in the observed waveform. A typical gain difference is illustrated in Fig. 12.2*b* and a lift difference in Fig. 12.2*c*. Other methods of achieving equality between channels are given below.

The final adjustments are carried out with a dynamic test using a standard card presented to the camera; these then take into account the complete transfer function of the camera from scene to output voltage including its colorimetric behaviour. Such a test card must not only be designed for the purpose but, what is even more important, it must be capable of being manufactured and reproduced to a very high degree of accuracy indeed. In the UK, and increasingly

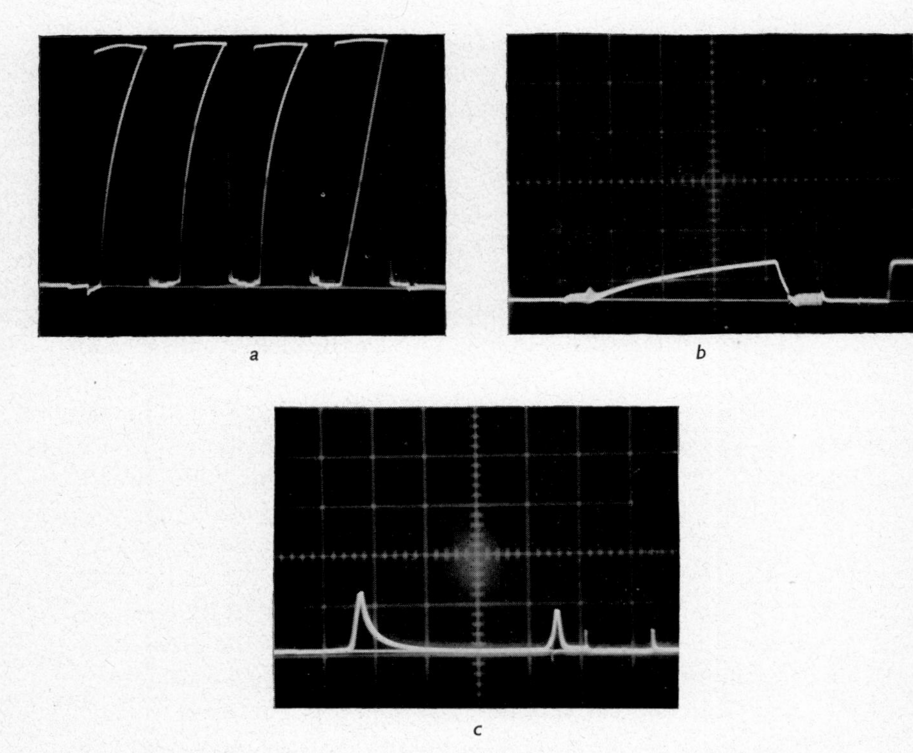

Fig. 12.2 Adjustment of gain and lift of colour-separation channels
a Test waveform through R, G and B channels sequentially
b Typical gain difference
c Typical lift difference

elsewhere, it is becoming customary to employ BBC test card 57 (Chapter 10.9), which fulfils these conditions.

Very briefly, it takes the form of a card containing a pair of step wedges, running in opposite directions, whose reflectances are to a good approximation the inverse of the required contrast law. The colorimetric neutrality (monochromaticity) of all the reflecting surfaces is maintained at a very high level. In the centre is a small area of extremely low reflectance known as the 'superblack', and is employed for setting black level and for the adjustment of the flare correctors.

This card is illuminated as evenly as possible with an incident light value typical of that used for studio work at whatever colour temperature is regarded as normal in the range 3000–3200 K. It is useful to insert a 1 MHz lowpass filter in series with the display device to reduce the effect of camera noise. This should have a linear delay characteristic and a slow rolloff to avoid unwanted and misleading transients on the waveforms under examination; a Gaussian type of filter is excellent. The display should have the appearance of Fig. 12.3*b*, where the close symmetry of the *R*, *G* and *B* channels is very evident. The symmetry of the waveforms corresponding to the two-step wedges shows that shading effects are small. For purposes of comparison, Figs. 12.3*c* and *d* demonstrate a faulty alignment of the red channel.

The flare correctors may now be set by watching the position of the superblack as the iris is opened and closed over a large range. The final setting of the exposure may then be made by noting that the white step of the step wedge, actually 60 %, is intended to correspond to a 700 mV output. From experience, this ensures an optimum reproduction of facial tones. The superblack should in principle be set to a point slightly above black level which corresponds to the ratio of its reflectance to the white step, but negligible error is introduced by setting it to blanking level.

The tracking, or grey-scale balance, is carried out by utilising the fact that, if, at any one of the steps of the grey scale on test card 57, the voltages of the colour-separation signals are equal as is desired, no change will be observed in the superimposed waveform if they are connected together. For this purpose, it is common to employ a relay which can be actuated by a push button conveniently located. When the relay is operated, any difference between the master and comparison channels can be detected and put right at the appropriate place. This is preferably supplemented by feeding the colour-separation signals to a correctly set up *RGB* colour monitor. The operation of the relay will then produce a coloration at any step of the grey scale where the channels do not track. Another version of the comparison method switches between the master and another channel at a 12·5 Hz rate. When the superimposed waveforms are viewed, any difference in level generates a very noticeable flicker at the point concerned.

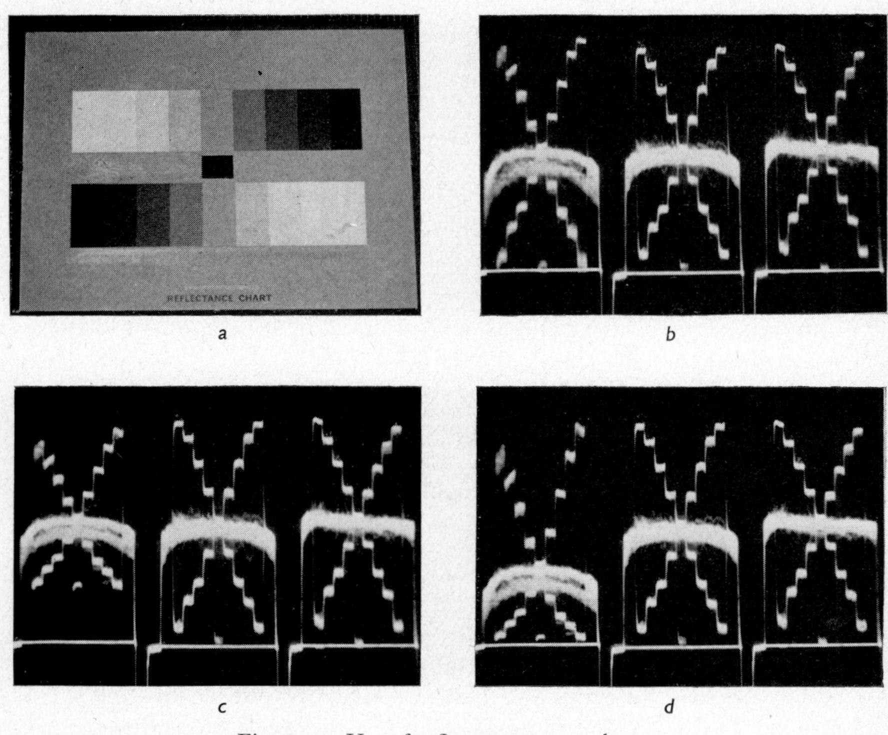

Fig. 12.3 Use of reflectance test card

a Prototype test card
b Correct adjustment of *R*, *G* and *B* channels
c, *d* Maladjustments of red channel

A technique now coming into widespread use is 'Clue' ('Colour lineup equipment'). The block diagram is given in Fig. 12.4. The master channel and the channel to be compared are fed to an electronic switch, the output of which is displayed on a monochrome picture monitor. The switching waveform is derived from line drive via a divide-by-four counter. Thus alternate blocks of four lines from each channel appear in sequence on the picture monitor, and move upward at the rate of one line per field.

At the start of lineup, the master channel is applied to both inputs of the electronic switch, and the gains are adjusted to give a display evenly illuminated in the vertical direction. The switch is then

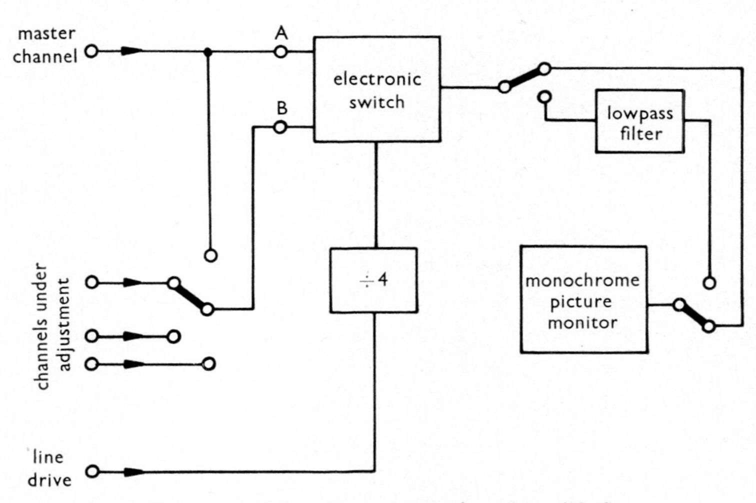

Fig. 12.4 Colour lineup equipment (simplified)

operated to bring in the channel under adjustment, when any difference in level is immediately obvious as a venetian-blind effect. The slow upward movement is no handicap; on the contrary, it aids discrimination.

However, even these refined techniques are not completely sufficient to ensure first-class rendering of the flesh tones in a scene, undoubtedly the most critical by far, on account of small residual errors in the cameras which are very difficult to take into account. It is accordingly widespread, at the time of writing, to carry out the final colour balance on a live model, usually a girl, under completely standardised conditions. Considerable work is in progress to replace the live model by some standard object, e.g. a dummy, a still-life subject or even colorimetric standards such as Munsell chips, but so far none of these has proved so reliable.

The problem of matching the colour-separation channels in the master camera is repeated when matching all cameras to be used in a given area one with another, and the above techniques are also applied to match the channels of the other cameras with those of the master. Not only does this achieve a closer alignment than would otherwise be possible, but the systematic approach with its clear indications to the operator speeds up the overall alignment, a matter of some importance since it increases the studio utilisation by reducing the nonproductive time.

The final subjective colour balance of cameras must be carried out with great accuracy, since even extremely small colorimetric differences show up when changing from one camera to another. A common practice is to group the cameras after individual alignment as close as possible together when viewing a standard scene, usually the lineup girl, so that, as nearly as can be achieved, each is producing the same picture. The cameras are then balanced against the master by employing a split-screen switch to display both the master-camera picture and that to be balanced on the same well aligned colour-picture monitor.

12.4 Contrast correction

12.4.1 General

As is well known, the relationship between the grid–cathode voltage E of a cathode-ray tube and the screen brightness B is given by $B = AE^{\gamma}$, where A and γ are constants. Hence, if the overall contrast law of the camera and display-tube combination is to be linear, a suitable nonlinear circuit must be included in the camera channel to equalise the combined transfer characteristics. This is known as the contrast corrector, γ corrector, or black-stretch circuit, the latter because its principal function is an increase of the gain in the near-black region.

Contrast correction is achieved by nonlinear circuit elements, usually diodes used over the curved parts of their characteristics. The overall contrast-correction curve is either continuous or composed of a number of approximately straight-line segments. The contrast corrector as a unit forms part of the processing amplifier.

Whatever the exact shape of the curve required, it must be accurately measured and set to the standard shape within close limits, not solely to ensure that the picture gradation is correct but, what is also important, to avoid noticeable differences between the pictures from a number of sources which may pass through a common mixer.

A number of methods of measuring contrast correctors are given below, some suitable only for investigational work and some more suitable for operational conditions.

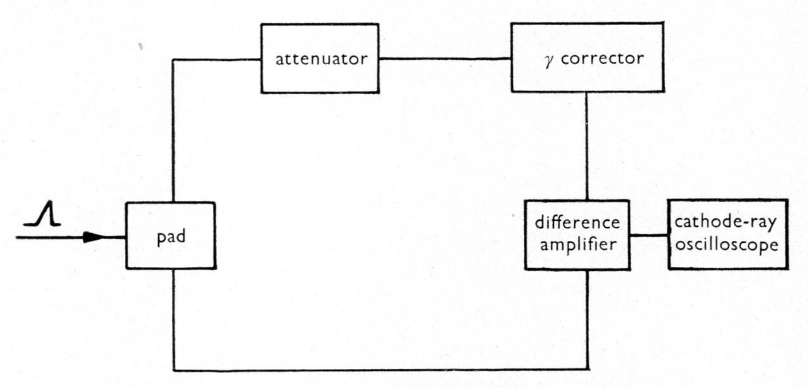

Fig. 12.5 Measurement of segmented γ corrector

12.4.2 Segmented curve

The measurement of the gain corresponding to the section of a corrector whose transfer characteristic comprises a small number of discrete segments is very simple if an accurately balanced difference amplifier is available such as is provided on certain waveform monitors.

The circuit is given in Fig. 12.5. A sawtooth waveform at line frequency is fed through a splitting pad to provide two 75 Ω outputs. One of these is taken through an attenuator to the input of the γ corrector, whose output feeds one of the inputs of the difference amplifier. The other output from the pad is taken to the second input of the difference amplifier. The output from this, i.e. the difference between the two input voltages, is taken to a waveform monitor whose sweep is externally triggered from the input sawtooth.

The variable attenuator in series with the γ corrector is adjusted until, say, the first segment becomes a horizontal line on the display. Then the attenuation introduced is obviously numerically equal to the gain of the first segment of the curve. The process is repeated with the other segments.

It is essential that the difference amplifier should be checked for balance before the measurement, and also that the sawtooth used should have the correct amplitude and be properly blanked. This latter is of particular importance since any malfunctioning of the clamp in the corrector will make it impossible to carry out the measurement.

12.4.3 Smooth curve

The method of Section 12.4.2 can still be applied even when the curve is smooth, by bringing a given point on the display down to coincidence with the horizontal axis. Then the attenuator reading gives the gain corresponding to the tangent at that point, i.e. the 'incremental gain' or 'slope gain'. Hence the curve can be reconstructed. However, if the correction curve is a true exponential, an elegant method, due to Potter (1958), can be applied.

A generator (Fig. 12.6) supplies a line waveform containing a full-amplitude line-width bar to two integrating circuits of the familiar *RC* type, one being fixed and the other having a time constant variable over a switch range. The output of the fixed integrator feeds the corrector under test, whose output is applied to one of the deflectional amplifiers of an oscilloscope. The output of the variable integrator passes through a gain-adjusting device and is then applied to the other deflectional amplifier of the oscilloscope.

The relative gains of the two sides and the time constant of the variable integrator are adjusted until the trace on the oscilloscope is a straight line at 45° to the axes. The constants of the circuit under measurement can then be calculated as follows.

Let the output from the fixed integrator be $A_1 e^{-\alpha t}$, and let the transfer function of the corrector under test be given by

$$E_{out} = B(E_{in})^{\gamma}$$

Then the output voltage applied to the oscilloscope input

$$E_{01} = A_1^{\gamma} B e^{-\alpha\gamma t}$$

Let the output from the variable integrator in the other branch of the circuit be $E_{02} = A_2 e^{-\beta t}$. Then if $E_{01} = E_{02}$, two equations are available. When $t \to 0$, $A_1^{\gamma} B = A_2$, and hence $e^{-\alpha\gamma t} = e^{-\beta t}$ or $\gamma = \beta/\alpha$. The constants α and β are known from the time constants of the integrators; and, consequently, γ can be calculated.

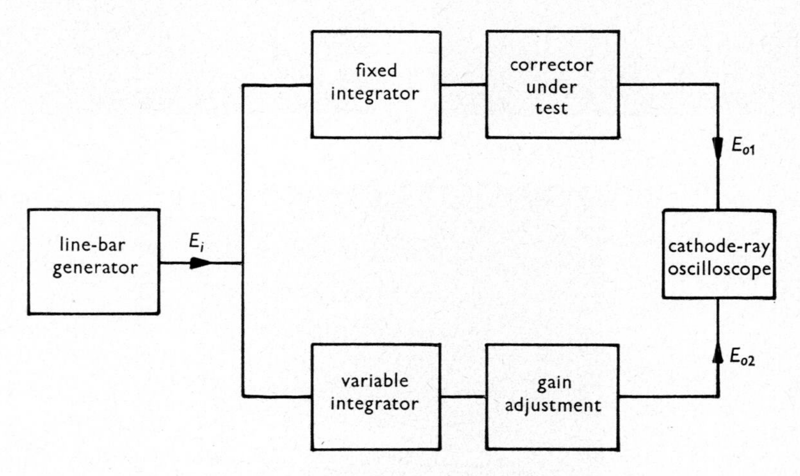

Fig. 12.6 Measurement of exponential curve

12.4.4 Operational method

In operational practice, a somewhat simpler technique is usually preferred. For this purpose the combined sawtooth and bar waveform mentioned in the Section 12.3.3 is recommended, but a plain sawtooth may also be employed. The test signal is inserted into the input of the processing amplifier, and the luminance-channel output in a 4-tube camera and the green channel output are examined individually on a waveform monitor provided with a suitable graticule. The sweep and the vertical-amplifier gain are set to standardised values. The channel gain and the 'sit' of the signal must previously have been correctly adjusted.

In the first variant, the graticule has drawn on it the shape of the sawtooth when distorted by the correct contrast law, and the corrector

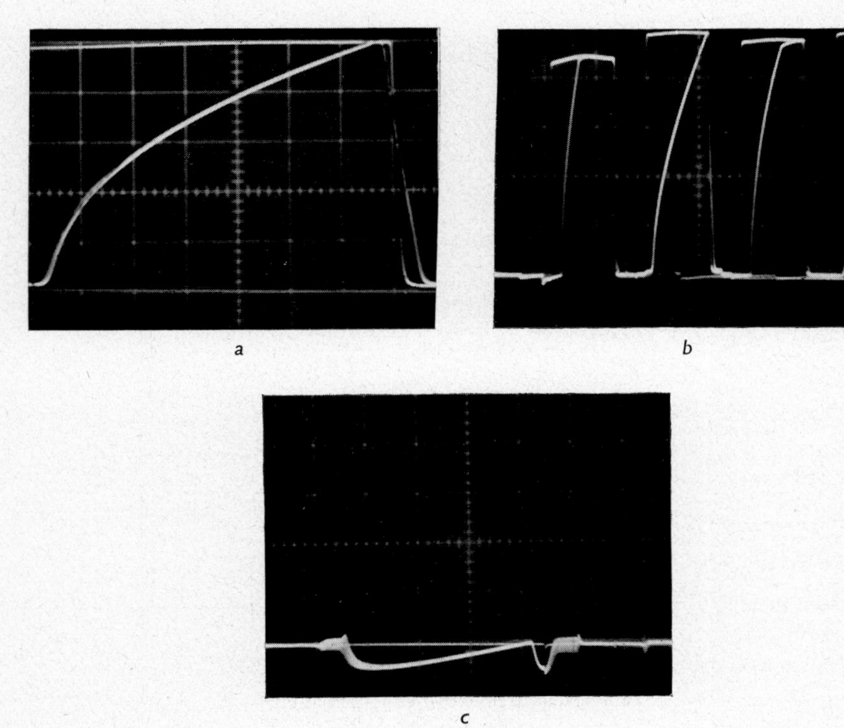

Fig. 12.7 Adjustment of contrast law of *R*, *G* and *B* channels

a Green channel against graticule
b Maladjusted red and blue channels
c Residual $R-B$ waveform

is adjusted to match this curve as well as possible. This process is shown in Fig. 12.7*a*, in which the match is so close that the graticule line is hidden except at the lower end of the curve, where the amount of correction is greatest.

In the second variant, a graticule is preferably also used, but may be dispensed with at the expense of some loss of accuracy. The output test signal is now not examined directly, but as the difference between the output and input signals. The result is a curve with a rounded hump, the magnitude and location of which provide some measure of the contrast law.

Another very successful method requires a 'preset law' staircase generator; i.e. one in which the relationship between the amplitudes of the successive steps can be chosen at will over a large range of values. Its great virtue lies in the fact that it may be adjusted so as to give the inverse law of the contrast corrector, this being easily checked against a prepared graticule. The resulting waveform at the output of the corrector should, of course, be a staircase with steps of equal heights, this condition being easily checked against the normal graticule; or, alternatively, the differentiating technique of Section 5.2.2 can be adapted. The network employed must be able to deal with ten steps, which seems to be a preferable figure for the present purpose, rather than the five steps standardised for general nonlinearity measurement.

Another approach for operational purposes utilises an external γ corrector, previously very carefully aligned by one of the methods given above, as a standard with which the corrector in the camera channel can be compared by feeding the two with the same sawtooth waveform from a distribution amplifier. The corresponding output waveforms are then subtracted in a difference amplifier, and the corrector under test is set for minimum residual.

With a colour camera, this operation is carried out on the green channel, and it is then necessary to adjust the red and blue contrast laws to match. Fig. 12.7*b* shows the condition, exaggerated for clarity, where the green channel is correct (second waveform) while the red and blue are out of adjustment. The matching may be carried by the same difference technique, and a typical residual waveform is shown in Fig. 12.7*c*. Otherwise one of the more refined techniques of Section 12.3 can be employed. It should then be checked that any common adjustment of the three channels does not upset their matching at any point.

12.5 Resolution

Resolution is a measure of the ability of an image-producing device to distinguish fine detail. It is objective, in contrast to picture sharpness, which is subjective.

It follows, therefore, that whatever criteria are used for the measurement of resolution must be capable of precise measurement, and must not depend on any visual estimate of picture sharpness, which can be very misleading since it depends on a number of factors having really nothing to do with the ability of a camera to render fine detail. An example is picture contrast, and it is very instructive to examine a number of pictures through a flying-spot scanner equipped with a large range of different contrast laws. It is found that certain pictures, which with normal gradation appear rather flat and lacking in detail, can have their apparent sharpness improved to a surprising degree by a suitable choice of contrast law, owing to the improved tone separation in areas of interest. Such techniques are used deliberately in aerial photography, and are found to be of great value.

Another example is linear waveform distortion of a suitable kind, such as the addition of the right amount of a differentiated version of the signal or the addition of a delayed negative echo of a suitable amplitude and delay. Better still is the use of a pair of small negative echoes. The result is to terminate each signal transition by a small overshoot of the opposite polarity which, in judicious amounts, has the effect of separating closely spaced transitions, and hence of improving the apparent resolution. Such crispening devices are frequently used to improve the picture rendition of cameras utilising tubes of the photoconductive type, and in this way they perform a valuable service. They are commonly applied also in television-camera viewfinders to reduce the uncertainty in finding the point of optimum optical focus. On the other hand, relatively low-frequency distortion, giving smearing or drooling effects, can markedly degrade the subjective sharpness of an otherwise excellent picture.

12.5.1 Horizontal resolution

Electro-optical convertors, such as cameras and picture monitors, resemble other signal-handling apparatus in that their linear transmission characteristics can be measured in terms of either frequency or standard waveforms. They differ, however, in that the necessity for conversion from, or to, an optical picture imposes an orientation on the resolution which may be, and with cameras generally is, different in the horizontal and vertical directions. These two quantities must therefore be measured separately when dealing with camera tubes. Certain types of picture tube, hitherto not extensively used, have similar limitations.

Although, with both picture and camera tubes, the effective overall resolution must of necessity be different in the two directions as a result of the line structure, the difference is more significant with camera tubes, since the quantised nature of the vertical readout of information affects the opto-electrical conversion process in a complicated manner, and measurements of vertical resolution must take this into account.

Measurements of horizontal resolution in the frequency domain are actually made in terms of spatial frequency, which for a camera tube would ideally be measured by test objects having a sinusoidal distribution of illumination in the horizontal direction, the spatial frequency itself being determined by the period of this sinusoidal distribution of illumination compared with the dimensions of a standard picture occupying the same plane as the test object. A series of such sinusoidal distributions with increasing numbers of periods per picture width would then show a progressive decrease in the amplitude of the corresponding output camera signal as a result of limitations or imperfections in the optical and electron-optical imaging systems and in the actual conversion process itself. The ratio of the output voltage at a given spatial frequency to that at low spatial frequencies is known as the 'modulation depth'; perhaps not a very apt term, but one widely used. For historical and practical reasons, it is very common to replace the sinusoidal distribution of the ideal test object by a square-wave distribution formed by successive light and dark bars of equal width, but sinusoidal test objects are to be preferred (Chapter 10).

The spatial-frequency characteristic, however, is not sufficient in itself, since one must be able to derive from it some simple and effective criterion for rating the resolution of a camera. In optical work, it is still common to use the limiting resolution of black and white test patterns, best defined by an arbitrary rule such as Rayleigh's criterion, but these have no relevance to modern camera tubes in which the decrease in modulation depth at spatial frequencies approaching the nominal limit of the videoband ought to be considerably less than limiting resolution, which in any event is not a very suitable criterion. Nevertheless, limiting resolution is still often used, even though the spatial frequency just resolved may be well beyond the nominal video bandwidth, and hence of little use as an indication of performance over the useful band. In such instances, the figure is usually quoted in terms of lines per picture height.

The other two criteria commonly used are the spatial frequencies at which the modulation depth falls by 3 or 6 dB, or alternatively the modulation depth stated in decibels at an arbitrary frequency close to the upper limit of the videoband; say, 5 MHz for a 625-line system. This last method is probably to be preferred. In any case, it is not advisable to define the resolution of a camera and its lens by a single figure, since the requirements are less stringent at points outside the central picture area, and it is convenient to divide the picture area into a small number of zones in each of which a separate measurement is made. This ensures at the same time that the resolution towards the edges of the picture is not allowed to deteriorate unduly, a situation which might occur is measurements of central resolution only were made.

The following basic rules for the measurement of horizontal resolution should be observed:

(*a*) The measurement of camera-tube resolution cannot be divorced from the resolution of the optical system used, so that, if the best measurement of tube resolution by itself is required, the lens chosen should have high resolution over the whole field and should be employed at the optimum stop and at the distance for best resolution if the lens corrections are not completely stable.

(*b*) Resolution measurements are easily confused by contrast-law limitations in the tube, as was pointed out by Brothers (1959). It is consequently very desirable for the test resolution patterns to have a low-contrast range and to be positioned in the upper-middle part of the grey scale.

(*c*) The pattern should be designed so that reference areas of low-frequency information are situated in the same line as the series of test-frequency gratings.

(*d*) The gratings should have an approximately sinusoidal shape, except perhaps for the very lowest-frequency gratings, to avoid the $4/\pi$ amplitude change arising from the removal of the harmonics of the grating waveform (Chapter 10).

For the measurement, the camera is set up in an optimum fashion for resolution; the aperture correction may be omitted or retained depending on whether the resolution of the tube itself or the overall practical resolution is required. The test pattern should be selected to conform to the principles stated above; test card 51 (Chapter 10) is highly recommended. It should be illuminated as evenly as possible and mounted in a plane exactly normal to the axis of the lens. The lens should be focused with the greatest care.

The test pattern is then located in the centre of the field of the lens and, by using a line selector on the output waveform from the channel, the modulation depth of the test patterns is measured with respect to the amplitude of the waveform corresponding to the lowest-frequency bars. The same measurement is then made at specified points in all four corners of the field. The corner resolution should not be allowed to fall below the central resolution by an amount depending on the use to which the camera will be put. It is useful to photograph this waveform, since, not only is a record of the measurement assured, but the integration of the random noise by the exposure time, provided that it is reasonably long compared with $\frac{1}{50}$ s, assists in obtaining accurate values of the modulation depth.

For quick operational checks of the horizontal resolutions of a camera, the exposure test pattern described in Section 12.3.2 is excellent. This preferably has approximately sinusoidal bars; but, even if this is not the case, the measurement is still valid for comparison purposes. The way in which the pattern provides its own reference amplitude for the sinusoidal frequency burst corresponding to the grating is very clear in Figs. 12.1*b* and *c*.

The alternative measurement of camera resolution by waveform-testing methods appears to offer not only a simpler criterion for the definition of resolution, but additionally the possibility of conducting the measurement in a manner similar to that used elsewhere in the signal chain, with all the advantages it would bring in the unification of measurement tolerances. Unfortunately, at the moment there are some practical difficulties to be overcome before this highly desirable result can be achieved.

The first reported attempt to test cameras in this manner appears to be due to Springer (1963), who investigated the spatial-pulse response by exposing a camera to a very narrow, illuminated, vertical slit. He did not propose any testing routine as a result, but noted that the output pulse from standard cameras bears a close resemblance to a sine-squared pulse. This result is not entirely unexpected, since the spatial-frequency response is normally quasi-Gaussian in shape, and it follows that the spatial pulse response in turn would be quasi-Gaussian, i.e. a waveform not dissimilar to a sine-squared pulse.

This method, however, is not very practicable, since, to obtain the true spatial pulse response of the camera, the slit must be extremely narrow. The output signal has consequently a very low amplitude and the high random-noise level makes accurate measurement difficult or impossible.

Another proposal came from Seyler (1963), who suggested a test card which has for its basis a slit with a width equal to the nominal $2T$ value for the television system in use (Chapter 6). This slit is disposed symmetrically between two large uniformly illuminated areas having the same brightness as the slit, which then ideally should produce two flat-topped bar waveforms which can be used as reference levels in the same manner as the bar height is used in sine-squared-pulse and bar testing. In such instances, the provision of two large illuminated areas is very desirable to take account of changes of sensitivity across the target, uneven illumination etc. The use of a $2T$ width for the slit suggests that it fulfils some basic relationship; in fact, it appears to be merely a suitable compromise value.

Potter (1964) extended Seyler's method with a means of allotting to the lens-plus-camera system a figure for effective bandwidth, which may in turn be used as a measure of the resolution. This is similar to a proposal by Lewis and Hauser (1967) for a definition of

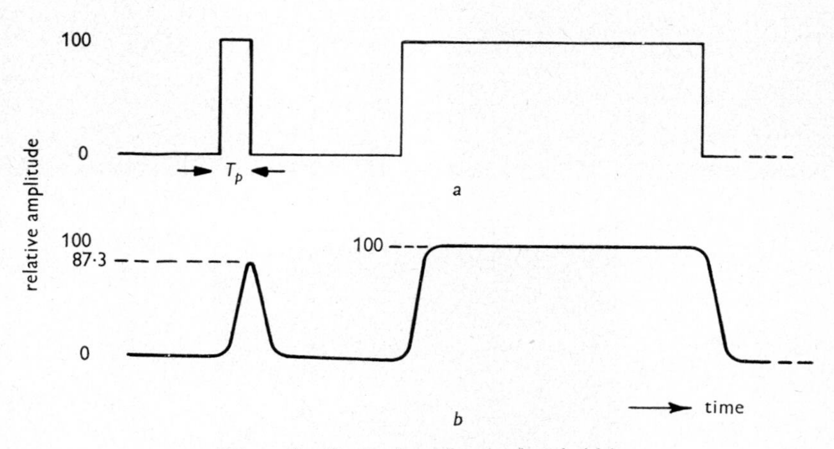

Fig. 12.8 Derivation of pulse bandwidth
a Input waveform
b Output waveform

bandwidth in lowpass systems, except that, whereas Potter assumed that the spatial-frequency characteristics of both the camera tube and the lens may be assumed to be of sine-squared shape, the definition suggested by Lewis is not restricted to a particular shape of characteristic.

The Lewis and Hauser definition of bandwidth is of considerable importance. It is based on the following consideration. Assume that the test signal applied to a lowpass filter has the shape of Fig. 12.8*a*, where the shorter rectangular pulse may have its width varied over a large range. When the durations of the two output pulses are both large enough, they are transmitted through the filter with their flat tops substantially unchanged, the distortion being confined to the transition areas. As the width of the smaller pulse is progressively decreased, the top of the output pulse narrows until it just disappears,

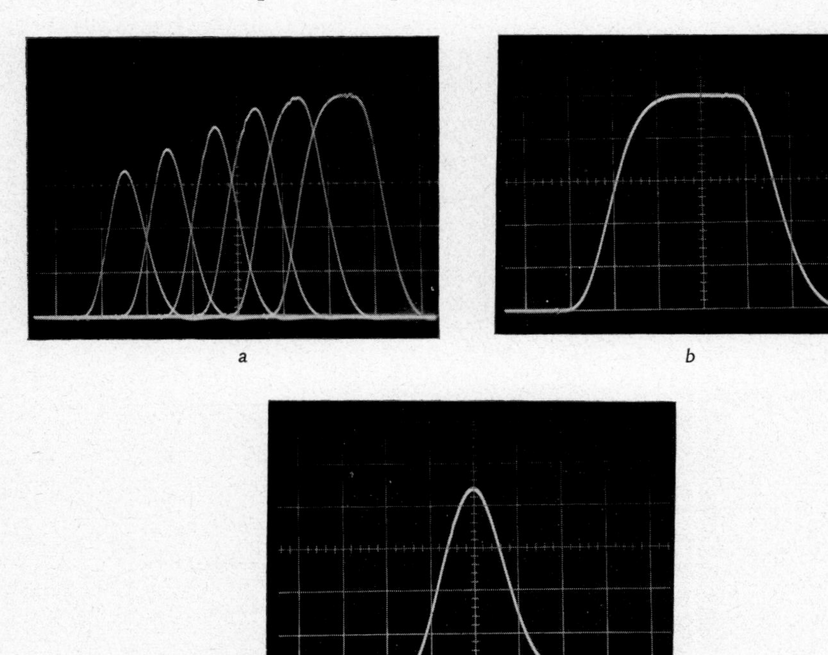

Fig. 12.9 Determination of pulse bandwidth

T_p = 320 ns
F_p = 1·56 MHz

a Effect of increasing pulse duration
b Reference flat-topped pulse
c Corresponding pulse of 87·3 % amplitude

and any further decrease in the duration of the narrow input pulse results in a progressive reduction of its height. A sequence of such pulses is shown in Fig. 12.9*a*.

A critical pulse duration T_p may then be found (Fig. 12.9*b*) for which the amplitude of the narrower output pulse is 87·27% of the amplitude of the wider pulse or, which is the same thing, of the amplitude of the step response. The 'pulse bandwidth' is then defined as

$$F_p = 1/(2Tp)$$

This particular amplitude ratio is chosen so as to equate the pulse bandwidth to the cutoff frequency in the special case of an ideal lowpass filter with an infinitely steep cutoff region.

Lewis also goes on to define a 'slot bandwidth' for a lowpass system, which is the frequency F_s at which the loss relative to zero frequency reaches 40 dB and subsequently never falls below that value. In other words, this is an arbitrary definition of the width of the maximum-frequency spectrum of any signals which have passed through the filter. The bandwidth ratio F_p/F_s can consequently be used as a figure of merit for lowpass filters, varying between the limits of 34% for a Gaussian filter (no ringing) and 100% for an ideal lowpass filter (maximum ringing).

The criterion used by Potter differed slightly from this since he defined the effective bandwidth in terms of the width of the rectangular pulse required to give an output pulse/step-response amplitude ratio of 81·75%, this value being derived, as has already been stated, from an approximation of the spatial-frequency responses of the lens and camera to sine-squared networks. In practice, it was proposed that a slit of fixed width could also be used and the critical pulse duration determined from the amplitude ratio of the output pulse by means of a calibrating curve provided.

Some experiments have been made by the author on a modification of Seyler's method employing a tapered slit. The test card, which is basically similar to Seyler's, is shown in Fig. 12.10. The slit is so dimensioned that, when it is correctly positioned with respect to the camera, the width of the top corresponds to 1 T for the system in use, i.e. 100 ns for 625-line systems, and the base if the slit were triangular would represent 2T, i.e. 200 ns. The centre of the slit corresponds therefore to 150 ns, which places the range of expected values near the centre of the pattern where the definition should be optimum. At the very bottom, the slit was provided with a short section having a very much more rapid taper so as to make a larger range of slit widths accessible without affecting the slower, linear change of slit width over the majority of the height of the test card.

The expected procedure was as follows. The selected line waveform displayed by the waveform monitor would be double-triggered so as to show the bar over the pulse. The selected line would be

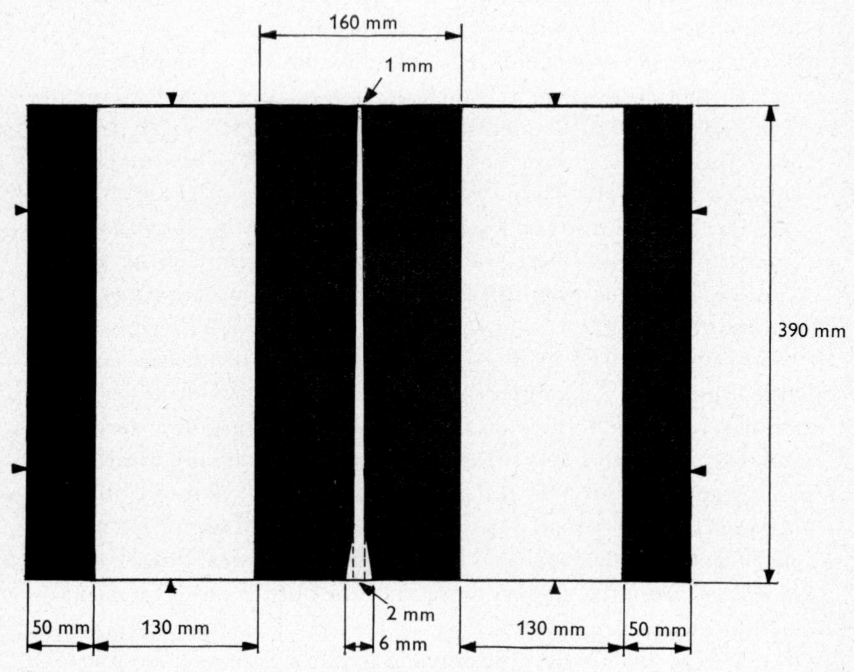

Fig. 12.10 Experimental camera-pulse test card
Slit is not to scale

moved along the field until the pulse amplitude was found to be 87·3 % of the bar height. This line would be identified with the line marker on the picture monitor, and hence the width of the effective input pulse found from the known contents of the test card. The waveform would have the appearance of Fig. 12.11 *a*.

In practice, it was found that the height of the bar could not be determined with sufficient accuracy for the critical pulse duration to be found with any reasonable degree of certainty, owing to shading effects, uneven illumination etc. (Fig. 12.11*b*). To overcome this uncertainty, a modified procedure was devised.

As the line under examination is moved downwards through the

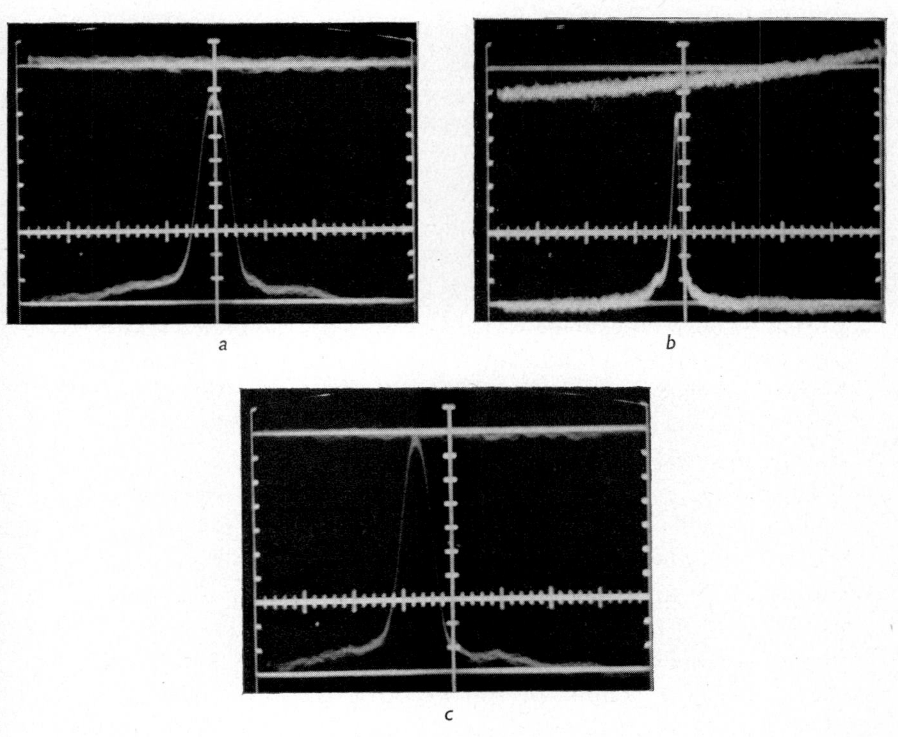

Fig. 12.11 Waveforms from experimental camera-pulse test card

a Pulse of critical amplitude
b Effect of image shading
c Determination of reference amplitude

field, so that the duration of the input spatial pulse increases, it is found that the amplitude of the output pulse increases until it reaches a limiting value, after which any further increase in the slit width merely broadens and flattens the top of the output pulse. Fig. 12.11*c* shows such an output waveform at the critical point just before any broadening takes place. In this instance, it was possible to achieve a reasonable agreement between the amplitudes of the pulse and the bar, which was not always found to be the case.

This limiting pulse amplitude replaced the bar height for purposes of measurement, and was found to be more reliable as a result of the use of the same restricted area for both the spatial-pulse response and the limiting pulse amplitude. To ensure that the latter could be measured without ambiguity, the original form of the test card was modified by adding the extra tapered portion at the bottom to give an increased range of slit width.

It would be further advantageous if a means could be found to identify the limiting pulse amplitude, using a further restricted portion of the total field, which would further reduce the residual errors. A proposed modification replaces the long, tapered slit by a short central slit whose width can be varied by means of a calibrated leadscrew. This could be evenly illuminated from behind. There would be no longer any need to try to match the illumination of the slit to the white bars, since only the area of the slit itself would be used for measurement.

During the experiments described above with the tapered slit, a crude but effective expedient was employed for measuring the effective duration of the input pulse. The displayed picture of the test card on the picture monitor was viewed while the point of a pencil was moved vertically along the length of the slit. When the point of the pencil was seen to be on the same line marker from the line selector of the waveform monitor, the portion of the slit corresponding to the displayed waveform had been identified. A small mark was made with the pencil at this point, after which a measurement of the distance of this point from the top or bottom of the card enabled the duration of the input pulse to be calculated.

The width of the rectangular input pulse for an output pulse-height amplitude ratio of 87·3 % being known, the pulse bandwidth as defined by Lewis could be calculated. The image-orthicon cameras tested were aperture-corrected, and, in this condition, a typical figure for the pulse bandwidth was 4 MHz, giving a figure of merit of 80 %, which is very creditable. On the other hand, with a vidicon camera intended for closed-circuit television, the bandwidth was 2·3 MHz, giving a figure of merit of only 46%. In each, the equivalent rectangular bandwidth of the 625-line channel has been taken as 5 MHz, corresponding to the figure used for sine-squared-pulse and bar testing (Chapter 6.2.2.)

It is not claimed that the pulse-testing methods described above, at least in their practical applications, have at all approached a final stage of development. They rather represent experiments which might stimulate further thought and further work in this direction. One likely avenue worth exploring is the use of a lower-contrast test pattern located approximately midway in the contrast scale between white and black, to avoid a number of the errors arising in camera tubes from high-contrast test objects. After all, a very large proportion of important detail in a picture lies in that range. With colour cameras, one might well apply the method to the investigation of the pulse bandwidth of the colour-separation channels as well as the luminance channel, the results of course depending on the type of camera.

12.5.2 Vertical resolution

The measurement of vertical resolution is much less often carried out than that of horizontal resolution, possibly owing to the instinctive but fallacious belief that the limited line structure must necessarily ensure that the vertical resolution is always adequate. In fact, the vertical resolution, measured in terms of the buildup of a horizontal edge, may well be appreciably more than one line, for reasons still not always understood.

Poor vertical resolution usually leads to a picture which subjectively does not seem unduly soft as long as the horizontal resolution is very good, but which rapidly deteriorates as soon as the horizontal resolution is impaired by any band limitation.

The standard measurement technique was first described by Theile and Pilz (1957); it is very simple and elegant. The camera is made to view a scene in which there is an abrupt transition between white and black, so arranged as to be at a very small angle to the horizontal (Fig. 12.12*a*).

The waveform monitor veiwing the video waveform is arranged to display a single field, and a pulse generator giving a pulse of, say, 1–10 μs long and synchronised to line frequency is connected to the grid or cathode of the cathode-ray tube in such a sense as to 'bright

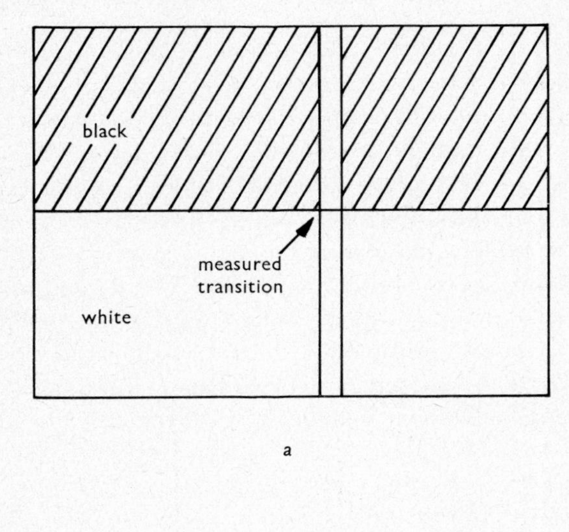

a

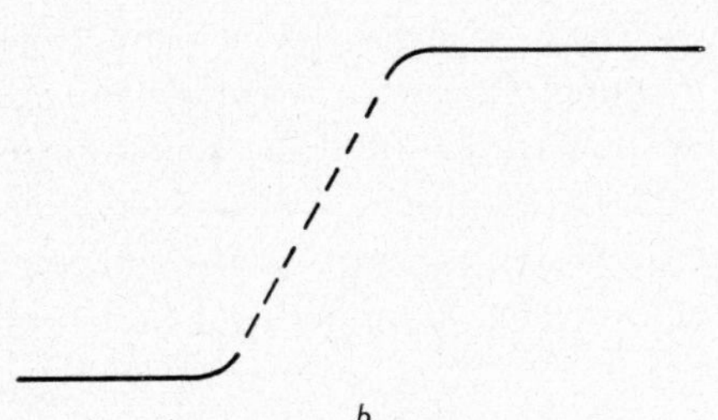

b

Fig. 12.12 Measurement of vertical resolution
a Test pattern
b Corresponding waveform

up' the display during each pulse. A variable delay with respect to the line synchronising pulse enables the brightup pulse to be located anywhere along a line. Such a device is known as a 'field strobe', because it makes visible on the waveform monitor the amplitudes existing over a whole field at points on all lines at an equal distance from the left-hand edge of the picture. In other words, it isolates and displays in sequence the line segments forming a thin, vertical strip of the picture.

The waveform displayed by the monitor takes the form of two horizontal levels joined by a transition marked out by a very small number of bright dots, corresponding to the vertical risetime of the picture (Fig. 12.12*b*). The object of the slight tilt of the test picture, combined with the variable delay of the test pulse, is to enable the test transition to start precisely on a single line; otherwise, there would be a certain ambiguity in the measurement. Ideally, the transition should be complete in a single line, but in practice this is usually exceeded. The waveform obtained during the measurement of a flying-spot scanner is given in Fig. 12.13*a*; the transition can be seen to take two lines, to a sufficient degree of accuracy.

12.5.3 Interlace

This is conveniently included here, although interlace errors are derived from the camera scanning circuitry or the synchronising pulse generator. The ideal interlace ratio is 50:50, since the spacing between each successive line in the complete picture is 50% of the spacing between successive lines in the same field. A typical limiting figure would be 45:55. The smallness of the deviation from the ideal makes the measurement somewhat difficult to carry out, although, fortunately, no high standard of accuracy is required.

The field strobe described in Section 12.5.2 can be adapted to measure interlace provided the vertical resolution of the tube is adequate. The black–white transition viewed by the camera is placed at such an angle to the scanning lines that several are crossed by it in the width of the picture, and the brightup pulse width is reduced to the minimum practicable. The resulting field-strobe waveform shows dots corresponding to the intersections of the transition with the scanning lines; and, by a suitable expansion of the relevant part of the waveform, a measurement can be obtained of the interlace ratio; e.g. in Fig. 12.13*b*, in which successive fields are superimposed, the ratio is 47·5:52·5.

A more commonly employed technique employs a black–white transition of the same kind which is viewed by the camera. At the same time, a grille-pattern generator or a vertical-line-pattern generator is connected to the input to the camera-control unit so that the waveform generated by the camera and the artificial pattern are viewed simultaneously on a waveform monitor.

The picture corresponding to the almost horizontal transition is quantised by the scanning lines – a phenomenon sometimes known as 'beading', since the transition itself takes on the appearance of a

Fig. 12.13 Use of field strobe
a Measurement of vertical resolution
b Measurement of interlace

series of beads, each corresponding to the intersection of a line with the transition. By careful adjustment of the angle of the transition with respect to the scanning lines, the successive beads can be spaced as far apart as is practicable. The artificial pattern may then be used to measure the intervals between the beads. This is particularly convenient if the number of vertical lines per picture can be varied smoothly, so that an integral number of lines occupies the space between alternate beads. Any error in the spacing of the central bead is very obvious, and can be estimated.

12.6 Linearity

Errors in the geometry of the picture may be due either to faults in the camera tube itself or to errors in the beam-deflection system. To separate these effects, there is really only one course available, i.e. to measure a number of tubes in the same yoke, from which it may be possible to distinguish between a systematic positional error common to all measurements, and hence probably, but not certainly, due to the beam deflection system and the more random elements which will be due to the tube.

It was pointed out by Brothers in his classic paper (1959) that the limits should be specified so as to take account, as far as possible, of the subjective picture impairment caused by positional errors. The following basic principles can then be laid down:

(*a*) The central area of the picture consisting of a circle whose radius is approximately 0·4 of the picture height yields the largest subjective impairment for a given error. When the largest tolerable error for the central area has been fixed, the remainder of the picture can be assigned a greater maximum error with a further relaxation for the extreme corners.
(*b*) The impairment caused by a given positional error is the same for all directions.
(*c*) The shorter the distance over which a given positional error occurs, the greater is the picture impairment. In other words, if a positional error Δh occurs in a distance Δr, the picture impairment is subjectively proportional to $\Delta h/\Delta r$.

The standard method of measuring these differential positional errors is not difficult to carry out. It requires a chart with a white grille pattern drawn on it, and an electronic grille pattern having the same number of squares in the picture height and width, which is added to the output of the camera channel. Either the distance of the chart, or preferably the constants of the electronic grille generator, are carefully adjusted so that, in the central portion of the picture, the two images can be made to overlap precisely as viewed on a good picture monitor.

In any given area of the picture where a measurement is to be made, some chosen point is made to coincide with respect to both signals. At some other point distant Δr from the first, two other points are found which should be coincident but, in fact, are displaced by Δh (Fig. 12.14). Then the positional error is defined as $\Delta h/\Delta r$.

In the simplest case, the measurement of the raster nonlinearity may be carried out by adjusting the camera or the optical test pattern

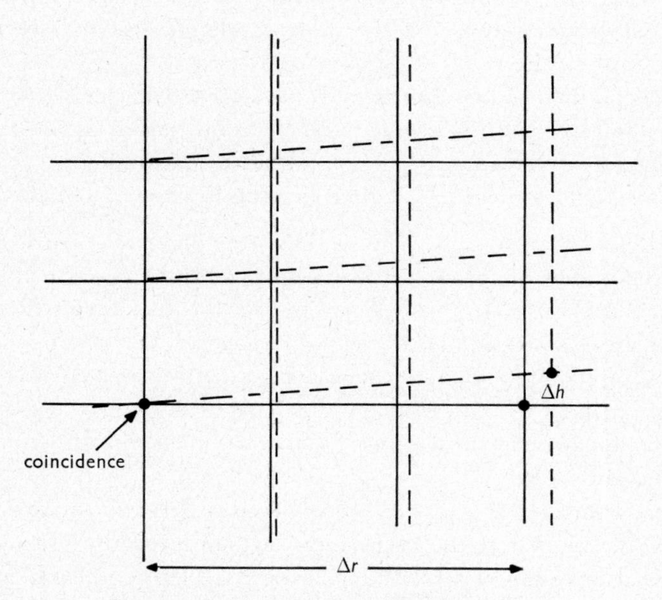

Fig. 12.14 Measurement of positional errors

to obtain coincidence of the camera picture and the electronic grille pattern wherever required, and estimating the error as a fraction of a square as seen on the screen of the picture monitor. This is crude, however, and a much more convenient and accurate method of aligning the patterns and measuring the coincidence error is to use an electronic pattern generator which has individual calibrated shifts applied to both the line and field synchronisation, so that the pattern as viewed on the monitor can be moved vertically and horizontally by voltages readily convertible into the corresponding spatial quantities. Some workers prefer to replace the electronic grille pattern by an electronic dot pattern, in which the dots correspond to the intersections of the horizontal and vertical lines in the grille pattern.

The above method of expressing and measuring raster nonlinearity is not universally accepted, however, in spite of its very logical and practical nature. The other common system is to measure the displacement of points on the displayed picture of a test pattern from the nominal position, and to express the shift as a fraction of the picture height. The use of the height instead of the width is a time-honoured convention. This type of measurement is quicker than the preceding one, and is therefore suitable as an operational check where the acceptance testing and maintenance testing are carried out by using the differential error.

For this purpose, the well known RETMA chart is very widely used. It contains a pattern built up from a number of identical, contiguous units, each of which is a rectangle of equidistant black circles. The thickness of the line defining each circle is 2 % of the picture height, and the white central area has a diameter of 2 % of the picture height. Black triangular markers are placed along the edges of the pattern so as to define the precise borders. This pattern may be obtained either in the form of an opacity or as a slide for use in a diascope.

As before, an electronic pattern generator producing a grille or an equidistant dot pattern must be used, and this should preferably be fitted with the shift controls already described. This is linearly added to the output of the camera channel or the input of the monitor displaying the camera picture. With the electronic pattern temporarily removed, the camera picture is carefully aligned so that the edge markers coincide precisely with the borders of the nominal picture area. The electronic pattern is then replaced and the dots or the intersections of the grille are moved by the shift controls until as many as possible fall within the circles on the RETMA chart. Where coincidence cannot be established, the error is read from the chart by using the circle thicknesses to estimate shifts of 2 % of picture height. This is only valid, however, where the error is measured with respect to the circle which the dot or intersection ought to occupy; it is possible in bad cases for a dot to move so far that it occupies a circle to which it has no right.

12.6.1 Colour-camera registration

The linearity and geometry of a colour camera must evidently comply with the kind of specification laid down for monochrome cameras, but in addition one has the responsibility of ensuring that all three colouring channels are in correct registration. This implies that, if one could reproduce the picture from the camera on an ideally converged picture monitor, the images formed by the three guns would be in coincidence within acceptable limits, these limits being very small, since the decreased chrominance resolution and colour fringing effects resulting from poor registration constitute a serious picture impairment. They are similar in appearance to convergence errors in a receiver; and, since it is evidently impracticable to ensure that domestic receivers are at all times impeccably converged, the onus is on the picture source to furnish a video signal with negligible distortion.

Although automatic cameras are now being introduced, camera registration is still an operational routine, and it is therefore important that the techniques employed should be such as to reduce as far as possible the time required and to make the task of the operator simple. Little seems to have been published so far on the subject; a good but rather brief account is given by Teear *et al.* (1970) of the methods used by the BBC. Since these are very representative of current practice they are worth mentioning here.

The basis of all registration methods is a test card containing a rectangular grille pattern of white lines on a black ground; normally it has 22 × 22 squares. When the camera has otherwise been correctly aligned, it is placed with the lens axis normal to the chart and the pattern focused correctly. With certain 4-tube cameras in which bandwidths of the colouring channels are limited by filters, on the general principle of reducing the noise bandwidth wherever possible, arrangements are made to switch out the filters during registration.

The resulting picture of the grille pattern is now observed on a monochrome monitor. As always during alignment procedures, the green channel is taken to be the master to which the others are matched. Accordingly, the blue channel, for example, is disconnected, and the red channel is adjusted to coincidence as far as possible with the green. Originally, this was achieved by superimposing the two sets of lines on the picture monitor, but with experience it was found much more satisfactory to displace the red image slightly by the centring control, after which the two sets of lines are brought as far as possible into parallelism. When the adjustment has been completed, the two images are superimposed by the same control. Finally, the red channel is replaced by the blue, and the process is repeated.

In a further version, the difficulty of the superimposed images is dealt with by reversing one of them, which cancels the resultant display wherever the two channels are in perfect registration. Another possibility is to make use of the Clue technique (Section 12.3.3).

When optimum registration is complete, it may be required to ascertain the magnitude of the errors, as is always the case during acceptance testing. Two methods are employed. One makes use of calibrated delay lines such as are described in Section 9.1.3. The polarity of the master channel is inverted and sufficient delay is inserted to ensure that all likely negative-delay errors can be catered for. A variable delay with a sufficient range is then included in the other path (or vice versa), and used to cancel out the horizontal components of the registration errors, the results being expressed in nanoseconds. The vertical components cannot be measured in this way, so an expedient is adopted. The horizontal component is cancelled, then the horizontal shift is found in 'equivalent nanoseconds', which is equal to the vertical displacement. Normally, 25 readings are taken spaced over the entire raster.

Another method depends on the fact that the shift produced by the centring control is linearly proportional to the current through the centring coil, and therefore to the potential difference across the coil. Accordingly, the centring control may be utilised to cancel the errors and the magnitude of the shift measured, preferably with a difference amplifier. This is calibrated directly by comparison with a shift equal to one of the squares of the pattern, not forgetting, with vertical shifts, to allow for the 4:3 aspect ratio of the picture.

With each method, two sets of results will be obtained, the misregistration of red to green and also of blue to green, and it is therefore not possible to specify the errors in any simple manner. It is customary during acceptance testing to ensure that the specification limits in terms of a maximum shift are not exceeded, after which the figures obtained on later testing are compared with the original figures to reveal any significant changes.

12.7 Positional hum

12.7.1 General

Positional hum on a camera arises from the effect of small hum voltages or currents in the camera scanning or deflection circuitry which cause the scans to become to some extent a function of the frequency of the power supply. At one time, the field frequency was firmly locked at all times to the supply frequency, which maintained any resulting geometrical distortion stationary, but the advent of colour television made it impossible to continue this practice, and it is now quite common to operate without field lock even with monochrome systems. The effect of positional hum is subjectively very disturbing, and it is common to apply severe tolerances to it. A typical value is a maximum movement of the raster by an amount equivalent to one-quarter of a picture element, say 25 ns for 625-line systems. The measurement consequently requires some skill.

12.7.2 Horizontal positional hum

The traditional method leaves the camera on test unlocked from the supply during the viewing of a suitable test object, such as a linearity chart. The resulting picture is displayed on a picture monitor having a very effective flywheel synchronising system which, unlike a hardlock time base, ignores any positional modulation on the synchronising information itself. If the camera synchronising-pulse generator is offset by a few hertz from the supply frequency, any horizontal positional hum becomes quite evident as a cyclic displacement of the image which may be measured with a travelling microscope or, more crudely, a graticule applied to the face of the monitor tube. However, such methods, although simple to carry out, are insensitive, and measurements to the standard of accuracy required nowadays are not easy to carry out.

An ingenious and very sensitive method has been described by Watson (1963). To understand its operation for horizontal positional hum, let us assume that we start from a stable frequency of 5·5 MHz. This is used in two ways, the first of which is to form one input to a mixing pad at the input of a picture monitor. The 5·5 MHz is also divided by 175, and the resulting 31·35 kHz signal is used to control a synchronising-pulse generator in place of the usual twice-line-frequency input. The output synchronising pulses are used to supply both the camera and the picture monitor (Fig. 12.5*a*).

The camera is used to view a test pattern, such as test card 51, which can be used either normally or in an unorthodox way by adjusting the pattern distance to provide an output frequency of 4·5 MHz from one of the gratings. One possible device consists in using a 405-line test card containing nominally 3 MHz bars with a camera operated on 625 lines. The output video signal is mixed with the 5·5 MHz directly from the oscillator.

The picture monitor is therefore displaying two signals simultaneously: 5·5 MHz directly from the oscillator and 4·5 MHz derived from the oscillator but indirectly through the scanning mechanism of the camera, and hence capable of being shifted in time by positional hum. If there is any positional hum in the picture monitor, both these patterns will be moved by an equal amount, and hence the measurement is unaffected.

However, any positional hum will cause one of the patterns to move with respect to the other, causing an obvious moving moiré pattern. This can be looked at in the light of a vernier, since nine of the 4·5 MHz bars occupy the same distance as ten of the 5·5 MHz bars, so that a movement of the pattern by an amount equal to the spacing of one pair of the 5·5 MHz bars is a movement due to positional hum of one-tenth of a picture element, i.e. 10 ns. To compensate for slight errors in the positioning of the chart, it is advisable to have a small range of variation of the 5·5 MHz which must, of course, itself be quite free from hum. By locating the frequency grating in various

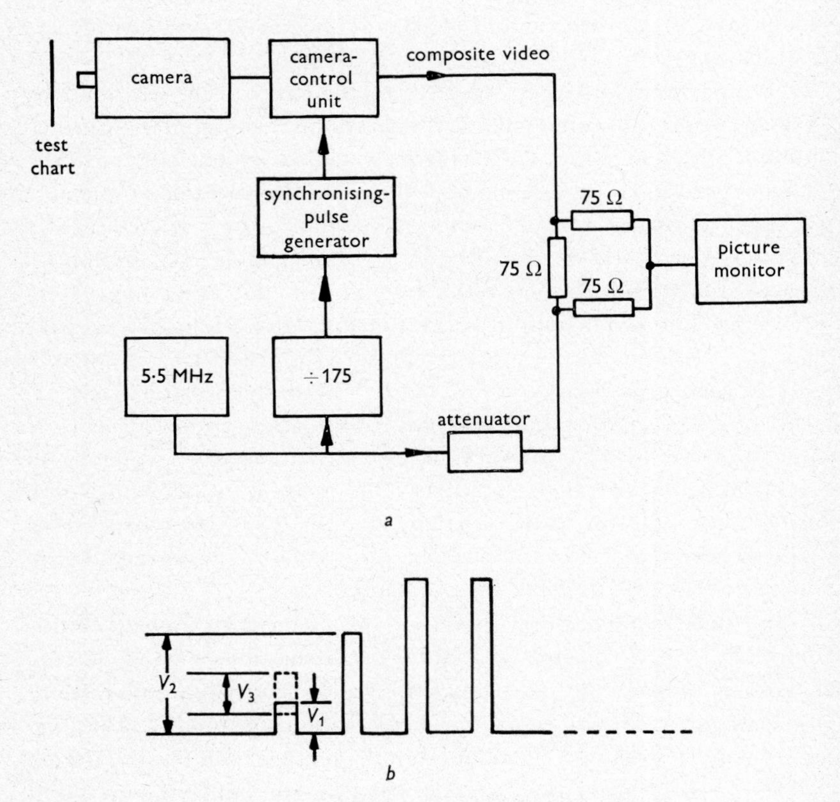

Fig. 12.15 Measurement of positional hum

a Apparatus
b Definition of positional hum

areas of the field, the variation of the positional hum with position can be measured.

The technique can be extended, if desired, for example, by using an optical test pattern providing 5·45 MHz, which will produce a moiré pattern scanning 100 periods of the 5·5 MHz, and thus increasing the sensitivity by a further ten times.

A word of warning: movement of the moiré pattern due to positional hum only should not be confused with movement due to a change in picture size. In the former, both ends of the pattern move by the same amount; but, if one end moves more than the other, a change in picture magnification is also taking place. Combined positional and magnification effects due to hum give a very unpleasant appearance to the resulting picture (jellyfishing). It is an advantage of this method that it is possible to identify the individual impairment.

12.7.3 Vertical positional hum

This can be measured with the field strobe (Section 12.5.8). Provided that no amplitude hum is introduced by the lighting of the test card or opacity, any cyclic vertical movement of the dots corresponding to the lines in the transition interval of the vertical step (Fig. 12.12*b*) is due to positional hum, and can accordingly be measured in terms of the line spacing.

If no field strobe is available, a line selector may be used if the test object is made a fairly narrow white area in a black field, say, not more than one-fifth of the width of the reproduced picture; it is conveniently between ten and 20 lines high. The triggering of the line selector must superimpose the two fields; then, provided that the edge of the test field is accurately parallel with the lines, a strobed waveform consisting of a series of vertical bars will be obtained. The first two or three represent the vertical buildup of the picture, and will be of lower amplitude than the bars corresponding to the evenly illuminated white area.

Watson (loc. cit.) suggests the following procedure, which may also be applied in principle when the field strobe is used. The synchronising pulses are first locked to the supply frequency so as to stabilise the display. The amplitudes of the first two bars are then measured; let these be V_1, and V_2 (Fig. 12.15*b*). The synchronising pulses are next unlocked from the supply, and the maximum displacement of the top of the first bar is measured. Let this displacement be V_3. Then the positional hum as a fraction of the line spacing is $V_3/(V_2 - V_1)$. Note that it is assumed that this first line has not such a small amplitude that it disappears into the baseline of the display under the effect of the hum. If this occurs, the camera or the test field must be very carefully repositioned so as to make the first pulse somewhat larger.

12.8 Unwanted signals

12.8.1 Random noise

Since no output signal can be obtained from a camera channel without the signal being blanked, the signal/noise ratio cannot be measured even at black level other than by a technique which allows the noise to be separated from the signal components; it is usual to measure the noise as at least one value of illumination of a uniformly lit field presented to the camera (Chapter 3).

The best method is undoubtedly the use of a gated-noise measuring set in which the random noise is separated from the signal components by time discrimination, to which the flat-topped bar signal resulting from a uniformly lit grey or white field lends itself admirably. It is additionally useful if the measuring instrument can also be provided with some means of determining the spectral distribution of the noise, since, at least with image-orthicon tubes, this allows the estimation of the amount of aperture distortion used and occasionally makes it possible to pick out an abnormal tube.

The selective method of Chapter 3 has been used very satisfactorily for some considerable time in the UK, both by broadcasting organisations and by television-equipment manufacturers. This method should require a number of measurements at frequencies spread over the whole of the video spectrum, but experience has shown that a measurement at 1 MHz only is sufficient for routine testing unless the value obtained appears to be anomalous. However, it is very simple to obtain the noise spectrum if required, by making a few extra measurements (Turk, 1966).

There is an unfortunate lack of uniformity in the practical measurement of the signal/noise ratio of cameras. The influence on the camera noise of factors such as illumination, aperture correction, lift and contrast correction is so large that a comparison between cameras is meaningless unless the precise operating conditions are known.

It is therefore desirable to measure the camera for random noise under conditions which simulate the normal very closely, and to standardise these conditions to facilitate comparison between cameras, whether of the same type or note. The exception may be the contrast-law circuit, which is commonly set to unity γ to maintain uniformity irrespective of possible changes in the law. With cameras employing lead-oxide tubes, the head amplifier is the major noise source, and it is usual to test this separately by feeding a sawtooth waveform through a high resistance into the input circuit, the voltage chosen so as to produce the standard signal current of 300 nA. The random noise is then measured in the presence of the sawtooth by one of the methods of Chapter 3.

With colour cameras, one known procedure is to measure the noise of each colour-separation channel with the camera adjusted to standard conditions when viewing test card 57 (Chapter 10 and Section 12.3.3); unity γ is used with no frequency enhancement, including vertical aperture correction if provided, and the linear matrixing is switched out. During the actual noise measurement, the camera lens is capped. The linear matrixing is then inserted, and any change in the random-noise level is noted. The noise in the luminance channel at the coder output with the chrominance switched out is also measured. If the signal currents for each channel are recorded, the noise performance at other illuminations may be calculated (Teear *et al.*, 1970).

12.8.2 Lag

All camera tubes suffer to some extent from lag; i.e. the response of the tube to changes in the incident light level is not instantaneous.

As is well known, it is most pronounced and important in tubes of the photoconductive type. The lag has two components: (*a*) lag arising from the incomplete discharge of the target by the beam during the course of a scan (capacitance lag) and (*b*) inertia phenomena in the photoconductive surface (photoconductive lag). The latter cannot be measured directly, but must be deduced, if required, from the capacitance lag and the overall lag, i.e. the sum of the two. For practical purposes, a measurement of the overall lag is sufficient, and the capacitance lag will not be considered further except to mention in passing that, for its measurement, the state of the target must be altered abruptly in the absence of illumination, usually achieved either by beam blanking or by modifying the cathode-target potential. Further information is given by Redington (1957) and Pilz and Schäfer (1963).

With the present very widespread use of photoconductive tubes, methods of measuring overall lag have become particularly important since abnormal lag characteristics give rise to 'ghosting' or 'print through' effects as a result of movement of camera or subject and, with colour cameras, to colour fringing behind moving objects.

Two forms of overall lag are recognised: 'buildup lag', the time taken for the tube output voltage to reach a specified percentage of its final value after a sudden application of illumination, and 'decay lag', the time taken for the output voltage to reach a specified small fraction of the initial value after the abrupt removal of the illumination. Any practical method must be capable of measuring both. The decay lag is especially difficult to deal with, since the typical decay curve has an initial very rapid drop in the amplitudes of the residuals followed by a long tail in which the amplitudes decrease quite slowly, and may persist for many fields. A measured curve of decay lag is shown in Fig. 12.16, in which the residuals are clearly discernible even after 16 fields. Even though they are apparently almost drowned in the noise level, they may be very visible on a picture monitor owing to the great ability of the eye to recognise patterns in the presence of noncoherent information. Hence it is desirable that any measurement equipment should be capable of dealing with amplitudes less than 1 % of the initial value even in the presence of high random-noise levels, and after a time interval of a considerable number of fields.

There seems to be no universally recognised method of defining the value of the lag in a given instance. Different criteria are employed by manufacturers, or specified by the user. Most usually, it is framed in terms of a minimum residual amplitude after a given number of fields for decay lag, and, with buildup lag, in terms of a minimum time to reach a specified fraction, e.g. 95 %, of the final amplitude.

The basic principle of the measurement of overall lag is evidently the use of an evenly illuminated surface which is arranged to expose the target and which may either be switched on at a selected instant for

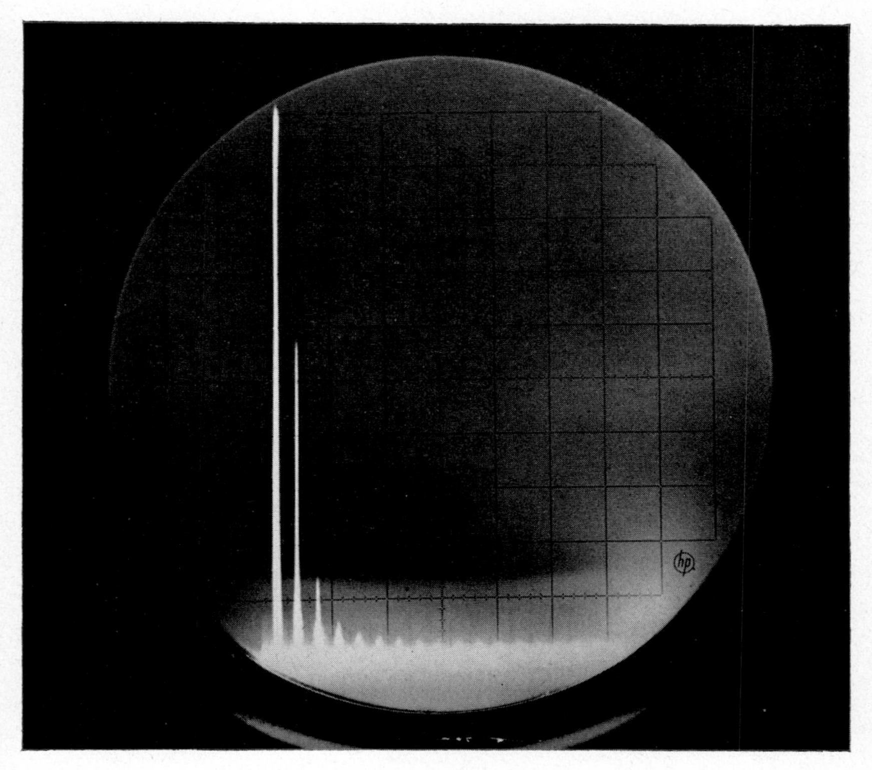

Fig. 12.16 Lag in photoconductive tube

the measurement of buildup lag, or conversely switched off for the measurement of decay lag. The standard method was formerly the use of a motor-driven disc, with a sector removed, rotating at a sub-multiple of field frequency, through which the camera could view the illuminated area (Brothers 1959; Pilz and Schäfer, 1963). However, this suffered from difficulties of synchronisation and was replaced by a cathode-ray-tube light source which can be controlled with great precision (Lubszynski *et al.*, 1960). A recent and refined version of this has been developed by the BBC Research Department, and is described by Sanders (1970). With this, residuals less than 1 % of the initial amplitude can be measured with satisfactory accuracy in any desired field and in the presence of high random-noise levels; the output reading is presented digitally. In spite of its refinement, the apparatus is rugged enough to be employed for routine operational tube testing as well as for investigational work.

The light source is a small cathode-ray tube with a short-persistence phosphor, operated without focusing or deflection; over an area of about 1 cm^2, it provides a good equivalent to an evenly illuminated, ideal diffusing surface. This is focused on the target as usual, the rest of the target being in darkness. The switching time of such a light source can evidently be considered as completely negligible as far as the measurement of lag is concerned.

The camera signal is measured at two points on the target, one within the illuminated area and one within the dark area; for each reading, the difference of the two is taken, thus eliminating errors such as flare in the optics and shading effects. The two areas of the target to be sampled are defined by two internally generated pulses, one of which is aligned with the illuminated area of the target by applying both the pulses and the tube output signals to a picture monitor, when, for correct adjustment, the sampled areas appear as white rectangles ('windows') superimposed on the camera output (Fig. 12.17). The on period of illumination may be chosen between 1 and 256 fields and the off period between 64 and 1024 fields. The precise switching instants are determined by the window waveforms, and sampling takes place between times t_1 and t_2 in each successive field (Fig. 12.18), i.e. over the total duration of each pulse.

For the measurement of decay lag, the illumination is switched off at instant t_1 in the selected field, i.e. at the start of a pulse; and, for buildup lag, it is switched on at instant t_2, i.e. immediately following the end of the pulse. The block diagram of the equipment is given in Fig. 12.19.

The reading is not taken from a single field but from the mean of

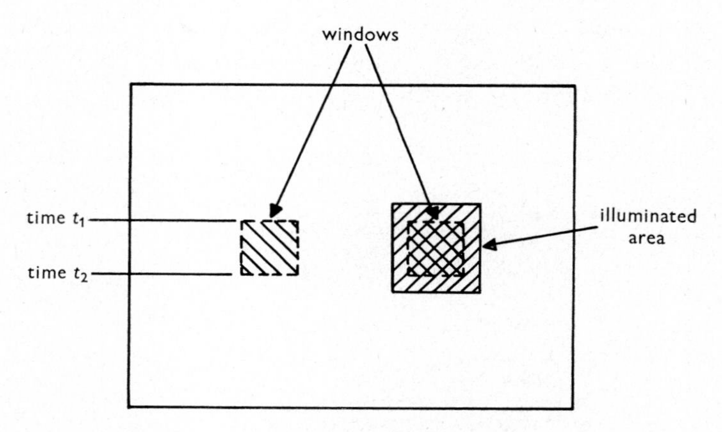

Fig. 12.17 Areas employed for lag measurement

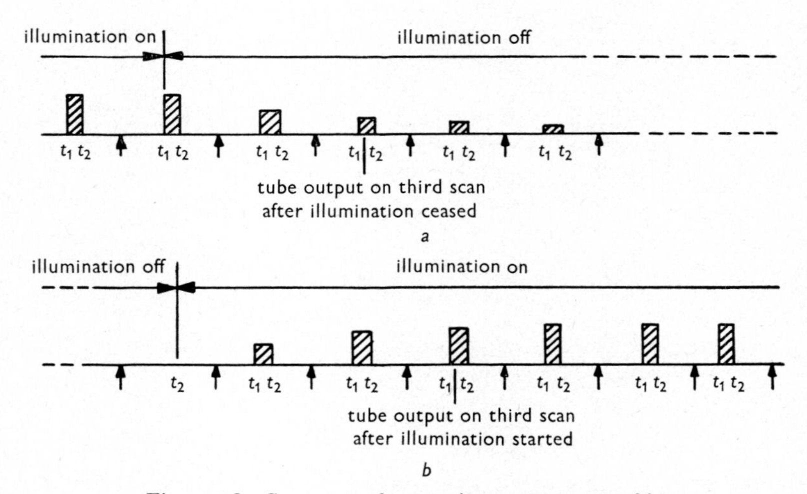

Fig. 12.18 Sequence of events in measurement of lag

a Decay lag
b Buildup lag

the samples taken in successive fields, whereby the signal/noise ratio is improved in proportion to the number of samples. This is achieved by the sample-and-hold technique (Chapter 5.5.), in which the samples are stored in a capacitor with a suitably chosen time constant. This makes it simple to add the refinement of a digital readout by a

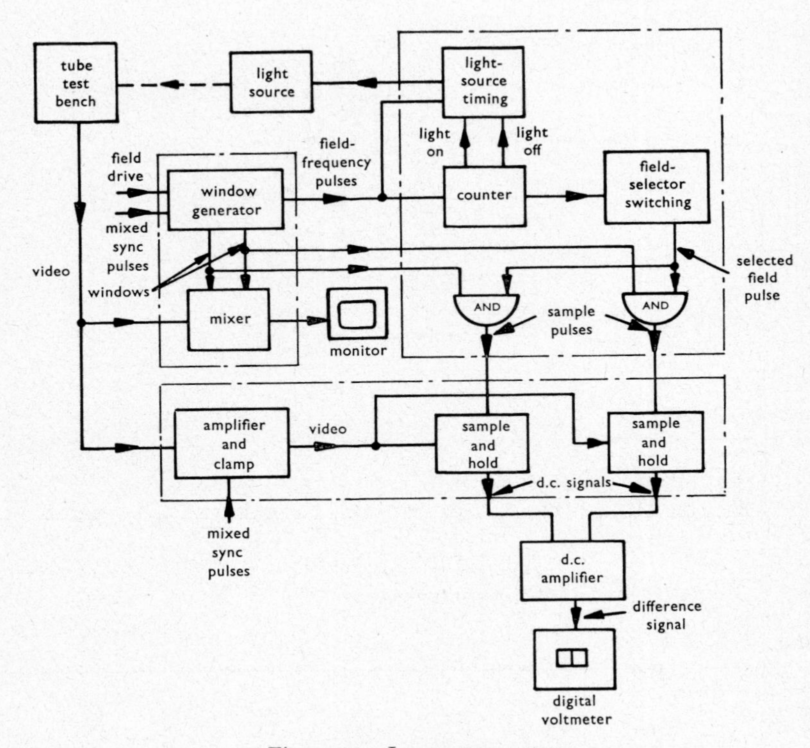

Fig. 12.19 Lag measurement

digital voltmeter reading the difference between the dark- and illuminated-area samples at the output of a difference amplifier. The digital voltmeter may also contribute a very useful degree of noise smoothing from its own internal integration. Because of the precautions taken, the lag signal can be measured to an accuracy within $\pm 1\%$ of white signal even in the presence of very high random-noise levels. The minimum of critical adjustments and the facility of an unambiguous digital readout on any chosen field make this equipment very suitable for routine or acceptance measurements.

12.9 Coder alignment

The coder, or encoder as it is also often termed, forms an essential complement to all colour-signal origination equipment; and its correct alignment must be considered as being of the greatest importance, since errors in signal coding can mar much painstaking work on the signal source itself. This is not merely a matter of adjustment, since, as will appear, specialised measurement techniques and apparatus have been found advisable. Although the coder has no more claim to be associated with the camera than with other colour-picture sources, it is convenient to discuss it at this point. For conciseness, only the Pal will be treated, although most of the techniques, with obvious modifications, can also be applied to NTSC equipment.

The coder converts the colour-separation signals from the picture source into the final coded video signal, so it seems logical to base the alignment on a standard input waveform in *RGB* form; in the UK, where a more or less standardised technique appears to be emerging, this is normally taken to be the 100–0–100–0 (100 %) colour-bar signal, although one of the other forms may also be employed.

The routine alignment of a Pal coder essentially requires

(*a*) carrier balance adjustments, i.e. the balance of the internal modulators to the subcarrier frequency and the elimination of subcarrier crosstalk
(*b*) the setting of the gains of the luminance and colour difference channels
(*c*) the adjustment of the angle between the $R-Y$ and $B-Y$ axes, and the angle between the two burst phases, exactly to 90°.

The procedure will vary in certain details according to the design of the coder in use, but the principles remain the same.

A reasonably high-gain waveform monitor which has small chrominance–luminance gain inequality is essential, and should be equipped with an equaliser of the type described together with the colour calibrator in Section 11.4.1 to maintain the chrominance and luminance gains of the waveform monitor at precise equality. Any leads connected to the input of the waveform monitor should be the same as will be later used during the coder alignment, as should likewise be the gain setting of the monitor at which equalisation is carried out. Only by taking such measures can one ensure that the alignment will not be vitiated by errors either inherent in the waveform monitor or introduced by the connecting leads.

This precaution need not apply to the high-gain setting at which the carrier balances are set, since, evidently, they are not affected by chrominance–luminance gain inequalities. This adjustment is made with both the luminance and the chrominance of the coder switched off, and the display must be such as to show a sequence of lines since the balance is not necessarily the same on successive lines (differential carrier-balance error).

The internal gains of the coder need to be set with great exactness, and experience has shown that measurements using the waveform-monitor voltage calibration are not adequate. Much time and trouble can be saved by a device known as the 'coder calibrator', which operates on the same principles as the BBC level-measurement equipment (Section 2.4). It provides square-wave calibrating waveforms, with an accuracy better than within $\pm 0{\cdot}1$ dB over long periods, which allow the following amplitudes to be set with precision:

(i) luminance	700 mV
(ii) synchronising-pulse and burst amplitudes	300 mV
(iii) $B-Y$ signal	610 mV
(iv) $R-Y$ signal	860 mV.

Calibrating waveform (i) also enables the components of the input colour-bar signal to be checked. Examples of the waveforms during calibration are given in Fig. 12.20*a*, *b* and *c*. For clarity, deliberate errors have been introduced, low gain on the $R-Y$ and $B-Y$ signals in Figs. 12.20*a* and *b*, denoted by the separation of the upper and lower traces, and high luminance gain indicated by the overlap of the two traces in Fig. 12.20*c*. It is also usual to check the white and black balances by short-circuiting the colour-separation inputs together, when the subcarrier should disappear, although this is not usually regarded as an operational adjustment.

The next procedure is the setting of the phase angles. The 90°

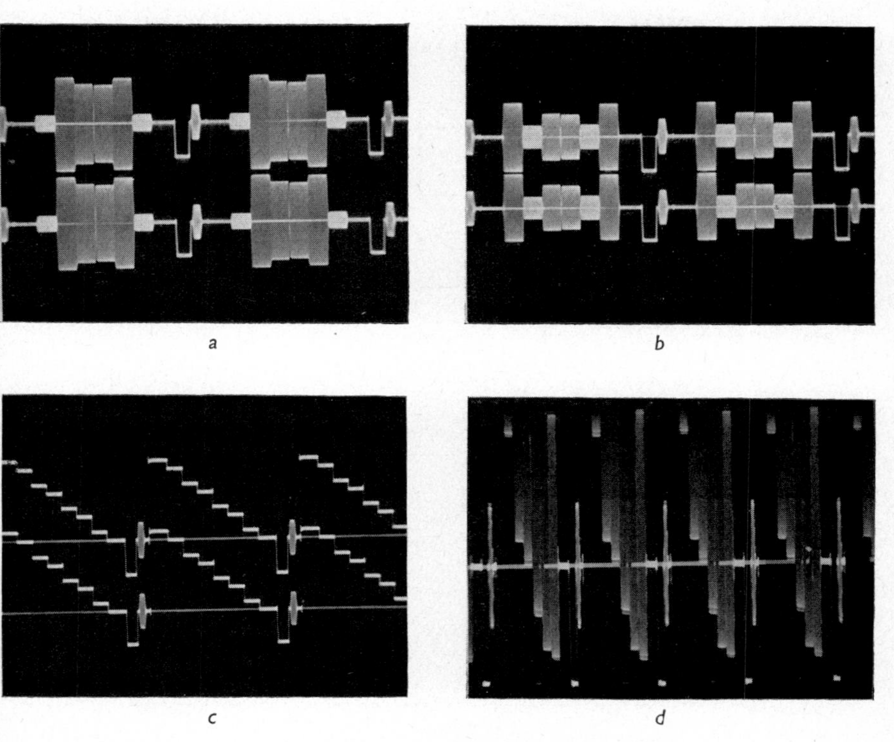

Fig. 12.20 Alignment of coder

a $R-Y$ signal, low gain
b $B-Y$ signal, low gain
c High Y-channel gain
d 'Twitter' on colour-bar waveform due to quadrature error

separation of the colour-burst vectors on successive lines is set with a vector monitor, which may be checked against a standard 90° phase shifter contained in the coder calibrator. The quadrature relationship of the $B-Y$ and $R-Y$ axes (or U and V axes) may be set in the same manner, but it is common to utilise the 'twitter' phenomenon. In the Pal system the V axis is reversed in polarity in successive lines, so that the resultant of any colour vector can only have the same length in both lines if the angle between the axes is precisely 90°. Any error in the angle can consequently be detected by triggering the waveform monitor so as to display successive lines superimposed, when a characteristic twitter is observed on horizontal chrominance edges.

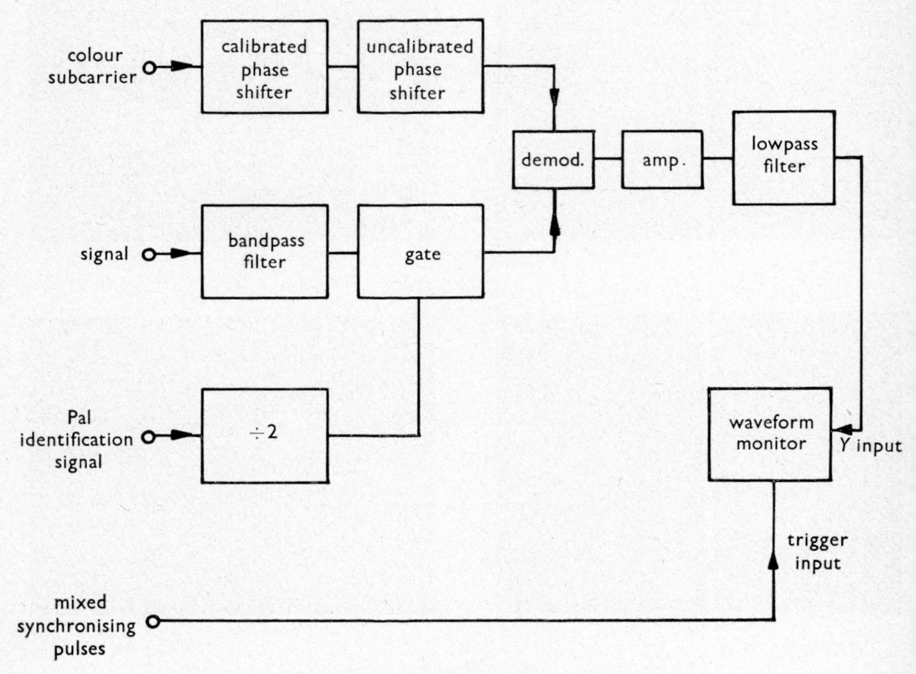

Fig. 12.21 Coder-analysis equipment

Fig. 12.20*d* attempts to show this effect; but, of course, a still photograph cannot give an adequate representation. The adjustment is performed by reducing the twitter to a minimum. Although in principle this should enable the quadrature to be set with extreme accuracy, in practice various secondary errors in the coder tend to decrease the discrimination somewhat.

Experience has shown that although this alignment procedure can hardly be bettered as far as the setting of the various amplitudes is concerned, the quadrature-phase adjustment is only just adequate, and the elimination of residual errors is made more difficult by the lack of a precise guide as to their origin. This has prompted the design of an instrument (Fig. 12.21), originally an adaptation of the line-display circuitry of the vector monitor described in Section 11.5.2, which overcomes these difficulties, and has proved useful for alignment and diagnosis of errors (Teear *et al.*, 1970).

The output signal from the coder is applied to a chrominance band-pass filter to remove the luminance component, after which the chrominance passes to a double-balanced synchronous demodulator via a gate controlled by the Pal 'ident' waveform after passage through a divide-by-two circuit. Thus the coder signal is applied to the demodulator for two successive lines and then removed for two successive lines. The corresponding zero output from the demodulator provides a reference line on the waveform-monitor display for the various measurements. A lowpass filter is connected in series with the output of the demodulator to remove both subcarrier leak and twice-subcarrier-frequency demodulation product.

The carrier supply for the synchronous demodulator, obtained from the local station subcarrier, passes through two phase shifters in series, one a precision 90° section with a calibrated vernier and the other a wide-range uncalibrated phase shifter for zeroing the selected subcarrier output component from the coder. The demodulator output is a direct current proportional to the cosine of the phase angle between the two subcarriers applied to it, which will appear on the display as a line of length proportional to the duration of the subcarrier vector. It follows that the various subcarrier components in the waveform from the coder will be separated on the waveform-monitor display by spacings proportional to their various phases and that the trace corresponding to any desired component can be made to coincide with the reference line by the use of the uncalibrated phase shifter.

As an example, to set the angle between the U and V axes, the coder is switched to $B - Y$ and the variable phase shifter is adjusted so as to bring the corresponding trace into coincidence with the reference line. The triggering of the waveform monitor is preferably made to overlay the pairs of successive lines. The precision 90° phase shifter is now switched in and the coder switched to $R-Y$. The precision phase shifter is now made to add 90° to the angle, when coincidence should again be obtained. If this is not the case, either the coder phasing control can be operated to bring it about, or, if desired, the error may be measured with the vernier phase shifter. Any differential-phase distortion within the coder will appear as a blurring of the trace, and can also be measured with the vernier.

When checking the residual subcarrier balance, it is now possible to isolate each axis. It is simplest to start by demodulating along the

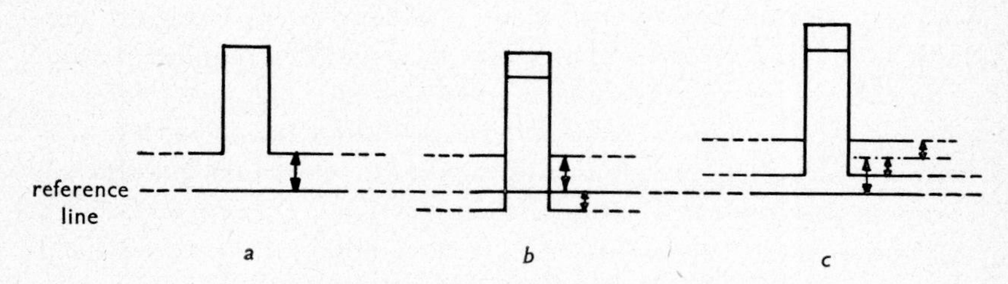

Fig. 12.22 Typical fault indications
Error voltages indicated by arrowheads

a $B-Y$ carrier unbalance, $R-Y$: differential carrier unbalance
b $B-Y$ differential carrier unbalance; $R-Y$: carrier unbalance
c $B-Y$ and $R-Y$: simultaneous carrier and differential carrier unbalances

$B-Y$ axis, a condition identifiable by the fact that both the demodulated burst components have the same polarity. Accordingly, the calibrated phase shifter being at zero, the uncalibrated phase shifter is operated so that these burst components are superimposed (Fig. 12.22*a*). No other demodulated colour vectors are visible, since, for this test, the chrominance is switched off at the coder. Any $(B-Y)$-axis carrier leak is now visible as a second line, and its extent may be measured by means of the separation from the reference line.

If the 90° phase shifter is brought in, the demodulation now takes place along the $R-Y$ axis and, because of the $R-Y$ switching in successive lines, a carrier leak will now have the appearance of

Fig. 12.22*b*. On the other hand, differential carrier balances will have the reverse appearance; i.e. a $B-Y$ differential-balance error will give the display of Fig. 12.22*a*, and an $R-Y$ differential-balance error the display of Fig. 12.22*b*. When all these errors are simultaneously present, the display of Fig. 12.22*c* is obtained.

White balance can be checked by switching on the chrominance signals and connecting all three inputs together. The error is indicated by a displaced trace, and the axes are identified as for the carrier balance.

12.10 Tube-test channel

The very high standard of performance demanded from modern cameras and the need for close matching of those used on the same production, not to speak of the tubes used in colour camera, necessitate a process of selection of tubes to offset the inevitable statistical spread of performance parameters and ensure the most favourable and efficient utilisation. Also, at some subsequent stage, it will be necessary to recheck a tube if it appears to have developed any anomalous characteristics or seems to be reaching the end of its useful life. If one can go so far as to check tubes at regular intervals, such data, if recorded and correctly analysed, can provide valuable information. Turk (1966) provides some useful remarks on the general subject of tube testing.

The equipment for such tests is known variously as a 'tube-test channel', 'tube-test rig', etc. There is no recognised standard, and it may take a wide variety of forms, ranging from a standard camera channel set apart for the purpose to the highly specialised and elaborate version described by Knight (1968).

In view of this and the differences occasioned by the different types of tube and the operational philosophy of the user, remarks are confined to a few general recommendations. Preferably, the tests should be arranged so as to conform very closely with operational procedures so that the results can be correlated directly with the day-to-day usage of the tube. Any reports of degraded performance can be checked similarly. On the other hand, as Knight (1968) points out, it should likewise be possible to repeat the tests laid down in the specification agreed with the manufacturer in case of questions of maintenance of quality of the product.

Nevertheless, some reasonable compromise must be found to minimise the number of tests and to obviate the expenditure of too much time and the accumulation of unduly large amounts of data. Above all, standardisation is very important, particularly with simple camera-test channels where there is a great temptation to improvise, with the result that the equipment is never quite the same on any two occasions; this applies also to ancillary apparatus, including test transparencies and their illumination. The adjustments of the tube itself should, of course, be maintained at consistent values throughout a series of measurements and from test to test. The influence of temperature is also not to be overlooked.

12.11 References

ANSTEY, H. G., and WARD, P. (1963): 'Operational control of television studio picture quality', *IEE Conf. Rep. Ser.* **5**, pp. 285–292

BROTHERS, D. C. (1959): 'The testing and operation of 4½ in image-orthicon tubes', *J. Br. I.R.E.*, **19**, (12), pp. 777–805

CAMPBELL, G. A., and FOSTER, R. M. (1931): 'Fourier integrals for practical applications', Bell Telephone System Monograph B584

DISHAL, N. (1959): 'Gaussian-response filter design', *Elect. Commun.*, **36**, (1), pp. 3–26

KNIGHT, E. R. (1968): 'A broadcaster re-tests his image-orthicons', *Roy. Televis. Soc. J.* **12**, (3), pp. 58–61

LEWIS, N. W. (1967): 'Optimisation of lowpass systems for analogue signals', *Electron. Lett.*, **3**, pp. 109–201

LEWIS, N. W., and HAUSER, T. V. (1962): 'Microcontrast and blur in imaging systems', *J. Photgr. Sci.*, **10**, pp. 288–301

LUBSZYNSKI, H. G., TAYLOR, S., and WARDLEY, J. (1960): 'Some aspects of vidicon performance', *J. Br. IRE*, **20**, pp. 323–334

PILZ, F., and SCHÄFER, H. (1963): 'Gerät zur Messung der Trägheitseffekte bei Fernseh- Kameraröhren', IRT Report 83

POTTER, G. B. (1958): 'Gradient measurement with an exponential', Australian Post Office Report 5597

POTTER, J. B. (1964): 'Use of an optical pulse-and-bar pattern for testing television cameras', *Proc. IEE*, **111**, (7), pp. 1219–1226

REDINGTON, R. W. (1957): 'The transient response of photoconductive camera tubes employing low-velocity scanning', *IRE Trans.*, **ED-4**, pp. 220–225

SANDERS, J. R. (1970): 'A camera tube lag meter', *IERE Conf. Rec.* 18, pp. 89–94

SEYLER, A. J. (1963): 'A pulse and bar TV test pattern'. *Proc. Instn. Radio Electron. Engrs. (Aust.)*, **24**, (3), pp. 321–322

SPRINGER, H. (1963): 'Die Vorteile des Impuls- und Sprung-Signales bei der Messung linearer Übertragungsverzerrungen', *Rundfunktech. Mitt.*, **7**, pp. 25–41

TEEAR, I. H., HUGHES, J. D., and HILL, M. E. (1970) 'Measurement techniques in television studios and outside broadcasts', *IERE Conf. Rec.* 18, pp. 417–443

THEILE, R., and PILZ, F. (1957): 'Conversion errors in image-orthicon camera tubes, *Arch. Elekt. Ubertragung*, **11**, pp. 17–32

THOMSON, W. E. (1952): 'The synthesis of a network to have a sine-squared impulse-response', *Proc. IEE*, **99**, Pt III, pp. 373–376

TURK, W. E. (1966): 'The practical testing of television camera tubes', *J. Soc. Motion Picture Televis. Engrs.*, **75**, (9), pp. 841–845

WATSON, G. R. (1963): 'Positional hum measurements in TV cameras', *Internat Televis. Tech. Rev.*, **4**, (11), pp. 406–408

13 VIDEOTAPE RECORDERS

13.1 General

The intensive development of videotape recorders since their early days, supplemented by work by the engineers of television administrations in many countries, has transformed the first relatively crude models into the very highly refined devices we know today. The developmental phase is still proceeding very actively, and newer types are coming into operational service. It is therefore only possible to discuss their testing in fairly general terms.

Although a videotape recorder is basically a signal transducer, as with so much other apparatus the inclusion of a mechanically operated electromagnetic conversion process introduces special features into the testing techniques. The input signal is converted into a specially processed f.m. intermediate frequency for recording, and this is subsequently demodulated after being recovered from the tape. It is possible to measure video–video in two ways, either by bypassing the tape and connecting the output of the recording side to the input of the replay side (electronics-only or *E–E* mode) or by recording on to the tape and then replaying the tape ('off-tape' mode). The latter mode is not unique, of course, since a tape recorded on one machine can be played back on any other suitable machine; but, for normal purposes it will be assumed that each machine is tested singly.

However, it is not easy to measure either of the component electronic systems singly, since such a measurement has to be carried out from video to the intermediate frequency, which is a specially processed frequency-modulated signal. In general, then record and replay are usually measured together, even though it is very important that each should be individually standardised to ensure interchangeability of tapes. The head system and the tape transport also introduce very special problems, largely owing to the need to reduce variations and instability in the mechanical system to extremely small amounts.

Some of the testing inevitably relates only to a particular model of machine, and it is impossible to give any general rule apart from referring the tester to the manufacturer's maintenance handbook. The overall transmission characteristics and certain other important aspects of performance can, however, be dealt with.

13.2 Linear waveform distortion

This can be measured with modern machines exactly as for other apparatus, except that a fullfield test signal, i.e. one containing complete field information, is mandatory. Any automatic equipment designed to minimise erratic positional errors in the output signal must be in use, otherwise the jitter on the test waveform makes readings difficult.

For monochrome machines, the 2*T* sine-squared-pulse and bar signal (Section 6.2) is the least stringent one should use. The augmented pulse-and-bar signal or the chrominance–luminance pulse-and-bar signal are preferable, however, since the amplitude of the chrominance components of the waveform adds a very useful item of information to that yielded by the 2*T* pulse-and-bar signal, i.e. the relative amplitude at the subcarrier frequency. The 1*T* pulse can naturally not be used in this instance, since the machine has a very definite upper limit to the acceptable frequency range.

For machines intended for colour, the waveforms and procedures of Sections 6.3.4 and 6.3.5 should be followed. The two important distortions here are chrominance–luminance gain and delay inequalities; distortion of the modulated chrominance envelope should be negligibly small.

13.3 Nonlinearity distortion

13.3.1 Luminance

The measurement of line-time nonlinearity distortion is carried out as described in Section 5.2, i.e. with the standard CCIR test signal consisting of one test line followed by either three black lines or three white lines. The test line in the CCIR signal may contain either a sawtooth with a superimposed sine wave at 1 MHz (for 625-line systems) or a 5-step staircase.

Even though, with a modern machine in good condition, the nonlinearity distortion is small, no especial difficulties should be encountered in measurement. In particular, the effect of random noise is greatly reduced by the filters used.

13.3.2 Differential phase and gain

The preferred test signal consists of a staircase with superimposed subcarrier at an amplitude of 140 mV pk–pk, although a sawtooth with superimposed subcarrier can also be used. This should again be utilised in the CCIR form with the three black and three white lines accompanying each line of test signal.

The measurement of the output signal is carried out as described in Section 5.4.2; but, although the signal/noise ratio of a videotape machine is not unduly high, this particular measurement is unusually sensitive to the effects of random noise, and, in the author's experience, the low values of differential phase obtainable with modern machines are not really measurable unless a sampling system of the type described in Section 5.5 is available. At least some method must be used of integrating out the effects of random noise if a reasonable degree of accuracy is required.

Differential gain may, in some cases, be proportionally somewhat greater than the differential phase, and, in addition, the chrominance-separating filter reduces the effect of random noise to some extent; so this measurement tends to be rather less dependent on the availability of an integrating device to improve the signal/noise ratio, but the use of such an adjunct to the measurement is nevertheless greatly to be recommended.

In each instance, it is advantageous if the measuring device can be gated from the head-switching pulses in a quadruplex machine, since it then becomes possible to estimate the differences between the individual heads, which is one of the factors contributing to head-banding effects.

13.3.3 Compact test signals

These may be defined as repetitive test signals in a form such that a multiplicity of parameters may be measured from it, simultaneously if desired. Such signals have already been mentioned in Section 6.3.10. Since the insertion test signal has been designed specifically to measure the important signal distortions over a very limited time, it seems sensible to take over its waveform and convert it into a compact signal by generating it repetitively, although special compact test signals may be designed for special purposes.

The particular advantage of such signals for testing videotape machines is as a means of assessing the total recording and replay distortions in an extremely short time, and in such a manner that the results obtained can be associated with a given replay of a certain tape. For this purpose, the compact test signal is recorded in the form of a short leader ahead of the programme signal, which means that the subsequently measured distortions are those corresponding to the total overall record and replay process at the moment when that tape was replayed.

The principal obstacle here is the extremely short time available for assessing all the signal distortions, less than 1 min in some cases. This has been overcome by replacing individual measurements by a photograph of a waveform-monitor display on which is multiplexed all the information necessary for analysing the distorted test waveform. This is achieved by the adapting equipment already designed for the routine evaluation of insertion test signals.

The technique is best explained by means of an example. In this instance, the test signal took the form of the two successive lines of the UK National i.t.s. repeated sequentially (Section 7.5), the only modification being the introduction of a 5% pedestal on the complete signal to ensure freedom from distortion arising from the action of the clippers. The multiplexed, analysed signal is given in Figs. 13.1*a* and *b*, the former corresponding to 'electronics only' and the latter to 'record–replay'. In each instance, the waveforms are, from left to right,

(i) luminance bar (time scale compressed)
(ii) $2T$ sine-squared pulse (time scale expanded)

(iii) $10T$ sine-squared pulse (time scale expanded)
(iv) top of bar, differential gain; superimposed inside bar, line-time nonlinearity
(v) colour burst
(vi) chrominance bar giving chrominance amplitude; central bright line corresponds to chrominance–luminance crosstalk
(vii) differential phase (initial spikes due to switching transient).

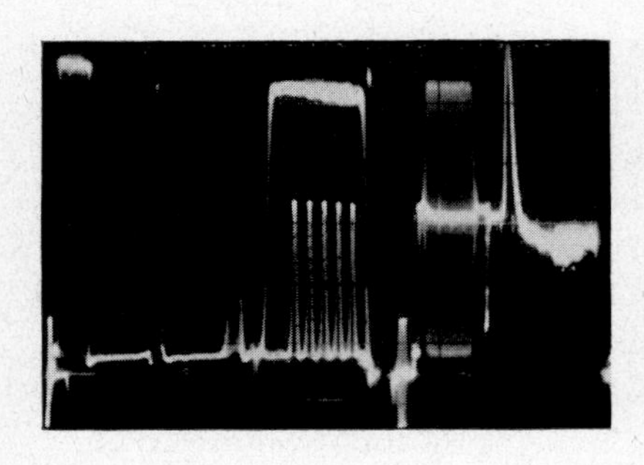

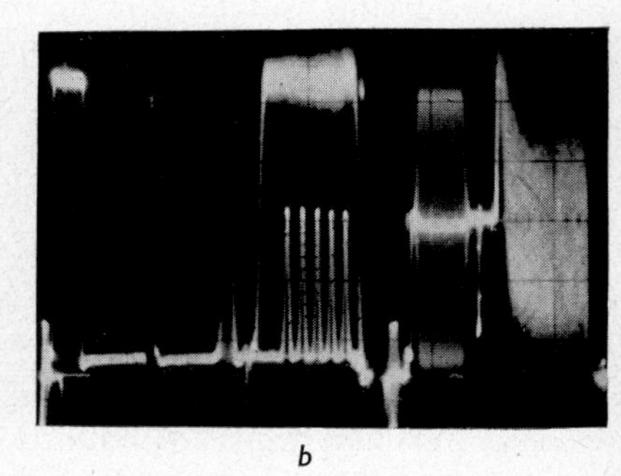

b

Fig. 13.1 Analysis of distorted compact test signal in multiplexed form

a Electronics only
b Record–replay for same machine as (*a*)

From the resulting photograph can be derived at leisure the values of:

(*a*) K_{bar}, K_{2T} and K_{pb}
(*b*) luminance–chrominance gain and delay inequalities
(*c*) differential phase
(*d*) differential gain
(*e*) line-time nonlinearity
(*f*) burst/chrominance ratio
(*g*) chrominance–luminance crosstalk.

A particular advantage of this technique is that a permanent record is available for later analysis. If the operator is properly prepared, the photograph requires only a fraction of the allowable time. No special difficulties have been experienced during the making of a considerable number of such records.

13.4 Phasing

13.4.1 Hue phasing

The tests for linear waveform distortion do not check the relationship between the subcarrier carried on the colour video signal and the subcarrier burst, which is reinserted at a late stage in the videotape machine. An adjustment is provided on the machine for achieving the correct relationship.

The setting is most easily carried out with a vector monitor providing a polar display of a colour-bar signal after passage through the machine. The phase adjustment on the machine and the zero-phase adjustment of the vector monitor are operated together until the burst vector lies along the graticule line and the subcarrier vectors fit as well as possible into their appropriate 'boxes' (Figs. 11.4*a* and *b*). Assuming that the colour-bar signal is correct and that the vector monitor is in good order, any errors are likely to be due to differential-phase effects. It should be borne in mind that the colour-bar signal cannot be used to measure differential phase in the true sense of the term, since the subcarrier amplitudes are too great; so it must not be supposed that differential-phase measurements can be obtained during a subcarrier-phasing check.

13.4.2 Source phasing

Where more than one videotape machine is available to a common mixer, the electrical lengths of the circuits between the machines and the mixer must be adjusted to equality so that no untoward effects appear when a cut is made from one machine to another.

A simple method of optical comparison has proved very effective, when the mixer, as is usual, is provided with a split-screen facility, preferably vertical as well as horizontal. All machines are supposed to have been synchronised previously with the local synchronising pulses. Then a colour-bar signal is transmitted through each of two machines and displayed on the same picture monitor using the split-screen facility. The picture-monitor phase reference may be derived from the local subcarrier, or some other arrangement is made for displaying the two sets of colour bars in synchronism depending on local circumstances.

Particularly when a vertical split is used, it is possible to match the two sets of colour bars visually very accurately, which then ensures that the two machines are phased. If there is an appreciable difference between the differential-gain and -phase characteristics of the two machines, a perfect match is not quite possible, and an estimated best match is sought. It is an advantage of this method that any slight compromise made does at least correspond to practical viewing conditions.

An objective, rather than a subjective, comparison can be made with a vector monitor. The accuracy is slightly restricted if only a polar display is available, but if the vector monitor is equipped with a line display and a means for precision phase measurement, a very close match is possible. As with hue phasing (Section 13.4.1) differential-phase effects may prevent a perfect match, and a best-fit condition must be sought.

13.4.3 Monochrome timing

The inclusion in a videotape recorder of an electronic device for reducing positional errors means that it is possible to shift the blanking of the reinserted synchronising pulses with respect to the original signal blanking. It is not normally required to measure the error but to reduce it to zero, which is easily achieved by displaying the line-blanking area of the signal on a waveform monitor and adjusting for minimum blanking width. A picture monitor may also be used for the same purpose. With machines for colour signals, the position of the colour burst is the most critical factor and will have to be set within the prescribed limits by means of a waveform monitor.

13.5 Noise

13.5.1 Random noise

The measurement of random noise with a videotape recorder presents certain difficulties owing to the fact that such machines normally blank the input signal; moreover, it is usually required to test at various levels of grey since the random-noise level is to some extent a function of the signal amplitude. These special requirements require suitable techniques.

One might measure the noise directly by estimation of the quasi peak-to-peak value on a signal containing fixed levels of grey, except that this measurement is not in itself very satisfactory (Chapter 3). A simple objective method is available if one merely wishes to measure through the electronics only. It is then possible to bypass the modulator and to replace its output by a steady frequency, from an external oscillator, corresponding to the level of grey at which the measurement is to be made. Under these conditions, there is no video-signal output from the machine and the random noise can be measured with a suitable random-noise meter, or this can even be improvised from a high-gain, wideband amplifier together with an electronic voltmeter suitable for the frequency range. Preferably, this should read true r.m.s. values, but an instrument reading mean values can be used if an r.m.s. meter is not available.

The measurement must always be made through a suitable lowpass filter so that out-of-band noise is eliminated, and a 10 kHz highpass filter is also required if there is any suspicion that the reading can be influenced by supply hum or other low-frequency effects which cannot be classified as random noise (Chapter 3).

The preferred objective method is the use of a gated type of random-noise measuring set (Chapter 3.9) which can operate in the presence of synchronising information and picture signals of certain simple forms, fixed levels of grey. It then becomes possible to include the tape noise, which forms the largest contribution to the total random-noise level.

This has proved extremely useful, and has facilitated investigation of a range of commercial tapes not only for signal/noise ratio as a function of luminance level but also for the noise spectrum. With a particular model of videotape recorder (Figs. 13.2*a* and *b*), there is a difference of 5 dB between the mean overall signal/noise ratios at 90% lift of the two tapes shown in the curves. However, the change in signal/noise ratio with luminance level was found to be quite small with all the tapes tested, in each case showing as a change in high-frequency, rather than low-frequency, noise.

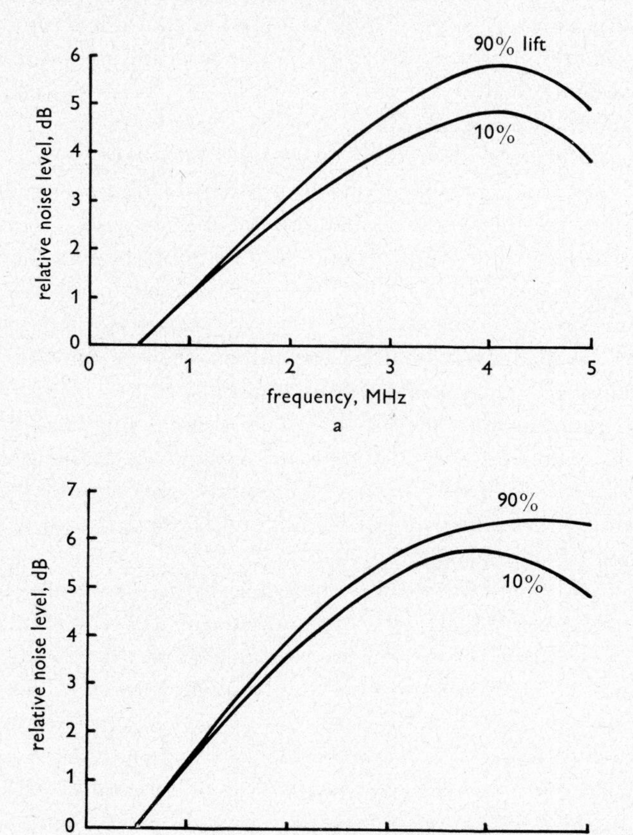

Fig. 13.2 Measured signal/noise ratio of videotapes

a Tape 1
43·5 dB at 90% lift
44·5 dB at 10% lift

b Tape 2
38·5 dB at 90% lift
42 dB at 10% lift

The spectrum of the random noise was measured with a useful auxiliary unit which operates on an output feed of separated random noise. The unit contains a range of simple bandpass filters whose midband frequencies are spaced conveniently over the videoband.

Any filter can be selected with a switch. The filtered output of random noise is amplified and finally measured on an output meter, so that the shape of the noise spectrum can be investigated rapidly and conveniently.

13.5.2 Moiré

This distortion, which takes its name from the 'watered silk' appearance it gives on a picture monitor, is due to spurious components arising from the f.m. process used in recording the tapes. The worst components arise from the colour subcarrier, since this is a high-energy signal component near the edge of the band.

It is generally accepted that the measure of moiré distortion is the ratio of the peak white signal to the peak-to-peak amplitude of the worst interfering frequency. Apart from that, there seem to be no standards of measurement.

The following is a subjective method found useful, although having a somewhat restricted accuracy. Colour bars are used as the test signal since measurements must always be made at a number of luminance levels, to explore the full range of f.m. deviation frequencies.

The colour-bar test signal is prerecorded, and is also available locally from a generator; in parallel with which is a video oscillator. The replayed signal is viewed on a picture monitor, and the appearance of the bar most affected by moiré distortion is noted. Then a switch is operated to interchange the replayed signal for the signal from the generator, and the frequency and the video-oscillator output are both adjusted until the appearance of the most disturbed colour bar is the same in both cases. It will be necessary to switch backwards and forwards a few times to establish the best match. The amplitude of the sine wave from the oscillator then enables the amplitude of the signal/moiré ratio to be calculated.

There are two principal weaknesses in this method. The first is obviously the need to switch one display to another to find the condition for equal impairment, which can be overcome if a split-screen device is available so that the two displays can be viewed simultaneously. The second weakness is that the visual interfering effect of a sine wave is not identical with moiré, and hence the two effects under comparison are not the same. For these reasons, the method is now falling into disuse.

The frequencies which produce the moiré effects show up in an unusual way on a polar display of the colour-bar signal on a vector monitor. The moiré frequencies are derived from the subcarrier vectors, and can be considered, in a sense, as sidebands of these

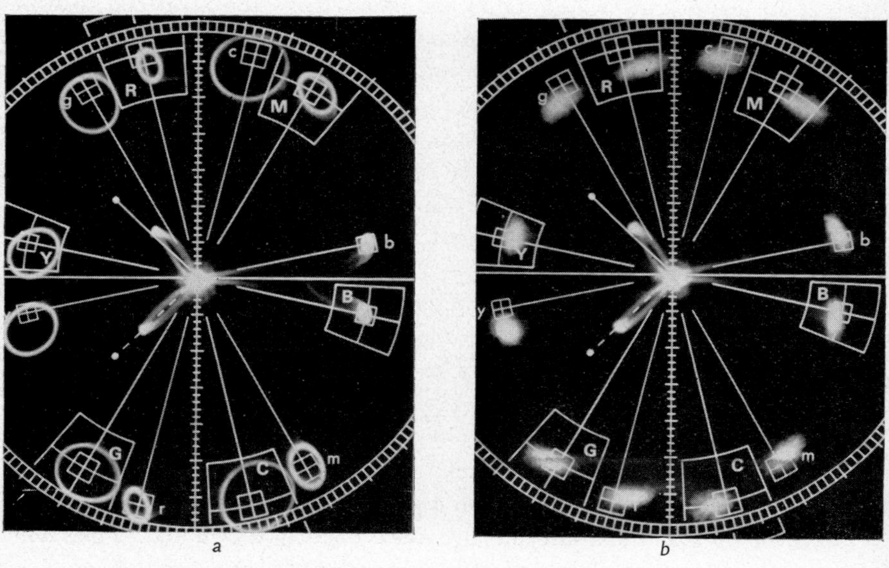

Fig. 13.3 Display of Pal colour bars showing moiré component
a Electronics only (moiré exaggerated)
b Pal colour bars offtape showing normal moiré

frequencies. Hence, in a polar display, the moiré frequencies will be found to be represented by circles around the tips of the subcarrier vectors, just as, in classical phasor theory, the sidebands of an amplitude-modulated wave are represented by phasors rotating about the tip of the subcarrier phasor, the locus of the tips of the subcarrier phasors being a circle.

These moiré circles, or 'doughnuts' as they are familiarly called, can clearly be seen in Fig. 13.3*a*, in which for clarity, the amount of moiré interference has been exaggerated; also the E–E mode has also been used to eliminate the tape noise. Some estimate can be obtained of the amount of moiré present by measuring the diameters of the circles and comparing them with the lengths of the corresponding vectors, although no great accuracy can be expected if the amount of moiré is small. Unfortunately the method is not effective in the offtape mode, since the tape noise and positional errors effectively obscure any small amount of moiré present (Fig. 13.3*b*).

Two methods which offer a greater accuracy and consistency of measurement require rather specialised instruments: in the one instance a spectrum analyser and in the other a gated type of random-noise measuring instrument, capable of measuring the heads of a quadruplex machine individually, if such a machine is under examination.

It is assumed that, before any measurement is undertaken, the machine under test will have been carefully aligned in accordance with the maker's instructions, particularly with regard to the correct f.m. deviation, playback gain and the response at the upper end of the videoband. The demodulator gain also needs to be calibrated.

In the method using the spectrum analyser, the carrier frequency of the internal modulator of the machine is detuned from its normal rest position so as to simulate the recording of a grey field without the necessity for synchronising and blanking information which would introduce spectral components masking those to be measured. A standard value for this 'lift' at the moment is 47%, which with the international highband standards corresponds to a frequency of 8·5 MHz; this can be set to the correct point with the spectrum analyser.

Subcarrier at a level of 700 mV is then applied to the video input, and the deviation checked to ensure that it corresponds to this value. Then, when the demodulator output is observed with the spectrum analyser, the subcarrier and the accompanying moiré components can be identified and their amplitudes measured. A typical display is given in Fig. 13.4; the colour subcarrier is somewhat obscured by the central graticule line, although the moiré frequencies at about 320 and 740 kHz are clearly visible at levels of -32 and -37 dB, respectively. It is customary with this method to take the amplitude of the largest moiré component present as the nominal value of the interference, i.e. -32 dB in this instance, even though it has been stated that the root sum square of all moiré components corresponds most closely to the subjective picture impairment. In many cases, of course, these will be nearly identical, i.e. when one component is predominant.

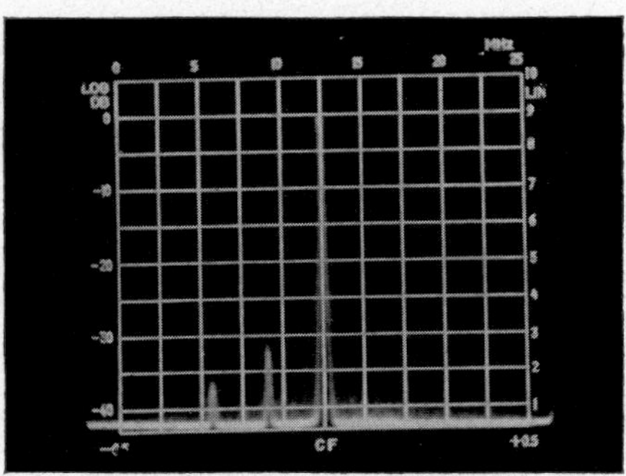

Fig. 13.4 Spectrum-analyser display of moiré components

The other method utilises equipment for the measurement of random noise in the presence of synchronising information and static signals or patterns; in the procedure described by Darby and Tooms (1970), the instrument is that dealt with in some detail in Section 3.12, one of the features of which is its ability to measure the heads of a quadruplex machine individually. The signal applied to the video input of the recorder is a coloured field with 100% amplitude and 100% saturation, and the measurement is made at the red-channel output of a decoder. Separate measurements are made for each of the primary colours and their complements, and the overall worst figure is taken as the value of the moiré interference. In this case, of course, the r.s.s. addition is performed automatically, but the result then has to be converted into a peak-to-peak value relative to 700 mV, while taking into account the $R-Y$ weighting introduced by the decoder. The measurement can be made in either the electronics-only (E–E) or the record–replay mode.

13.6 Tape dropouts

Dropouts are microscopic blemishes in the tape coating which cause an interruption either of the record or of the replay process and so give rise to short pulses, typically of durations between 3–25 μs, which appear as either black or white dots according to the modulation system of the videotape machine. If they occur at all frequently, they represent a major impairment of the reproduced picture, and, although their effect has now been greatly reduced, owing to improvements in tapes and to the widespread use of efficient dropout correctors, it is nevertheless still necessary to monitor tapes and to discard those which show an unduly great dropout activity.

It is not sufficient just to record the number of dropouts, since visible impairment produced is not the same for every dropout. Very short dropouts and those which do not cause a complete interruption of the f.m. carrier have a very small effect, whereas long dropouts or groups with only a short interval between successive members of the group cause a large impairment. A very good discussion of this point and the results of subjective tests to determine the relationship between dropout width and picture impairment are given by Geddes (1965). It was found that a simple *CR* circuit with a correctly chosen time constant forms an adequate weighting network for dropouts, preferably also with an amplitude-limitation criterion built into the system so that any dropout which does not reduce the f.m. carrier to less than, say, 10% of the normal value is excluded.

The information from which the occurrence of a dropout can be determined can be derived from the video signal on replay or from the f.m. carrier. In the first, the tape is recorded with either black level or a peak white signal according to the modulation conditions. The dropouts are then detectable as pulses occurring during the constant video signal.

This simple method suffers from a number of practical drawbacks. One of the most important is the need for an individual measurement of each tape, by recording a special signal and then replaying it. The degree of amplitude limitation and the operating conditions of the f.m. demodulator also affect the result. Nevertheless, it has been successfully used (Goldberg and Hannah, 1961; Waechter, 1961).

In the second, the information is derived from the f.m. signal itself, since a dropout consists of a complete or partial interruption of the f.m. carrier, and a simple a.m. detection following amplitude limitation will reject the f.m. modulation and furnish the dropout signals directly. In practice, it is rather more complicated than the first method, but the advantages are considerable, in particular the possibility of monitoring tapes during normal operational service without the need for special measurement. A good example is given by Altmann (1964) and Stübbe (1964).

The simplified block diagram is shown in Fig. 13.5. The machine with which it was intended to use the dropout monitor was equipped with a monitoring point at the head switcher which provided a very convenient input signal for the device. Some such intermediate-frequency output point is usually available. The f.m. signal is first amplified by the automatic-gain-controlled amplifier which maintains the output level constant, and is then limited to 10% or less of its original amplitude, thus ensuring that only those dropouts are included which reduce the carrier amplitude to the limited value or less.

The next stage is an a.m. demodulator which detects the dropouts and the fast edges from the switcher pulses, which are reduced to insignificant proportions by a lowpass filter connected between the demodulator and the pulse amplifier. The Schmitt trigger circuit which follows is amplitude-sensitive, so that it delivers an output pulse of standard amplitude and a duration corresponding to that of the dropout, but only for input pulses whose amplitude exceeds the critical value. These pulses are turned into corresponding current pulses and used to discharge an *RC* circuit dimensioned, as has already been explained, so as to act as the analogue of the subjective picture impairment due to dropouts.

At this point, the signal processing has been completed and the analogue signal may be used to operate a pen recorder to obtain a record as in Fig. 13.6. It has been found that a paper-tape speed of 12 in/h is convenient for most purposes, and the 18 in paper length obtained with a 90 min tape is not too unwieldy for storage.

Further data may be obtained from the instrument by driving electronic counters from the output pulses of the Schmitt trigger, first to give a total count of all pulses and secondly to totalise the periods of all the pulses to provide the total dropout time.

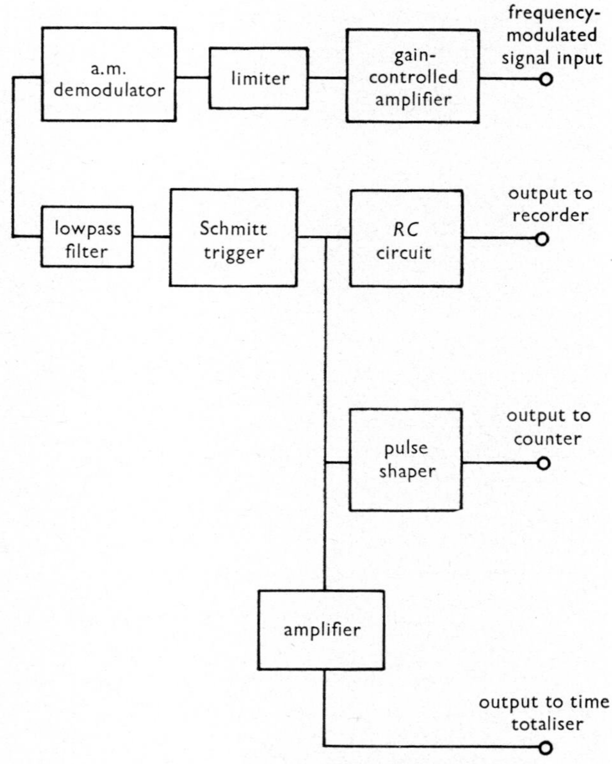

Fig. 13.5 Dropout recorder (simplified)

13.7 Tape drive

13.7.1 Servosystem

The mechanical-drive system of a videotape machine is normally locked, not only to the incoming field pulses, but also to the line synchronising pulses to minimise the positional errors on a picture monitor due to the inevitable random variations of the drive mechanism, i.e. of the tape transport and the head drive.

The means adopted to produce the desired effect vary with the type of machine, and it is not possible to give any detailed information on the type of measurements to be used on the servosystem. However, there must be at least one servoloop in charge of this type of synchronisation, and this requires checking.

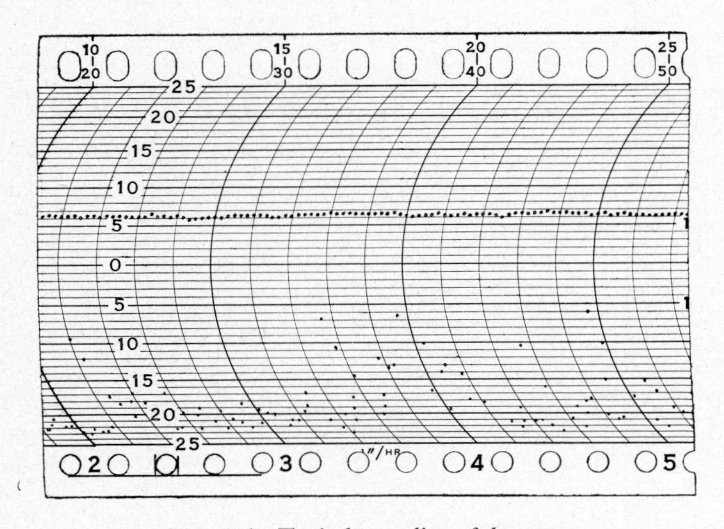

Fig. 13.6 Typical recording of dropouts

At the moment, a very common method for checking the correct operation of the main mechanical servoloop seems to be to introduce a sudden deceleration into the mechanical system by lightly touching the recording-head drum and watching the output waveform from the comparator. This should be very highly damped and not exhibit oscillatory tendencies. Not only is the method crude, but the checking of the head drum can lead to harm to the machine as well as to the tester unless it is very carefully and judiciously carried out. Also it does not, in fact, test the whole of the servosystem.

A more scientific and less risky method is to add a step waveform to one of the two frequencies being compared in the phase comparator of the servosystem. This should be long compared with the transient response of the output waveform. The input and output waveforms can now be standardised and working tolerances laid down.

The method of testing one particular machine which has a combination of two servoloops, one acting on the main mechanical system and the other electronically on the reproduced signal is described in detail by Vollenderweider (1965). He was particularly concerned with the effect of a phase step resulting from the sudden, small positional error caused by a tape splice.

Operationally, it is common to check the servosystem by measuring the time taken to recover from an abrupt discontinuity in the synchronisation of the signal on replay, and the maximum time taken to recover from a similar type of disturbance in the input video signal during recording.

13.7.2 Velocity errors

The correction in the machine of positional errors does not of itself ensure that the velocity of the head across the tape is uniform. Velocity errors are particularly obnoxious in colour television, because a velocity change is equivalent to a phase shift of subcarrier, with the result that what should be an area of uniform hue across the screen of the picture monitor will show in most instances a progressive change of hue from left to right.

Velocity errors in fact are no longer considered as a problem in modern machines since correctors are incorporated, but it is still worth discussing how the magnitude of the inherent errors may be obtained.

This hue change suggests a method of detecting the presence of significant velocity errors; it cannot be called a measurement, since the magnitude of the error is not derived. A colour-bar signal is recorded and replayed. The original signal and the replayed signal are displayed together on a picture monitor, preferably using a split-screen-effect facility with a vertical split. If this is not available, the two signals can be switched successively on to the same monitor; this is less convenient and less accurate.

Care must be taken to phase the two bursts accurately, so that the two sequences of colour bars both start with the same hue; then, if a velocity error is present, the corresponding sets of bars show progressive errors, with the red bars normally showing the greatest difference.

An alternative method is to use a red bar extending over the whole width of a line; this can conveniently be combined with a colour-bar signal. Red is chosen because the eye appears to be especially sensitive

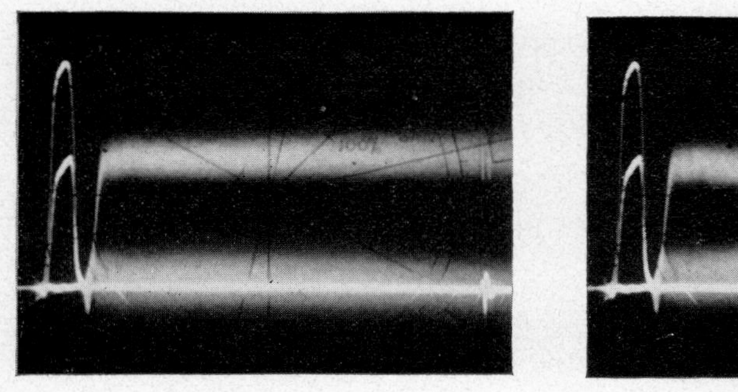

a

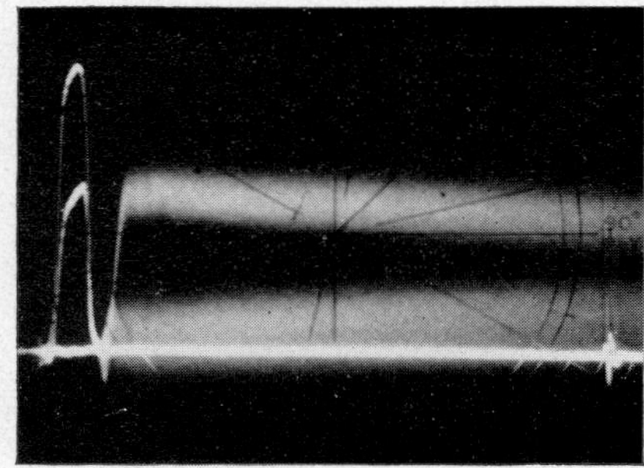

b

Fig. 13.7 Display of velocity errors

a Line display of phase of red patch
b As (*a*), but showing 10° velocity error

to slight hue changes in that region. When such a signal is replayed, any hue changes along the line are immediately obvious. However, the colour monitor must be checked beforehand to ensure that no visible purity errors are present which could be mistaken for a velocity error.

Even better is to use the same bar, which need no longer necessarily be red, and to display it on a vector monitor as a line display. If no velocity errors are present, the trace corresponding to the bar should be parallel with the baseline of the display, as in Fig. 13.7*a*, where, in fact, a return trace is also visible. When a velocity error is present, the trace slopes, since there is a steady change in the subcarrier phase, and the maximum departure gives the phase error corresponding to the velocity error. In Fig. 13.7*b*, it was measured as a trace displacement of 1 cm corresponding to a phase error of 10°.

In Figs. 13.7*a* and *b*, the noise on the trace limits the accuracy of the measurement, much more so when a visual observation is made than when the trace is photographed, since the noise integrates fairly well to a rectangular bar. The accuracy can be improved considerably by using a waveform sampling device (Chapter 5), which effectively exchanges the rate of presentation of the display for signal/noise ratio.

Incidentally, such a display is excellent for checking that the tip penetration of the heads into the tape is optimum from the point of view of velocity errors. A change in penetration is clearly visible as a tilt of the trace corresponding to the subcarrier phase, and a useful practical correction is possible within reasonable limits for existing errors.

13.8 References

ALTMANN, K. (1964): 'Drop-out Registriergerät', Institut für Rundfunktechnik Report 89

DARBY, P. J., and TOOMS, M. S. (1970): 'Colour television studio performance measurements', *IERE Conf. Proc.* 18, pp. 1–13

GEDDES, W. K. E. (1965): 'Drop-out in video recording', BBC Engineering Division Monograph 57

GEDDES, W. K. E. (1966): 'The subjective impairment produced by dropouts in video tape recordings', *J. Televis. Soc.*, **11**, pp. 154–160

GOLDBERG, A. A., and HANNAH, M. R. (1961): 'Video-tape analyser', *J. Soc. Motion Picture Televis. Engrs.*, **70**, pp. 85–89

STÜBBE, M. (1964): On the measurement and valuation of dropout errors in video-tape recording', Proceedings of the IEE international conference on magnetic recording, London, pp. 9–11

VOLLENDERWEIDER, E. (1965): 'Untersuchung an Ampex- Bildaufzeichnungsanlagen bei Wiedergabe von Magnetbändern mit Schnittstellen', Institut für Rundfunktechnik Report 99

WAECHTER, D. (1961): 'Einrichtung zur Erfassung der Drop-out Stellen auf Video-Magnetbändern', *Rundfunktech. Mitt.*, **5**, pp. 295–296

14 TELECINE MACHINES

14.1 General

The measurement and testing of telecine machines suffers from the restriction inherent in all optoelectrical conversion systems that much of the work must of necessity involve a comparison between optical test objects and the corresponding output waveform. These objects are always difficult to produce to a sufficiently high accuracy and consistency, all the more so when they must be available on small-format cinefilm. A further complication is the existence of two principal types of machine, those employing flying-spot devices and those based on camera tubes, which necessitates a different approach in certain directions. Further problems are added by colour operation, not solely by the need for colour analysis, but also as a result of the wide variety of film material which must be dealt with, each more or less individual from the colorimetric point of view.

The most important measurements on the electrical side will be covered briefly, the optical side being beyond the scope of the present work, with particular reference to colour telecine machines. The application of the techniques cannot be given in any great detail since proprietary makes of machine differ considerably in construction. However, a large proportion of these techniques are also applicable, with certain obvious modifications, to caption scanners, since, apart from the film-transport mechanism, their principles have much in common.

In general, before any measurements are attempted, it is necessary to ensure that the standard lineup procedure has been gone through; in particular, it is essential to allow sufficient time for warmup, i.e. with camera-tube machines for the yokes to reach correct temperature and with flying-spot machines for the normal phosphor temperature to be attained.

14.2 Gamma

Under operational conditions, the principal aim is to adjust the γ characteristic to the correct shape, with special reference to the near-black end of the scale, where a considerable degree of black stretch is usual. This has an important effect on the noise performance, since noise in the near-black region is subjectively a serious impairment, and the amount of black stretch must be carefully controlled. The most satisfactory method is to insert in the gate a series of accurately calibrated neutral-density filters, but a film strip containing a standardised tone wedge with, say, ten steps of density is more commonly used in maintenance areas. The output waveform can then be compared with a standard curve, or a direct measurement can be made. For this, the step wedges should be situated in the central area of the field only, to avoid errors due to vignetting. The ideal is on the lines of test card 57 (Section 10.9).

Such an overall measurement ensures that the amplitude–amplitude characteristic of the conversion system and also the light scatter in the optical-imaging and light-transfer systems are taken into account. However, it is also desirable to check the electronic γ correction, usually by inserting an appropriate signal into the input of the processing amplifier. If a preset-law staircase generator is available, a very convenient check is obtained. This generator has the property that the output waveform can be modified in such a way that the successive steps of amplitude follow a desired law, so that the inverse waveform corresponding to the desired black-stretch circuit can be generated; and, ideally, the output waveform should then be a linear staircase. Adjustments can then be made to correct any deficiencies. Other methods of measuring γ are discussed in Chapter 12.

With colour-telecine machines the correct balance of the γ-correction circuits of the three colouring channels is of great importance; and it should be set up only against a reference of colour-film stock, otherwise serious errors can be introduced.

In all operations concerned with the balance of the colour channels in any colour-signal source, it is usual to take the green channel as the reference. This is set up to the nominal γ curve, then each of the other two channel outputs is compared with the green-channel output in a difference amplifier, so that the waveform difference is displayed, and should then be reduced to the smallest possible amount. If an overall gain control is provided, the balance should also hold to changes in gain.

14.3 Resolution

The resolution of a telecine machine, as with a camera, is measured in terms of the modulation depth of the output signals corresponding to a series of resolution gratings, i.e. the amplitudes of the output signals expressed as a percentage of the output signal for gratings having a relatively wide spacing. These gratings are most often rectangular because of their simplicity, even though a correction has to be made for the $4/\pi$ effect arising from the removal of all the harmonics (Chapter 10). Better gratings would be sinusoidal, although even these cause difficulties from the distortion of the waveform in the γ corrector. The measurement is always artificial to some extent, since the behaviour of the pickup tubes is not consistently linear, independently of the signal areas which it is handling. It is all the more important, therefore, to use the same method on all occasions so as to ensure consistency of the results. Usually, unity γ is specified.

The test object is most often a film of a standard test card, since this is readily available, but the characteristics of available small-format films are such that it is not possible to manufacture such test strips with full modulation of the resolution gratings for sizes under 35 mm; and, even with 35 mm film, the utmost care has to be taken if good results are to be achieved. Hence it is necessary in the general case to standardise the film beforehand with a microdensitometer.

Better than a film of a standard test card is a specially prepared film in which the grating blocks are of a much greater length, since the waveforms then correspond more closely to a square wave or a sine wave as the case may be. The fact that the gratings have a finite length means that mathematically speaking the grating waveform is multiplied by the pulse waveform corresponding to a pulse having the width of a grating block. This is a modulation process and gives rise to a symmetrical train of sidebands about the grating frequency.

As a rule, extremely little trouble is experienced in the lower half of the video-frequency range. At the lower video frequencies, it arises from afterglow correction.

14.4 Afterglow correction

It is not normally required to measure afterglow effects, since they are capable of correction by means of circuitry built into the machine. The chief problem is to provide a suitable test object to allow the adjustment to be carried out with the greatest precision. The standard is a film containing in each frame a number of bars having a range of lengths. The bars must have a thickness of at least, say, the equivalent of ten lines. A choice of lengths is needed, because the effects of insufficient correction may well show up much more clearly with a certain length of bar.

The adjustment is carried out using the means provided while the picture on a monitor is examined. The reason for this is that correction is not usually ideal, and it is much more satisfactory to judge on a picture monitor when the optimum setting has been reached. With a waveform monitor, it becomes much more difficult to assess the importance of any residual errors.

14.5 Signal/noise ratio

The measurement is normally made as a routine on a colour machine on all three channels, and it is important to ensure, before starting, that the adjustments of γ, resolution etc. are satisfactory since these have an important influence on the signal/noise ratio.

The noise also depends on the signal to some extent, and it is necessary for routine purposes to select a grey level for the measurement low enough to take account of the increased noise in the blacks, but not so low that an undue preponderance is given to this area. An opacity of unity below the opacity corresponding to peak white has been found most suitable, since this approximately corresponds to the region of unity gain.

Since the noise spectrum, particularly with camera-tube types of telecine machine, tends to be sharply peaked towards the highest video frequencies, it is highly advisable to make the measurement through a suitable lowpass filter to ensure that the noise bandwidth is compatible with the nominal signal bandwidth. Such filters may be included in the R, G and B channels of a colour machine, either explicitly or implicitly.

Since a video signal is always present, the measurement is best made with a gated-type random-noise meter, preferably including a means for measuring the spectral distribution of the noise, since this provides a comparison of the noise performance of different machines. With monochrome machines, it is usual to make unweighted measurements, even though this does not seem altogether logical. When measuring the colouring channels of a colour machine, weighted measurements are not possible owing to the subsequent matrixing of the channels. It would seem that a useful single figure should be obtained by measuring one of the colour-difference channels at the output of a coder, since, at this point, the matrixing has been completed.

14.6 Registration

Errors in registration in a colour machine may arise from deficiencies or lack of alignment in the scanning of optical systems, apart from differences in the times of transmission of the signals through their respective channels. This means that the errors are unlikely to be identical at all parts of the scanned picture area, and it is usual to measure them at a number of points.

For laying down tolerances, it is customary to divide the picture area into zones of different importance as regards registration errors. A typical scheme is two zones formed by tracing an ellipse on the picture area, with major and minor axes equal to the picture width and height, respectively. The inner elliptical area is termed zone A and the outer B. The allowable tolerance is 1/2000 of picture width in zone A and twice that figure in zone B, say, 25 ns and 50 ns for 625-line standards, so that measurements ought to be made to an accuracy of a few nanoseconds. The use of nanoseconds in this connection is widespread, although a spatial separation is actually meant.

14.6.1 Shift method

The simplest method, which requires the minimum of specially designed apparatus, is versatile but somewhat inconvenient as a routine, since it depends on a calibration of the shift currents, achieved by the insertion of a meter. In general, the accuracy is improved if the bandwidths of the three colouring channels are made absolutely identical by means of three closely matched lowpass filters, which nevertheless should have the widest bandwidth possible, but it is not usual to go to these lengths; and, indeed, it may, in many instances, not be necessary.

After the usual alignment of the machine, a film loop is run of some convenient test card with suitable patterns in the various regions of the picture area. The channel outputs are taken in pairs to the inputs of a high-grade difference amplifier, and the resulting difference signal is displayed on a picture or waveform monitor. Adjustment of the relative levels should then result in cancellation of the central portion of a particular pattern, the residual effects at the edges being due to registration errors. Adjustments are then made to obtain optimum registration over the whole of the picture area, the picture monitor display being particularly suitable for this purpose.

At a previous stage in the measurement, arrangements have been made to monitor the shift currents of the deflection system; and, by inserting a known small delay and readjusting to the previous condition, a measure is obtained of the current stage per nanosecond. This facility is then utilised to measure the registration errors by adjusting the shifts, so as to minimise the edge effects of the pair of channels under test. Two other similar sets of measurement must, of course, also be undertaken so that all the relative errors are known. The requirement for the largest possible bandwidth of the output filters should now become apparent; the greater the channel bandwidth, the more precise is the setting for minimum registra tion eror, since the risetime of the edges is decreased.

14.6.2 Calibrated-delay method

This follows the same broad lines as the previous measurement. However, the adjustment for minimum registration error is achieved by inserting into the appropriate channel a precision calibrated time delay capable either of adjustment in steps of a very few nanoseconds or preferably with a continuously variable fine adjustment. The error can then be read directly from the dials. This is capable of greater accuracy than the previous method, and is certainly much more convenient. Although in principle this is only suitable for measuring horizontal registration errors, it is possible to obtain a satisfactory estimate of a vertical error by comparing it with a calibrated horizontal shift. The accuracy is of the order of 5 ns, which is adequate for the purpose.

14.7 Flare

Flare is the spurious signal resulting from the scattering of light in the optical system, which takes the form of a fairly uniform low level of illumination in areas which would be black. It is measured in terms of the amplitude of the spurious signal found in a black patch surrounded by a uniform white area.

The test object is a specially prepared film or slide having as high a contrast ratio as practicable between the central black area, usually taken to be 30 picture-elements square, and the surrounding clear region. Suitable test transparencies are possible by known photographic techniques, but need to be looked after very carefully to avoid degradation from handling. The corresponding output signal is a fullwidth line bar, the amplitude of which is taken to be 100%, having in its centre a narrow negative-going pulse, the amplitude of which is required to be measured. The only unknown now is the amplitude of the signal corresponding to true black or 0%, which is easily found by capping the lens. The difference between this and the peak of the negative pulse represents the flare signal, usually expressed as a percentage of the white level.

The reverse measurement using a white patch in the centre of a black area is not usually carried out as a routine, but may be required for investigational purposes.

14.8 Lag

The accurate measurement of lag with telecine machines employing camera tubes can be carried out very accurately by, for example, an adaptation of the method of Section 12.8.2. However, for operational purposes, such a procedure is unduly time-consuming, and simpler methods are preferred. Typical is the use of a test film containing a white pendulum oscillating against a black background, which provides a very quick assessment of the magnitude of the lag. This has the practical advantages that the figure thus obtained, i.e. the apparent angle of persistence of the pendulum is more nearly related to the subjective picture impairment than is the size of the residual signal after a given number of fields. Moreover, it allows operational adjustments to be carried out to optimise the apparent lag against other parameters, such as signal/noise ratio and resolution.

14.9 Shading and vignetting

A certain variation of light intensity across the picture area is inevitable, even with a perfect lens system, but this is always aggravated to some extent by defects in the optical system and possibly also by some vignetting. This latter differs essentially from shading, in that it is caused by a progressive cutoff of the extremal rays in the transmitted light, but in practice the two defects may be difficult to separate. Also, with a camera-tube machine, further shading effects may arise from the operation of the tube.

The amount of shading and vignetting is easily measured by examining the output waveform corresponding to an open gate, and is specified as the amount of deviation of the flat top of the bar expressed as a percentage of white. During this measurement, it is as well to examine the whole of the waveform, using a line selector for other spurious camera signals resulting from defects in the photocathode.

In a twin-lens machine, vignetting can give a 25 Hz flicker at the top and bottom of the picture, readily detectable on a picture monitor.

With both telecine systems, it is also possible to find dichroic tilt produced in the colour-analysis unit of a colour machine. Precise measurement of this is not simple, but a quick assessment of its seriousness can be gained by using a neutral test object and looking at the resulting picture on a picture monitor. Any dichroic tilt is shown as coloured shading.

14.10 Colour analysis

A check of the colour analysis of a machine on acceptance may well be desirable, but without lengthy colorimetric measurements. Again, it may be thought prudent to monitor the colour analysis at intervals while the machine is in use, to ensure that any deterioration of performance due to aging of the optical elements is noticed. Such a check can now be carried out very conveniently and rapidly with a proprietary spectral test-transparency provided with markers at fixed intervals of wavelength. When this is scanned using a γ of unity, the outputs from the three colouring channels can be displayed on three of the channels of a 4-channel oscilloscope, and then compared either with the theoretical curves or with those obtained on previous occasions.

15 PICTURE MONITORS

15.1 General

Black-and-white picture monitors are required in operational areas for a variety of purposes. The strictest standards of performance are demanded from monitors associated with picture sources, where they are required for judging picture quality. These must possess first-class resolution combined with extremely accurate synchronisation, and must be provided with stable black-level control from an efficient clamp. Since defects in performance of all varieties in the picture sources must be made visible, the geometry and linearity of the raster will also have to be as good as possible. Above all, first-grade picture monitors must possess the stability to be able to function for very long periods at a stretch without noticeable changes in performance; any residual small variations have to be such that all monitors of the same type change as far as possible in the same manner so as to minimise difficulties in picture matching.

Second-grade picture monitors are required more for informational purposes than for judging picture quality; accordingly, the standards may be relaxed somewhat in most respects. They usually need much larger screens, e.g. for relaying programme information to studio staff, for presenting tape or film inserts to studio audiences etc.

The above also applies very largely to colour-television practice, since it is advisable to confine the colour monitors to the small number of positions where colour information is essential. This is principally due to the difficulty of maintaining an accurate match between a number of colour monitors over long periods of operation, but the associated saving in cost is also an understandable motive.

15.2 Contrast and brightness

It is extremely important in operational work that a standardised technique should be used for the adjustment of the brightness and grey scale of picture monitors; in particular, it must permit a very close match between a number of monitors to be reached with the minimum of effort.

In principle, this can be achieved with a test card containing black and white areas combined with a suitable step wedge; but, in practice, although there is no difficulty in setting the peak white brightness to a standard value with a suitable photometer, it is extremely difficult to adjust the picture blacks with anything like sufficient accuracy for the purpose.

By far the most satisfactory method is the use of a special waveform of the Pluge type (Chapter 10.6), which provides a partly objective and partly subjective means of setting the black level in the form of a signal containing, amongst other things, a standard near-black level with a very small percentage increase in brightness and a very small decrease in brightness provided in the shape of narrow 'ribbons'. It is found that the adjustment of the brightness control of the monitor to the point at which extinction of the darker ribbon is just observed is an extremely sensitive and reliable criterion for the precise setting of the black level, which at the same time takes account of reasonable variations in the ambient lighting. The complete procedure is described in Chapter 10.6.

Colour monitors are also preferably adjusted by using Pluge after the decoders have been properly aligned or, in the case of *RGB* monitors, after the channel gains are correctly set. In either case, a luminance signal ought then to produce a white of the correct colour temperature. However, with a colour monitor, the black levels on the *RGB* signals from the decoder tend from experience to vary somewhat, depending on the presence or absence of the colour burst, so that a balance carried out without the burst present does not necessarily hold when the burst is introduced into the input signal. It is consequently prudent to ensure that the burst is present whenever the Pluge and step-wedge signals are in use with colour monitors.

For operational work, it is necessary to have some reliable means of setting the white point to the correct value. This is quite impossible without an objective reference, since the eye adapts very quickly to the ambient lighting conditions, and a correct white field may appear subjectively to be distinctly incorrect; in particular, when the ambient lighting has a low colour temperature, the true white seems much bluer than one would expect.

A tristimulus colorimeter, a number of good examples of which are commercially available, can be used. An alternative, operationally much more convenient, is a standard comparison field. One that has proved very useful consists of a tungsten light source in a small box, at the end of which is an aperture provided with a suitable coloured filter. The lamps are supplied from a regulated current source, which, to standardise the instrument, is adjusted to set the colour of the comparison field against a standard colour-temperature meter. Reliable devices employing fluorescent tubes are also available. In the simplest application, the comparison source is placed on the picture monitor as close as possible to the white field on the screen. A worthwhile complication is a mirror device by means of which both light sources may be viewed simultaneously. Both sources must be reflected from the same number of mirror surfaces of the same type, otherwise errors are introduced.

A more sophisticated device is described by Sanders, Gaw and Wyszecki (1968). In this, the comparison field is generated by means of a small tungsten-halogen lamp in conjunction with suitable optical filters, which is powered by a regulated d.c. supply. A surface-silvered halfmirror is used to locate a portion of the white field to be adjusted so that, as in a photometer, both these areas are in close juxtaposition and in the field of view at the same moment. The adjustment of the two colours to a perfect match then becomes extremely precise. To reduce the effect of the ambient lighting, the comparison device may be used in direct contact with the protective front glass of the picture monitor; in this case a diffusing screen is required in front of the monitor field to minimise the effect of the screen dot structure.

The photometric type of comparison used in this device suggests that a refinement in which the luminance of the comparison field is standardised could be added so that one could set both the colour temperature and the luminance of the picture-monitor white field with the same instrument. This would be a decided operational convenience.

Unfortunately, with a colour monitor, the adjustment of picture black and white, although necessary, is not sufficient, since the grey-scale tracking must simultaneously be correct; i.e. ideally the colour temperature should not change with luminance-signal amplitude over the whole range of input levels from black to white. Such an ideal adjustment can only be approximated to, and difficulty may be experienced in deciding on the optimum condition. Some operators prefer to have available close to the monitor a grey scale of the correct colour temperature which has the same luminance steps as the reproduced grey scale used for tracking purposes, so that the two can be compared.

It might be appropriate to point out here that, if separate decoders are used with *RGB* monitors, it is very important to ensure that the colouring-channel gains have not been adjusted to compensate for any deficiencies in the decoder in use. If such a situation arises and the decoder is changed, the colour temperature may also change. A very simple and immediate test for the independence of the decoder and the monitor is to arrange for the *R*, *G* and *B* inputs to be simultaneously connected together while a luminance signal only at white level is applied to the input of the decoder. Any change in the colour of a white field indicates a maladjustment. This condition does not necessarily hold good at intermediate grey levels, owing to likely errors in grey-scale tracking.

15.3 Resolution

15.3.1 General

There is as yet no universally accepted criterion for measuring the resolution of picture monitors. The three principle criteria used to denote the ability of the monitor to render picture detail are as follows:

(*a*) limiting resolution, i.e. the frequency corresponding to the set of frequency gratings in which the pattern is just discernible. The performance of picture monitors for professional use is now universally so good that this criterion is no longer meaningful. The frequency of limiting resolution, which is the point at which the response has fallen by a large amount, say 20 dB or so, is so far above the videoband that it gives no useful information about the performance within the band. This is not necessarily true for receivers, because of the need to remove the sound channel, but the concept of limiting resolution is in any case too vague to be useful

(*b*) 6 dB frequency, i.e. the frequency corresponding to the set of frequency gratings at which the effective modulation depth falls to one-half of the modulation depth at the lower video frequencies

(*c*) relative modulation depth, expressed in decibels, or as a percentage of frequency gratings corresponding to some nominal upper video band-limiting frequency.

Some broadcasting authorities replace frequency by 'line number', which is the number per effective picture height of alternate black and white rectangular bars of equal width corresponding to the grating in use. Although this convention is fairly widespread, there seems little to commend it for operational television, for which it is more useful to think in terms of frequency directly, without having to relate the line number to the video standard in use.

Indeed, the discussion of the quality of a picture monitor in terms of 'lines' makes it clear that the instrument is still being treated as though it were an optical instrument instead of as one item only in the complex signal chain. This is illogical since, in respect of resolution at least, the picture monitor is not unique. Any transducer in the picture chain, and even the signal source itself, can equally affect the resolution. If one were presented with a picture monitor displaying an image of poor definition and were not allowed to examine the input signal, one would not normally be able to distinguish between waveform distortion of the signal of the type which provides a poor sine-squared-pulse/bar ratio and a monitor having poor resolution.

It would be far more appropriate to test a picture monitor as though it were for this purpose on a par with the other elements of the total signal chain. This cannot be done by examining the drive voltage to the cathode-ray tube, since resolution applied to a picture monitor must include the degradation of the overall resolving power introduced by the image-forming means of the electron beam and the phosphor, which in practice is far from negligible. Very interesting overall measurements have been carried out by Grosskopf (1966 and 1967), which are described below. They are analogous to the camera-resolution measurements described in Section 12.5, in which a further discussion of the problem of resolution is given.

With colour monitors, it is customary to measure resolution on a luminance signal, even though, for special purposes, the three colouring channels may be measured independently. The resolution is consequently impaired by any convergence errors which exist, although from experience, they affect the resolution somewhat less than one might suppose. Nevertheless, the monitor should be re-converged very carefully before any measurements are made, and if, as very occasionally seems to occur with shadow-mask tubes, a particular small area of the screen seems to have abnormally high convergence error when the rest of the screen is very well adjusted, this area should be avoided as far as possible for resolution measurements.

15.3.2 Simple methods

The time-honoured method of measuring picture-monitor resolution is to set up the monitor to standard contrast and brightness conditions on a test card such as test card D, and then to estimate the modulation depth of the higher-frequency gratings visually.

In this form, it is very crude and hardly to be dignified by the name of measurement, not the least because the estimation of modulation depth visually is highly inaccurate. It is true that some engineers are more consistent in their judgements than others, but the existence of the 'calibrated eyeball' is a myth. Any attempt to estimate the bandwidth in terms of limiting resolution is equally highly fallible.

If the reduction in modulation depth at the highest frequencies is small, say a few decibels, the following procedure can be adopted. The picture monitor is first carefully aligned for contrast and brightness, preferably by the Pluge technique or the equivalent. This is important, because the resolution of the tube is a function of these quantities.

A test card or pattern containing resolution gratings is then displayed on the monitor. For best results, the gratings should be sinusoidal frequency bursts derived preferably from an electronic source rather than from a camera or caption scanner, and the higher-frequency bursts should be located not too far from the centre of the screen. It is also helpful if the bursts do not extend up to full white level, and if the pattern is such that the lower-frequency bursts, say 1 MHz or so, are located reasonably close to the high-frequency bursts.

In series with the input of the picture monitor is a variable-aperture corrector which is set to the 'flat' position during the preliminary adjustments. This corrector is then adjusted until it is estimated visually that the contrast of the 5 MHz bars, say, is the same as the contrast of the 1 MHz bars. A measurement of the aperture corrector will then give the approximate droop at 5 MHz. A useful aid here is to turn down the brightness control and observe whether the bright bars in both gratings disappear simultaneously.

This method does not pretend to be highly accurate, but it is a stage more scientific than any purely visual estimate can be. Unfortunately, it cannot be used for large droops at the higher frequencies owing to the possibility of overloading the picture-monitor amplifier. Care should be taken that any limiter in the monitor is disabled before the measurements are made.

A variant of this method giving reasonably good results, when the reduction in modulation depth at the higher frequencies is not large, requires the use of picture-inlay equipment. The pattern used should contain physically rather larger frequency gratings than in, for example, test card D, preferably inverted in sequence in successive horizontal groups, so that each highest-frequency burst is enclosed vertically by two groups of the lowest-frequency burst.

Two independent feeds of the same monochrome test pattern are applied to the two sides of the electronic switch in the inlay equipment, one being taken through an amplifier and attenuator to allow the relative gain to be varied. A fairly small square of grey field from the latter feed is inserted into the corresponding larger area of grey field handled by the other side of the electronic switch. The gain of the smaller insert is then varied until a photometric match is obtained.

One of the sets of highest-frequency gratings is then replaced by the variable feed, the gain of which is increased until the contrast of the high-frequency grating appears to be equal to that of the immediately adjacent low-frequency gratings. The brightness control may again be used in judging this. The increase in gain required gives the reduction in modulation depth at the frequency of measurement. The limitation of the method, as before, is the signal-handling capacity of the equipment, especially towards the top of the videoband.

15.3.3 Photometric methods

The introduction to Section 15.3 pointed out that it would be much more logical and useful if, instead of treating a picture monitor as an optical instrument, it were tested as though it were a signal-handling element in the overall picture chain. This has been carried out by Grosskopf (1966 and 1967), who, in a number of his measurements, utilised a sine-squared-pulse and bar signal to measure the linear waveform response of the monitor from video input signal to output visual response, thus aligning the measurements with those made elsewhere in the signal chain. One aspect of these measurements corresponds to resolution.

The input signal used was a standard fullfield $2T$ sine-squared-pulse and bar signal except for measurements on television receivers in which the signal was appropriately modulated on to a high-frequency carrier. The measuring device consisted of a very sensitive, wide-range photometer which measured the mean brightness of a very narrow vertical slit placed in contact with the screen of the tube. For monochrome monitors, the slit was only about 0·15 mm wide. The signal was traversed past the slit electrically in synchronism with a pen recorder, which then drew the output waveform of the sine-squared-pulse and bar signal.

The results of a number of such tests are shown in Figs. 15.1 and 15.2; in each case, it should be noted that care has to be taken in interpreting the pulse/bar ratio, which is the analogue measurement of the resolution, since the ordinates are in logarithmic units. Figs. 15.1 *a*, *b* and *c* show the overall $2T$ pulse-and-bar response of a video monitor with three different settings of the peak screen luminances,

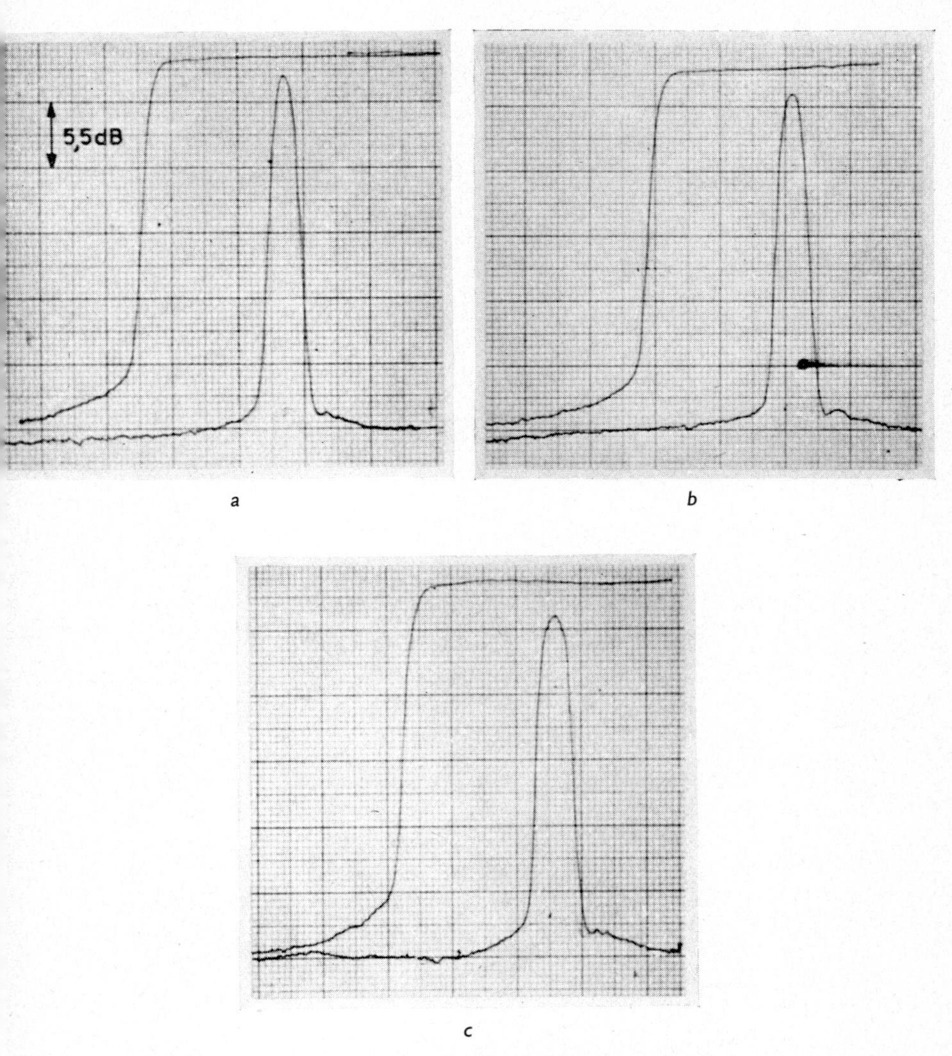

Fig. 15.1 Overall sine-squared-pulse and bar response of picture monitor at three values of brightness

Contrast range = 40:1
a 200 asb
b 400 asb
c 800 asb
(3·14 asb = 1 cd/m²)

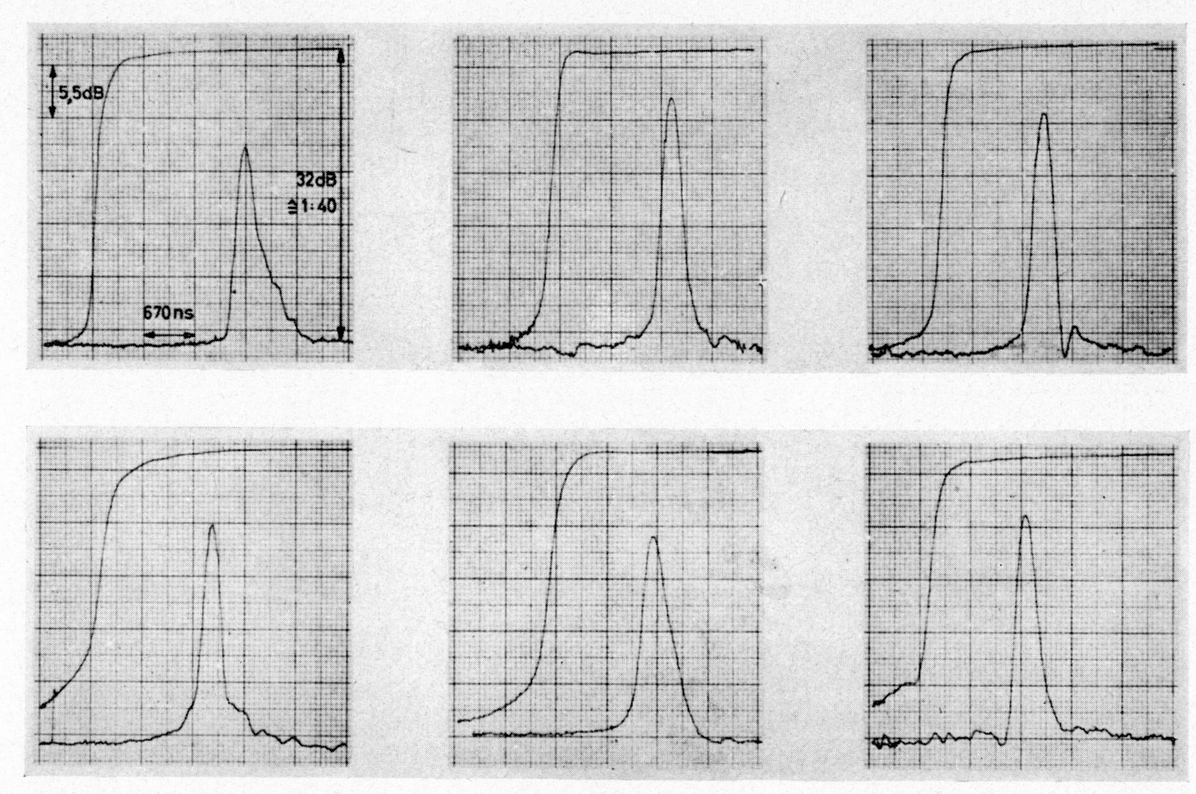

Fig. 15.2 Overall sine-squared-pulse and bar response of six television receivers

Peak brightness = 150 asb
Contrast range = 40:1
3·14 asb = 1 cd/m²

200, 400 and 800 asb, (3·14 asb = 1 cd/m^2) but the same contrast in each case. The decrease in resolution with increasing screen luminance is quite clear, and would have been considerably more marked if a 1*T* pulse-and-bar signal could have been used.

Fig. 15.2 shows a number of commercial television receivers measured as above. The resolution is in every case considerably inferior to the video monitor even at the lower peak screen luminance of 150 asb (48 cd/m^2) used, but this would be expected since, first, the receivers would not be built to the standard of quality of a professional video monitor and, secondly, the inclusion in the receivers of the h.f. circuitry necessarily shapes the upper part of the videoband.

With colour monitors and receivers, the slit had to be appreciably wider than that for monochrome, to include in each measurement at least one complete triad of phosphor dots. The aperture distortion thus introduced falsified the measurement of resolution, although comparative measurements could be obtained very accurately. For these, Grosskopf used a 4·5 MHz sinusoid in addition to the sine-squared pulse-and-bar waveform of the previous monochrome measurements, which furnished a very useful comparative figure for resolution.

It appeared from these measurements (Grosskopf, 1967) that a colour monitor or receiver with optimum convergence has at least the same resolution as the equivalent monochrome version. Particularly interesting also are the separate resolutions and waveform responses of the three guns of the shadow-mask tube, obtained by feeding the *R*, *G*, and *B* channels of the monitor separately with the test signals. The shapes of the 4·5 MHz sinusoids as measured at different screen luminances with a colour monitor are also very revealing of the loss of resolution brought about by high luminance levels.

15.3.4 Measurement of spot profile

The two principal factors which govern the resolution of a picture monitor are the waveform response of the amplifying system as far as the grid or cathode of the cathode-ray tube, and the resolution of the cathode-ray tube itself. In any well designed monitor the latter is the limiting factor. The inability to resolve all frequencies equally well arises from the aperture distortion caused by the finite size and the nature of the profile of the scanning spot; and, if this aperture distortion can be measured, the resolution of the monitor can be deduced, allowance being made if necessary for any effect on the resolution due to the amplifier.

This has been investigated by Woodbridge (1965), who measured the aperture distortion in the following manner. The stationary spot generated by the tube when the scans are stopped is focused with a microscope on to a graticule which contains a series of frequency gratings, each consisting of a set of equally wide opaque and transparent strips. The successive sets have reduced spacings so as to represent sets of square waves with frequencies increasing in steps. The light from the spot which traverses the graticule is diffused by a piece of opal glass, and is then picked up by a photomultiplier.

The graticule is arranged, by means of a cam drive, to slide backwards and forwards across the stationary spot, so that the series of gratings passes over it in sequence. Coupled to the mechanical movement is a potentiometer or resolver, by means of which a direct voltage proportional to the linear movement of the graticule can be applied to the X deflection input of an oscilloscope. The Y input is connected to the output of the photomultiplier.

When the grating spacing is large compared with the spot, the oscilloscope trace will be a square wave with sloping sides; and, as the spacing decreases, the flat top becomes shorter, and finally disappears at the narrowest spacings, leaving an approximate sine wave with reduced amplitude. The aperture distortion can be deduced from this response and combined, if necessary, with the amplifier response to yield the overall monitor resolution.

There is actually a redundancy in this procedure, since the aperture effect can be derived from the shape of the flank of one of the square waves alone, so that the grating could be replaced by a stationary slit across which the spot is deflected electrically. The technique then approaches that of Brown, described below.

The Woodbridge method has two principal drawbacks compared with the Grosskopf technique: the resolution is not given directly, but must be derived from the measurements; in addition, it is not easily practicable to relate the readings obtained with the peak brightness of the monitor display. In fact, to avoid phosphor burn, the tube will almost certainly be run at a lower brightness than under normal running conditions, which will tend to give an optimistic answer. There is also reason to believe that the results obtained with a purely static spot are not the same as those obtained with a spot under normal scanning conditions (Sandor, 1960).

An alternative method was employed by Brown (1967), which makes the measurement under more nearly normal operating conditions. The picture monitor is supplied with a fully composite video signal containing a white bar, and the contrast and brightness are set up to standard conditions. A low-power microscope is used to focus a portion of the centre of the field on to a very narrow horizontal slit behind which are mounted a spectral filter and a photomultiplier, the purpose of the filter being to make the overall spectral response approach that of the eye.

The picture-monitor display is made to move vertically at a slow rate by means of a variable amplitude of a sawtooth at a frequency of 2–5 Hz, which is inserted in series with the vertical-deflection yoke, so that a given line is moved vertically across the slit in a series of successive scans at the rate of one per picture. The same waveform is applied to the horizontal-deflection circuit of the oscilloscope, which then draws a series of vertical lines, the envelope of which is the line profile. From this profile, the aperture distortion of the picture tube itself can be calculated, and this can be combined with the amplitude–frequency response of the amplifier of the picture monitor to give the overall resolution.

Apart from the fact that the measurement of resolution is not direct, as it is with Grosskopf's method, the chief drawback of Brown's method is the measurement of the vertical line-profile instead of the horizontal line-profile, which is the quantity required for the estimation of the picture-tube aperture distortion. In the central region of the screen, it is probable that the spot profile will be reasonably symmetrical, but this certainly does not necessarily hold good elsewhere, owing to the effect of the various electron-optical aberrations.

Mention must also be made of one of the most widely used methods, the so-called 'shrinking raster' technique, commonly employed for both picture and waveform monitors. Its great advantages are simplicity and a minimum of special apparatus.

A white area is displayed on the screen of the picture monitor, preferrably as a white bar or 'window' located in the central area of the screen to ensure optimum spot symmetry. The vertical dimension of this area contains, say, precisely N lines. The vertical-scan amplitude is then progressively reduced until the individual lines begin to merge and finally just disappear, leaving an apparently uniform white area. At this point, the measured height of the white area is, say, L centimetres giving a critical line spacing of $L/(N-1)$ centimetres. From this, the spot profile in the vertical direction is deduced on the assumption that the brightness distribution transversely to the direction of scan is Gaussian.

It is often believed that the critical line spacing, as just defined, is equal to the half-amplitude width of the spot, but this is not true. Alternatively, it is supposed that the Gaussian distributions at the critical condition are spaced by two standard deviations; i.e. the curves intersect at the 60·7% amplitude points, which is much nearer the mark. This latter is derived from the very easily proved property of two identical Gaussian curves that the result of adding them linearly has a region of zero slope at the centre when they are separated by two standard deviations; it is an oversimplification, since the luminance distribution in the raster during the measurement is the sum of a much larger number of individual Gaussian distributions.

In view of this, it was decided to carry out an investigation of the problem, which fortunately lends itself very readily to solution with a computer. It was assumed that each individual luminance distribution can be expressed as $B = \exp(-x^2/2\sigma^2)$, where σ, the standard deviation, is now a convenient shape factor for the curve; x therefore becomes the distance variable expressed as numbers of standard deviations. To simplify the results of the analysis, the variation or 'ripple' generated by a sequence of such equidistant distributions is defined as a percentage given by $\{2(A_{max}-A_{min})/(A_{max}+A_{min})\}100$.

The calculated percentage ripple as a function of the spacing is shown in Fig. 15.3. It will immediately be seen that as the spacing is progressively decreased from a large value, the ripple decreases very rapidly until a transition region is approached, in the neighbourhood of two standard deviations, after which the curve quickly becomes asymptotic to the axis. This behaviour, of course, justifies the use of the method. However, the choice of the spacing corresponding to the

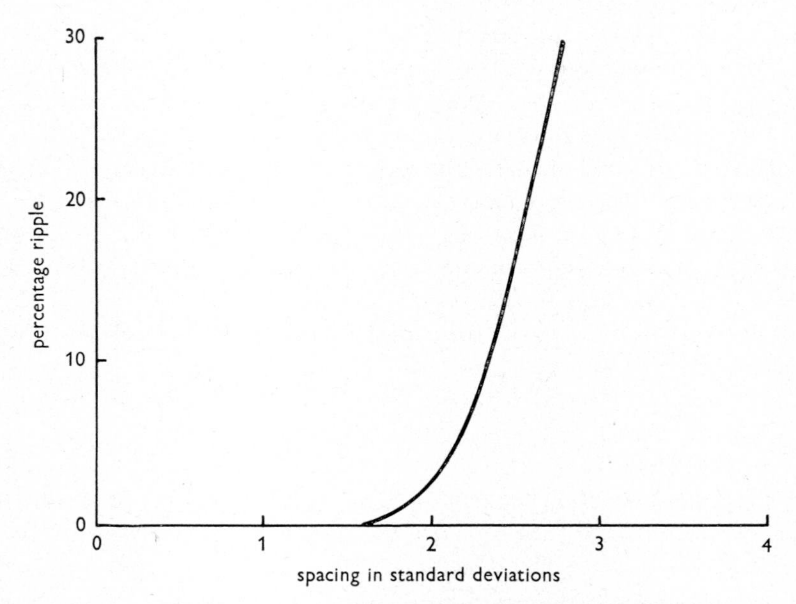

Fig. 15.3 Sum of a number of identical Gaussian distributions with equal spacing: peak-to-peak ripple as function of spacing

$$\text{percentage ripple} = \frac{2(A_{max} - A_{min})}{A_{max} + A_{min}}$$

critical spacing as estimated visually presents some difficulty, since it depends to such a large extent on the smallest luminance variation which can be perceived under the experimental conditions, apart from the influence of external effects, such as multiple reflections in the face of the tube and stability of the trace. It is not difficult to show, by adding calibrated sinewaves to the display, that the eye is sensitive to brightness variations considerably smaller than 1 %; but, allowing for practical difficulties, one might arbitrarily take this figure as the minimum perceptible luminance variation. This corresponds in round numbers to a spacing of 1·8 σ, so that the spot profile in the vertical direction is then defined by $\sigma = L/\{1{\cdot}8(N-1)\}$ centimetres.

The above assumes, of course, that the scan interlace is perfect. Although the interlace can be very good, it is unlikely to be so close to perfection that its effect can be ignored, unless one is content with a fairly rough measurement. The difficulty can be overcome by arranging for the interlace to be removed from the display, which means that steps must be taken to ensure that the vertical-scan amplitude can be reduced sufficiently to enable the measurement to be carried out; this in turn may entail a temporary modification to the monitor circuitry. The simplest way of achieving a single, noninterlaced display is to generate the white signal on one of the two fields only. A drawback is the 25 Hz flicker which some observers find irritating. The alternative would be the provision of a 625-line noninterlaced display, which is less easy.

15.4 Interlace

The interlace ratio is defined as $\{100d_1/(d_1+d_2)\}:\{100d_2/(d_1+d_2)\}$, where d_1 and d_2 are the smaller and larger spacings, respectively, between a given line of the raster and the lines immediately adjacent to it. An alternative criterion is the interlace factor defined as $2d_1/(d_1+d_2)$. Thus, if the interlace ratio is 45:55, the interlace factor is 0·9 or 90%.

The simplest method of measuring the quality of the interlace is by measurement from the screen when the picture monitor is fed with a line bar on standard synchronising pulses of the correct amplitude, so that an evenly illuminated field is available, with the synchronising and scanning circuits operating completely normally.

If a low-power microscope is available with an eyepiece incorporating a measuring graticule, or preferably a travelling microscope with accurately calibrated movements, the measurement presents no great problems. Alternatively, a reflex camera provided with extension tubes for photographing at close distances can be pressed into service. 35 mm or 6 × 6 cm negatives can be mounted directly in transparency mounts and projected on to a screen, on which the line spacings may be measured with an ordinary ruler. Since only the ratios of the measured spacings are required, there is no need to measure the magnification. The exposure should be kept to the minimum, and, whichever method is employed, the monitor brightness should not be

Fig. 15.4 Incorrect scan interlace
a Interlace ratio 29:71
b Interlace ratio 47:53

increased beyond the maximum value for best line definition. Some assistance can also be gained by increasing the picture height as much as possible. Typical photographs of incorrect scan interlace are given in Figs. 15.4*a* and *b*.

A very accurate measurement of the interlace can be obtained by a simple modification of Brown's method of measuring the line profile (Section 15.3.4). This consists in increasing the amplitude of the very low-frequency sawtooth in series with the vertical scan to the point where three successive lines are included. The spacings may then be measured from a photograph.

15.5 Waveform distortion

Defects in resolution are not the only picture impairments which can be considered as linear waveform distortion. The two commonest are streaks and ringing.

Under certain picture conditions, surprisingly small amounts of streaking can be seen; indeed, it may be quite difficult to measure the amount on a waveform monitor. It is accordingly usual to check on the picture monitor itself for any streaking, since the indication is so much more sensitive. Professional types of picture monitor should not show any perceptible amount.

The test signal should be purely electronic, since it is not easy to ensure that a picture from one of the usual sources is itself completely free from streaking. The signal preferably consists of both white bars on a black field and black bars on a white field. The bars should be 10–20 lines in height and variable in length. It is also useful if the surrounding field can be varied from black or white to grey.

The screen of the picture monitor, which must have previously been put into correct normal adjustment, is then examined with this signal under different conditions of bar length and polarity, and any trace of streaking is noted. If there is more than a trace, the amount can be measured with an oscilloscope with a high-impedance voltage-attenuating probe connected to the grid or cathode of the cathode-ray tube.

The streaky noise which may be generated by black-level clamps when the video signal contains a fair amount of random noise is, however, a nonlinear phenomenon. This is an intermittent and random change in black level due to the action of the random noise peaks on the clamp. This can be investigated with the same video test signal as for normal streaking, to which is added varying amounts of flat random noise. If a random-noise generator is not available, a receiver can be used as a noise source. That the noise spectrum produced by a receiver is not usually completely flat is not important for this purpose. The signal/noise ratio at which streaking starts should be measured. There is no need to go beyond a ratio of about 22 dB, say, very roughly the point of which the noise level on a waveform monitor appears to have about the same amplitude as the synchronising pulse.

The other picture impairment is ringing, best measured on a $2T$ sine-squared-pulse and bar signal applied to the input terminal of the monitor. The output voltage from the drive amplifier is displayed on a waveform monitor with an attenuating probe of very low capacitance. With a high-quality picture monitor, no perceptible ring should be found.

In the case of an *RGB* colour monitor, the individual colouring channels can be checked similarly. If a decoder is incorporated and the measurement is made from the decoder input, care should be taken to disable any subcarrier-trap circuit which may be included.

15.6 Return loss

Picture monitors are usually provided with the facility for either a 75 Ω termination or a loop-through connection, so that the monitor may be bridged across a coaxial cable terminated in another piece of equipment.

When the terminated connection is in use, the return loss required is usually fairly high, and should be measured by the methods of Chapter 8 to ensure that this is so.

With the loopthrough connection, there is the possibility of a mismatch from the input impedance of the picture monitor in shunt with the cable, which is well situated to generate a delayed echo of the signal if the sending-end termination of the cable is not ideal, and in any case may well modify the signal travelling beyond the bridging point.

Precautions are normally taken to reduce distortion of this type to negligible proportions by one of two means: either the bridging impedance is made sufficiently high compared with 75 Ω, or the input capacitance of the monitor, which is usually the cause of poor impedance, is absorbed into a lowpass network connected between the input and output sockets. Nevertheless, it is a wise precaution to check the impedance with a return-loss measurement. For this, the output socket is terminated with a good 75 Ω resistor, using the minimum cable length for the purpose, and the measurement is made looking into the input socket. The measurement may then be repeated with the sockets interchanged.

15.7 Display nonlinearity

15.7.1 General

If an electronic pattern of simple geometrical form, e.g. equidistant dots or a grille, is displayed on a monitor with the correct picture height and aspect ratio, the equidistance of the dots or of the crossovers of the grille should be maintained accurately at all points over the picture area; otherwise, deflectional nonlinearity is present.

Since the screen of the cathode-ray tube is not completely flat, we therefore have to decide if we wish to be completely unambiguous about whether the equidistance of the features of the electronic pattern refers to distances measured along the surface of the screen or to distances along a plane normal to the axis of the tube. Although the differences are not great, they are significant, and affect measurement.

The normal assumption is that the image on the screen is measured when projected geometrically on to a plane normal to the axis of the tube and in contact with the centre of the face of the screen. A point on the axis of the tube is chosen as the viewing point and a straight line is drawn from each significant point on the displayed pattern to the viewing point P (Fig. 15.5a). The intersection of each such line with the plane gives the projection of the corresponding point on the screen, so that A corresponds to C and B to D, the difference between CD and AB being the effect of tube curvature.

The distance d from the centre of the screen to the viewing point P must be standardised. A convenient value is five times the picture height, which is a comfortable viewing distance for critical examination of the whole of the screen.

15.7.2 Definition of distortions

There are eight principal distortions of the raster which can be defined. In general, it is preferable to measure the deflectional nonlinearity of a picture monitor without reference to particular forms of distortion; with high-quality monitors, this is usually the only course available, since the errors are all very small. However, it sometimes happens that one or two forms of distortion predominate, and it may then be appropriate to specify the amount of each form individually.

A convenient classification is the following:

(*a*) Trapezoidal distortion, in which two opposite sides of the raster are parallel but unequal in length. Fig. 15.5*b* shows vertical trapezoidal, sometimes called 'keystone' distortion. A convenient measure of this is the difference between the lengths of the parallel sides divided by their sum, i.e. $(DC-AB)/(DC+AB)$.

(*b*) Pincushion distortion, characterised by an inward displacement, or bowing of the sides. It usually occurs symmetrically; most often, all four sides are affected to some extent. Fig. 15.5*c* illustrates pincushioning of the horizontal lines only. It may be measured by $2(h_1+h_2)/(AB+DC)$, if h_1 and h_2 are the maximum displacements of the sides from the straight lines joining the corners. For this measurement, the corners A, B, C, D of the raster are adjusted to coincide with the corners of the hypothetical ideal raster on the face of the tube.

(*c*) Barrel distortion. This is the inverse of pincushion distortion, and is measured in an exactly similar manner. Barrelling of the horizontal lines only is shown in Fig. 15.5*d*.

(*d*) Parallelogram distortion. The raster is correct except for a shear through an angle θ, which measures the magnitude of the distortion.

(*e*) Waviness distortion or ripple distortion, in which one or more of the edges shows a stationary ripple-like departure from the ideal straight line. It is conveniently specified by the ratio of the peak-to-peak amplitude of the ripple to the length of the side of the raster in which it occurs. If more than one side is affected, the worst value is taken. The measurement of this distortion should always be made with synchronising pulses which differ slightly from mains frequency, to observe whether the pattern moves. If it does, it originates in hum and is considerably more serious than a purely geometrical distortion.

However, the distortion of a professional type of picture monitor should be very small; in any case, it is usually a mixture of very small

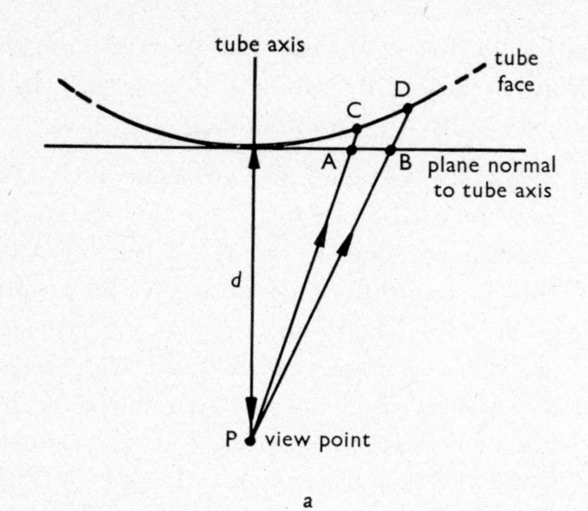

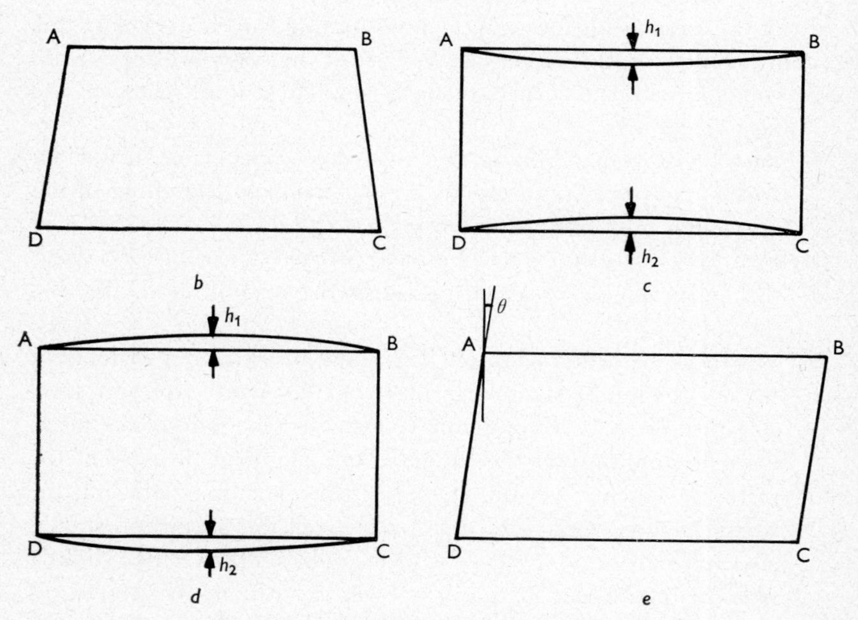

Fig. 15.5 Measurement of scan nonlinearity

a Effect of screen curvature
b Trapezoidal distortion
c Pincushion distortion
d Barrel distortion
e Parallelogram distortion

amounts of several distortions. It is then preferable to define the nonlinearity in a similar manner to that used for cameras (Chapter 12), in which the distortion is measured as the ratio of the displacement of a picture point from its true position to the distance between that point and the nearest picture point which coincides with its ideal position. The area of the raster may be divided into zones which are allowed successively increasing errors with increasing distances from the centre of the tube. Such a definition is independent of the type of the distortion, and takes into account the way in which deflectional nonlinearity distortion affects the observer.

15.7.3 Measurement

The basis is always an electronic grille or dot pattern displayed by the monitor. If the size of the raster is at all affected by the setting of the brightness control, this should be suitably reduced.

Probably the most generally suitable method of obtaining a comparison pattern is to have prepared photographically a transparency which can be projected on to the face of the tube. The lens of the projector is placed at the standard viewing point, say five times the picture height and on the axis of the tube, and the size of the image on the transparency is preset so that the spacings of the grille intersections or of the dots are equal to the spacings in the projected image. Slight inequalities in size can be corrected by adjusting the projector focus without destroying the definition.

The adjustment of the projected comparison pattern is somewhat fussy. A professional type of cinecamera tripod is useful for mounting the projector. The rigidity of the mounting and the movements provided on the tripod make it much simpler to bring the patterns into alignment, but care has to be taken that the axis of the projector is maintained precisely on the axis of the tube.

A very useful facility is the provision, on the pattern generator, of a range of calibrated delay on both line and field, so that the pattern can be moved over distances on the face of the tube corresponding to known intervals of time; this is not unduly difficult to achieve. Thus, not only can the last stage of the superposition of the patterns be carried out with great ease, but any errors due to nonlinearity can be put right and the corresponding error read from the calibrated delays. Otherwise, the errors are not easily measured with a projected pattern, with any great accuracy. Another, far less convenient, device is the calibration of the horizontal- and vertical-shift controls.

A modification of the above technique is especially convenient for acceptance checking of batches of picture monitors or for factory checking of production units. The electronic pattern consists of an equidistant-dots structure, and the projected pattern is a series of circles arranged so that an ideal monitor would place an electronic dot exactly in the centre of each. The diameters of the circles correspond to the acceptable limits, and consequently the division of the raster area into zones can very easily be catered for. It is then possible to see at a glance whether or not a given monitor meets the specification limits on a 'go or no-go' basis.

An alternative to the projected image is a thin but rigid sheet of transparent material on which the comparison pattern has been drawn or photographically reproduced. This graticule is then placed in contact with the face of the tube, but is not constrained to follow the curvature of the tube face. It and the displayed picture are viewed together from the correct distance. A convenient device is a hole consisting of a ring of wire, or a hole in a metal sheet, which is located at the current point on the axis of the tube. One advantage of the graticule suggested is that fairly precise measurements of errors in the superposition of the two patterns are possible with a cathetometer, i.e. a low-power telescope with calibrated movements. A photograph may also be taken with the camera lens at the standard viewing point.

15.7.4 Measurement by moiré pattern

A very ingenious and sensitive method for measuring the difference between a pattern displayed on the screen of the picture monitor and a comparison pattern has been suggested by Henk (1966). A brief description only will be given here; for a detailed account and for a method of manufacturing the graticules, reference should be made to

Henk (loc. cit). The moiré-pattern method has since been considerably extended by Sobolowski (1970).

Moiré patterns are optical interference patterns produced by the superposition of two regular fine structures, the transmitted or reflected light from which can reach the eye. Such patterns are produced, for example, when two transparencies of similar black and white grating patterns are superimposed with a slight error in coincidence of the two patterns. An almost unique and very convenient feature of such a moiré pattern is that the smaller the error in coincidence the larger is the pattern, so that even minute errors are detectable. In fact, one has to suppose mathematically that, when coincidence is perfect, the moiré fringes are still present but have retreated to an infinite distance.

Assume that the display takes the form of a series of alternate, parallel black and white bars, as in a frequency grating, and this is examined by superimposing an identical pattern in the form of a transparency. If the display and pattern are identical, it will be possible to position the two so as to obtain identical patterns; or, when a white bar coincides with an opaque bar in the transparency, there will be zero transmission. If, however, at some point in the display, the spacing between all bars of the first pattern is in error by 1%, say, moiré fringes will be seen displaced by distances corresponding to every 100 cycles of the pattern, from which the error can be determined with greater accuracy than would be possible by direct measurement.

Henk recommends that a comparison graticule should have alternate transparent and opaque bars which are slightly convergent instead of parallel. When such a transparency is in front of a displayed series of alternate parallel black and white bars whose spacing is adjusted to coincide with the graticule at the centre of it, the resulting moiré pattern is a family of rectangular hyperbolas whose asymptotes intersect at the centre of the monitor screen.

When the horizontal linearity is measured, i.e. when the two patterns are both vertical, the horizontal asymptote, which is also the horizontal line of symmetry through the centre of the pattern, is a line of zero error. Any error in the horizontal linearity causes a vertical displacement of this line from its correct position, which is directly proportional to the magnitude of the error. The vertical linearity can be measured similarly when the comparison pattern is turned through 90°. Here, it is advisable to suppress alternate fields to avoid movement of the pattern arising from any slight interlace jitter, bearing in mind the magnification introduced by the moiré pattern. In fact, the presence of any such movement should be a sensitive test for stability of the field synchronisation.

The use of moiré patterns is an elegant device capable of great sensitivity, but it seems best adapted for simple and regular errors in linearity. Otherwise, the patterns become rather difficult to interpret.

15.8 Synchronisation

The synchronising performance of a professional-type picture monitor should be unaffected by picture content and capable of holding a steady picture up to the point at which the signal/noise ratio makes it virtually unusable because of the high degree of impairment. Such measurements as are made are consequently qualitative rather than quantitative.

The efficiency of the synchronising-pulse separator is best assessed with a test card such as test card D, which is framed by castellations, i.e. alternate blocks of black and white intended for just this purpose. Those on the left-hand side of the picture subject the synchronising-pulse separator to sudden transitions of picture content which can alter the triggering point of the line sawtooth generator if any defect exists in the synchronisation.

For the test, both scans should be reduced slightly, so that the borders of the picture are clearly visible. With a video signal of correct level and picture/synchronising-pulse ratio, no perceptible shift of the vertical castellations should occur. The horizontal-synchronising control, if any, should be adjusted over its entire range to avoid any critical dependence on its setting.

Particular attention should be given to the first few lines at the extreme top of the picture. No horizontal displacement is allowable. It is additionally useful to replace the test card by a moving picture containing rapid changes of picture content, as a further test of the field-synchronising pulse regeneration. As before, it should be possible to vary the field-synchronisation control, if provided, over its complete range with negligible effect on the synchronisation except at the ends of the range, at which control is lost.

Finally, the signal/noise ratio of the video signal should be progressively degraded, either by the linear addition of white noise from a generator or, if such should not be available, from a receiver used as a noise generator. Usually, the synchronisation breaks down quite suddenly, often within a 1 dB step of the attenuator. This value should be noted. The tolerable value depends on the type of use for which the picture monitor is intended: but, with modern monitors, a ratio of 24 dB signal/noise should normally be attainable, and is often bettered quite appreciably. During this test, particular attention should be paid to any movement of the verticals, as a function of the picture content, or any movement of the top of the picture, since failure in this region is possible before synchronisation breaks down completely.

15.9 E.H.T. generation

The high-voltage supply to the tube is capable of causing two principal defects of the image. The first, generally termed an 'e.h.t. ring', consists of a velocity modulation of the scan as a result of the operation of the e.h.t. generator. It comprises parallel dark and light bands across the left-hand half of the picture, which may only be visible under certain background conditions, and are therefore best sought by displaying a uniform grey field variable between black and white. Usually, the e.h.t. ring will be noticeable, if present, on a fairly low value of grey.

The other defect arises from a poor effective internal resistance of the generator. It is an increase in the spot size with an increase in brightness, or even of a change in the picture size with a change of average picture level. The former, of course, degrades the definition, and will be tested separately.

The difficulty here is that too low an internal resistance is undesirable on safety grounds, so that an optimum value of internal resistance is often specified, say 1 MΩ, which must be adhered to within reasonable limits to maintain the best compromise. This is usually measured by the standard procedure of noting the voltage change when the brightness control is varied, so as to change the current taken by the tube between limits corresponding to the extreme demands occurring in practice. To obtain better accuracy, an external laboratory-type high-voltage generator may be used to back off a voltmeter of limited range. Precautions must be taken to avoid errors due to leakage or corona effects. The greatest care must also obviously be taken to avoid danger to personnel, since, even with a nominally nonlethal supply, the secondary effects of the muscular contractions resulting from a shock can be very unpleasant, and even dangerous.

A further test is also desirable to ensure that any feedback stabilisation of the e.h.t. generator does not give rise to low-frequency oscillations on sudden changes of current from the generator. For this, a 'bump test' signal is applied to the picture monitor, which must be in a normal and correct state of adjustment. This signal cuts as abruptly as possible from a full-black field to a full-white field without any disturbance of the synchronising pulses. Under these conditions, no transient change in the picture dimensions should be discernible. Above all, no continued, damped variation in the picture size at a slow rate can be tolerated.

15.10 References

BROWN, E. A. (1967): 'A method for measuring the spatial-frequency response of a television system', *J. Soc. Motion Picture Televis. Engrs.*, **76**, (9), pp. 884–887

GROSSKOPF, H. (1966): 'Die elektro-optischen Übertragungseigenschaften von Fernseh-Heimempfängern', *Radio Mentor*, (5), pp. 394–399

GROSSKOPF, H. (1967): 'Vergleich der elektro-optischen Übertragungseigenschaften bei Schwarzweiss und Farbe', *Rundfunktech. Mitt.*, **11**, (6), pp. 286–293

HENK, A. J. (1966): 'Scan linearity measurements without tears', *Wirelesss World*, **72**, (9), pp. 449–454

SANDERS, C. L., GAW, W., and WYSZECKI, G. (1968) 'Colour calibrator for monitors in television studios', *J. Soc. Motion Picture Televis. Engrs.*, **77**, (6), pp. 622–623

SANDOR, A. (1960): 'Effective spot size in beam scanning tubes', *ibid.*, **69**, pp. 735–738

SOBOLEWSKI, V. C. (1970): 'Using moiré patterns to determine the distortion of graphic displays and graphic input devices', *Proc. Inst. Elect. Electron. Engrs.*, **58**, (4), pp. 567–583

WOODBRIDGE, L. A. (1965): 'High resolution cathode ray tubes', EMI symposium on television, London, EMI Publication R/C001

16 APPENDIXES

16.1 Colour-bar-signal amplitudes

16.1.1 Pal system

Given below for each colour are the amplitude of the luminance signal measured from black level and the peak-to-peak values of the U and V subcarrier components and of the resultant chrominance vector. The angle of each vector is relative to the $B-Y$ axis. The angles for line n correspond to the odd lines of the first and second fields and the even lines of the third and fourth fields. Those for line $n+1$ correspond to the even lines of the first and second fields and the odd lines of the third and fourth fields.

Table 16.1 100–0–100 (100%) *colour-bar signal*

Hue	Luminance	U	V	Resultant chrominance	Angle: Line n	Angle: Line $n+1$
	V	V	V	V	deg	deg
White	0·700	0	0	0	0	0
Yellow	0·620	0·612	0·140	0·627	167	193
Cyan	0·491	0·206	0·861	0·885	283·5	76·5
Green	0·411	0·405	0·721	0·827	240·5	119·5
Magenta	0·289	0·405	0·721	0·827	60·5	299·5
Red	0·209	0·206	0·861	0·885	103·5	256·5
Blue	0·080	0·612	0·140	0·627	347·0	13·0
Burst	0			0·300	135·0	225·0

Table 16.2 100–0–100–25 (95%) *colour-bar signal*

Hue	Luminance	U	V	Resultant chrominance	Angle: Line n	Angle: Line $n+1$
	V	V	V	V	deg	deg
White	0·700	0	0	0	0	0
Yellow	0·640	0·459	0·105	0·470	167	193
Cyan	0·543	0·155	0·646	0·664	283·5	76·5
Green	0·483	0·304	0·541	0·620	240·5	119·5
Magenta	0·392	0·304	0·541	0·620	60·5	299·5
Red	0·332	0·155	0·646	0·664	103·5	256·5
Blue	0·235	0·459	0·105	0·470	347·0	13·0
Burst	0			0·300	135·0	225·0

Table 16.3 100–0–75–0 (*EBU*) *colour-bar signal*

Hue	Luminance	*U*	*V*	Resultant chrominance	Angle Line *n*	Angle Line *n*+1
	V	V	V	V	deg	deg
White	0·700	0	0	0	0	0
Yellow	0·465	0·459	0·105	0·470	167	193
Cyan	0·368	0·155	0·646	0·664	283·5	76·4
Green	0·308	0·304	0·541	0·620	240·5	119·5
Magenta	0·217	0·304	0·541	0·620	60·5	299·5
Red	0·157	0·155	0·646	0·664	103·5	256·5
Blue	0·060	0·459	0·105	0·470	347·0	13·0
Burst	0			0·300	135·0	225·0

16.1.2 NTSC system

Below are given the amplitude of the luminance signal measured from black level and the peak-to-peak values of the I and Q subcarrier components and of the resultant chrominance vector. The angle of each vector is given relative to the $B-Y$ axis.

Table 16.4 100–0–100–0 (100%) *colour-bar signal*

Hue	Luminance	*I*	*Q*	Resultant chrominance	Angle
	V	V	V	V	deg
White	0·700	0	0	0	0
Yellow	0·620	0·450	0·437	0·627	167·0
Cyan	0·491	0·834	0·295	0·885	283·5
Green	0·411	0·385	0·732	0·827	240·5
Magenta	0·289	0·385	0·732	0·827	60·5
Red	0·209	0·834	0·295	0·885	103·5
Blue	0·080	0·450	0·437	0·627	347·0
Burst	0			0·300	180·0

Table 16.5 100–0–100–25 (95%) *colour-bar signal*

Hue	Luminance	*I*	*Q*	Resultant chrominance	Angle
	V	V	V	V	deg
White	0·700	0	0	0	0
Yellow	0·640	0·337	0·328	0·470	167
Cyan	0·543	0·625	0·221	0·664	283·5
Green	0·483	0·289	0·549	0·620	240·5
Magenta	0·392	0·289	0·549	0·620	60·5
Red	0·332	0·625	0·222	0·664	103·5
Blue	0·235	0·337	0·328	0·470	347·0
Burst	0			0·300	180·0

Table 16.6 77–7·5–77–7·5 (*US*) *colour-bar signal* (*Fig.* 10.20)

Hue	Luminance	*I*	*Q*	Resultant chrominance	Angle
	V	V	V	V	deg
White	0·525	0	0	0	0
Yellow	0·465	0·337	0·328	0·470	167
Cyan	0·368	0·625	0·221	0·664	283·5
Green	0·308	0·289	0·549	0·620	240·5
Magenta	0·217	0·289	0·549	0·620	60·5
Red	0·157	0·625	0·222	0·664	103·5
Blue	0·060	0·337	0·328	0·470	347·0
Burst	0			0·300	180·0

INDEX